Thrasyvoulos Spyropoulos
Karin Anna Hummel (Eds.)

# Self-Organizing Systems

4th IFIP TC 6 International Workshop, IWSOS 2009
Zurich, Switzerland, December 9-11, 2009
Proceedings

Springer

Volume Editors

Thrasyvoulos Spyropoulos
ETH Zürich, Computer Engineering and Networks Laboratory
ETZ G 96, Gloriastr. 35, 8092 Zürich, Switzerland
E-mail: spyropoulos@tik.ee.ethz.ch

Karin Anna Hummel
University of Vienna, Department of Distributed and Multimedia Systems
Lenaugasse 2/8, 1080 Vienna, Austria
E-mail: karin.hummel@univie.ac.at

Library of Congress Control Number: 2009940141

CR Subject Classification (1998): C.2, F.1.1, C.2.2, C.1.3, I.2.2, I.2.8, I.2.6

LNCS Sublibrary: SL 5 – Computer Communication Networks
and Telecommunications

ISSN        0302-9743
ISBN-10     3-642-10864-4 Springer Berlin Heidelberg New York
ISBN-13     978-3-642-10864-8 Springer Berlin Heidelberg New York

springer.com

© IFIP International Federation for Information Processing 2009
Printed in Germany

Typesetting: Camera-ready by author, data conversion by Scientific Publishing Services, Chennai, India
Printed on acid-free paper      SPIN: 12810772      06/3180      5 4 3 2 1 0

# Lecture Notes in Computer Science     5918

*Commenced Publication in 1973*
Founding and Former Series Editors:
Gerhard Goos, Juris Hartmanis, and Jan van Leeuwen

## Editorial Board

# Preface

We welcome you to the proceedings of the 4th International Workshop on Self-Organizing Systems (IWSOS 2009) hosted at ETH, Zurich, Switzerland. IWSOS provides an annual forum to present and discuss recent research in self-organization focused on networks and networked systems. Research in self-organizing networked systems has advanced in recent years, but the investigation of its potentials and limits still leaves challenging and appealing open research issues for this and subsequent IWSOS workshops.

Complex and heterogeneous networks make self-organization highly desirable. Benefits envisioned by self-organization are the inherent robustness and adaptability to new dynamic traffic, topology changes, and scaling of networks. In addition to an increasingly complex Future Internet, a number of domain-specific subnetworks benefit from advances in self-organization, including wireless mesh networks, wireless sensor networks, and mobile ad-hoc networks, e.g., vehicular ad-hoc networks. Self-organization in networked systems is often inspired by other domains, such as nature (evolution theory, swarm intelligence), sociology (human cooperation), and economics (game theory). Aspects of controllability, engineering, testing, and monitoring of self-organizing networks remain challenging and are of particular interest to IWSOS.

This year, we received 34 full paper and 27 short paper submissions. The high quality of the submissions allowed us to provide a strong technical program. Based on the recommendations of the Technical Program Committee and external expert reviewers, we accepted 14 full papers from the full-paper submissions and invited 3 as short papers. Of the 27 short-paper submissions we accepted 10 for presentation for a total of 13 short papers. All full papers received three to four reviews and all short papers three reviews, with a few exceptions that were clearly out of scope. A number of papers were shepherded toward publication by the Technical Program Committee and external expert reviewers.

Our technical program consisted of full-paper sessions on Ad Hoc and Sensor Networks (3 papers), Services, Storage and Internet Routing (3 papers), Peer-to-Peer Systems (3 papers), Theory and General Approaches (3 papers), and Overlay Networks (2 papers). Additionally, there were three short-paper sessions on Peer-to-Peer Systems and Internet Routing (4 papers), Wireless Networks (5 papers), and Networking Topics (4 papers). To complement the technical program we invited a discussion paper on design approaches for self-organizing systems. Finally, we were delighted to have two keynote addresses by Dario Floreano and Martin May.

We are grateful to all Technical Program Committee members and additional reviewers who provided thorough reviews that made the selection of the papers possible, and the ones who additionally helped with the paper shepherding process. Special thanks go to our IWSOS 2009 General Chair, Bernhard Plattner, for his outstanding support in all the phases of the workshop organization.

Our thanks go to Georg Carle and Guy Leduc from IFIP TC6 for supporting the workshop with IFIP sponsorship. We are also indebted to the Euro-nf Network of Excellence for supporting IWSOS 2009 by sponsoring travel grants for their members attending the workshop. We thank Caterina Sposato from ETH Zurich for help in the local organization and the side program, and finally ETH Zurich for providing us with excellent facilities free of charge.

December 2009                                                    Thrasyvoulos Spyropoulos
                                                                            Karin Anna Hummel

# Organization

IWSOS 2009, the IFIP 4th International Workshop on Self-Organizing Systems, was organized by the Communication Systems Group, ETH Zurich, Switzerland, December 911, 2009.

## Steering Committee

David Hutchison      Lancaster University, UK
Hermann de Meer      University of Passau, Germany
Randy Katz      UC Berkeley, USA
Bernhard Plattner      ETH Zurich, Switzerland
James P.G. Sterbenz      The University of Kansas, USA
Karin Anna Hummel      University of Vienna, Austria
Georg Carle      TU Munich, Germany
     (IFIP TC6 Representative)

## General Chair

Bernhard Plattner      ETH Zurich, Switzerland

## Technical Program Co-chairs

Thrasyvoulos Spyropoulos      ETH Zurich, Switzerland
Karin Anna Hummel      University of Vienna, Austria

## Program Committee

Sandford Bessler      Forschungszentrum Telekommunikation Wien, Austria
Christian Bettstetter      University of Klagenfurt, Austria
Taric Cicic      University of Oslo, Norway
Alexander Clemm      Cisco Systems, USA
Costas Courcoubetis      AUEB, Greece
Anwitaman Datta      Nanyang Technological University, Singapore
Simon Dobson      University College Dublin, Ireland
Wilfried Elmenreich      University of Klagenfurt, Austria
Stefan Fischer      University of Luebeck, Germany
Michael Fry      University of Sydney, Australia
Matthias Hollick      Technical University of Darmstadt, Germany
Amine Houyou      University of Passau, Germany
Merkourios Karaliopoulos      ETH Zurich, Switzerland

| | |
|---|---|
| Alexander Keller | IBM Global Technology Services, USA |
| Wolfgang Kellerer | DoCoMo Lab Europe, Germany |
| Rajesh Krishnan | Scientific Systems Company, Inc., USA |
| Guy Leduc | University of Liege, Belgium |
| Marco Mamei | University of Modena e Reggio Emilia, Italy |
| Andreas Mauthe | Lancaster University, UK |
| Paul Mueller | Kaiserslautern University, Germany |
| Masayuki Murata | Osaka University, Japan |
| Ben Paechter | Napier University, UK |
| Dimitrios Pezaros | Lancaster University, UK |
| Christian Prehofer | Nokia Research, Finland |
| Lukas Ruf | Consecom AG, Switzerland |
| Susana Sargento | University of Aveiro, Portugal |
| Marcus Schoeller | NEC Laboratories Europe, Germany |
| Caterina Maria Scoglio | Kansas State University, USA |
| Mikhail Smirnov | Fraunhofer Fokus, Germany |
| Paul Smith | Lancaster University, UK |
| Dirk Staehle | Würzburg University, Germany |
| Burkhard Stiller | University of Zurich and ETH Zurich, Switzerland |
| John Strassner | Motorola Labs, USA |
| Zhili Sun | University of Surrey, UK |
| Kurt Tutschku | Würzburg University, Germany |
| Patrick Wuechner | University of Passau, Germany |
| Albert Zomaya | University of Sydney, Australia |

## Local Organizing Committee

| | |
|---|---|
| Bernhard Distl | ETH Zurich |
| Simon Heimlicher | ETH Zurich |
| Bernhard Plattner | ETH Zurich |
| Caterina Sposato | ETH Zurich |

## Additional Reviewers

| | |
|---|---|
| Helmut Hlavacs | Giovanni Neglia |
| Karin Anna Hummel | Thrasyvoulos Spyropoulos |

## Sponsoring Institutions

ETH Zurich
Euro-NF
IFIP TC 6

# Table of Contents

## Ad Hoc and Sensor Networks

## Services, Storage, and Internet Routing

## Peer-to-Peer Systems

## Theory and General Approaches

## Overlay Networks

## Short Papers

## Peer-to-Peer Systems and Internet Routing

## Wireless Networks

## Networking Topics

# Self-management of Routing on Human Proximity Networks

Graham Williamson[1], Davide Cellai[1,2], Simon Dobson[1,2], and Paddy Nixon[1,2]

[1] Systems Research Group, School of Computer Science and Informatics,
University College Dublin, Dublin, IE
[2] Lero, School of Computer Science and Informatics,
University College Dublin, Dublin, IE

**Abstract.** Many modern network applications, including sensor networks and MANETs, have dynamic topologies that reflect processes occurring in the outside world. These dynamic processes are a challenge to traditional information dissemination techniques, as the appropriate strategy changes according to the changes in topology. We show how network dynamics can be exploited to design a self-organising data dissemination mechanism using only node-level (local) information, which detects and adapts to periodic patterns in the network topology. We demonstrate our approach against real-world human-proximity networks.

## 1  Introduction

Most networks in society and technology present a time dependent topology. Networks of friendships, phone calls, but also mobile devices, probes and satellites, are dynamic. Of course, networks often change slowly and can be effectively represented by a static model, or they change in a precise direction (e.g. they grow) and their dynamic behaviour is well understood. There are cases, however, of highly dynamic networks, where change consists in a complex rewiring of the edges. A good example is the network formed by mobile objects (such as mobile phones) with short range communications. This example is particularly interesting, because the fast growing complexity and diffusion of mobile devices is envisioned to lead to a scenario where a number of applications will be used and shared by people living in the same city (Pocket Switched Networks). If these devices can communicate with each other ad hoc, without infrastructure, the result is a proximity network. For these reasons an interesting question is to understand when such networks can support stable applications such as information dissemination. In recent years, there has been growing interest in information propagation on dynamic networks. In particular, delay tolerant networking (DTN) protocols are designed to work in challenged or sparse networks. Recent papers have approached the problem of formulating efficient protocols for these types of networks [1,2,3]. Hui et al. [4] have proposed an efficient algorithm named BUBBLE, which exploits the popularity of a node to provide a quick and parsimonious way for a message to reach its destination.

T. Spyropoulos and K.A. Hummel (Eds.): IWSOS 2009, LNCS 5918, pp. 1–12, 2009.

However, most of these methods rely on either the particular characteristics of the considered experiment, or global information which will not presumably be available in real applications. Given the nature of this type of network, it appears sensible to design protocols with significant self-management capabilities.

In this paper we explore different data sets and investigate the importance of local properties for data communication. Then we propose a dissemination protocol which is defined locally at node level and does not imply global knowledge of the network. Finally, we define a self-managing mechanism allowing the nodes to adapt their dissemination strategy based on the detection of periodic patterns.

## 2 Analysis of Experimental Data

### 2.1 Data Sets

In this paper we take advantage of two very interesting data sets: the *Reality* experiment, performed at MIT [5,6], and the *Cabspotting* data set [7,8].

In the Reality experiment, 103 smart phones are assigned to 97 people (mostly among undergraduate and graduate students, but also staff at MIT), who carry them along every day. The smart phones detect other smart phones or any discoverable bluetooth device every 5 minutes. In this way, a network of proximity based encounters is built at any time. The experiment lasts 9 months, covering the terms of the academic year 2004-05.

In the Cabspotting experiment, the positions of 536 taxi cabs where tracked for about a month in the city of San Francisco. The positions were recorded as GPS coordinates at intervals of approximately 10 $s$. The proximity of cabs can be calculated on the basis of their movement patterns. The network formed by connecting cabs at a distance of $10m$ or less is already quite dense for the purpose of testing our protocol, and thus we will not consider larger communication ranges in this paper.

### 2.2 Network Connectivity

For communication purposes, it is clear that some nodes may be more important than others. For example, in a static network nodes can be considered more important if many shortest paths between node pairs pass by them. This concept was rigorously formulated about 30 years ago with the definition of *betweenness centrality* [9]. Moreover, it is quite well known that social networks are characterized by a community structure, where nodes within a community are very well connected, whereas few edges link different communities [10]. Therefore, it is clear that nodes communicating among different communities have an important role in information dissemination.

In this paper, the *degree* of a node is defined as the number of distinct nodes encountered in a certain time interval. This quantity is therefore different from the number of encounters, because many edges can come and go between the same pair of nodes within a given time. The importance of the time scale in calculating the degree is addressed in Section 5.

We now show and comment results from the analysis of the Reality data set. Fig. 1(a) shows the daily behaviour of the node degree (the number of distinct nodes encountered in a day) over the duration of the experiment. We observe a significant difference in the activity, with lower degrees during public holidays, as well as strong oscillations for different working days. Since the global activity

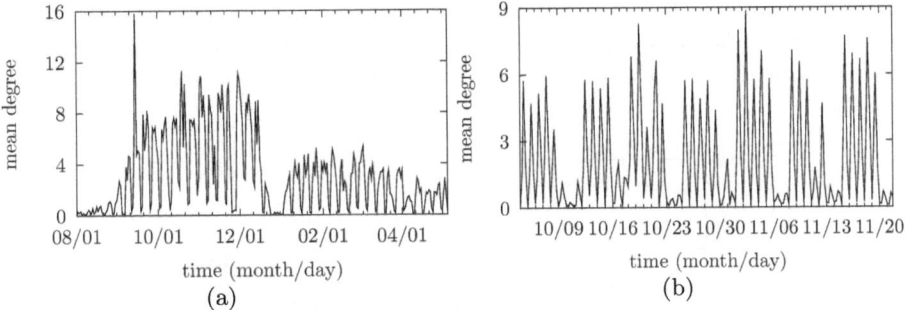

(a)                                                    (b)

**Fig. 1.** (a) Daily degree vs time. (b) 6-hour degree vs time in a 7 week interval. The trace can be highly variable, but there are long (several weeks) periods with regular behaviour.

of the trace is so heterogeneous, it is useful to focus on some regular parts such as the one represented in Fig. 1(b), where node degree calculated in time slots of 6 hours is plotted vs time. As shown by the plot, week-ends are clearly recognizable as well as the alternation between day and night.

Most of the time the examined network is sparse, as different individuals are involved in different activities, usually far apart. It is then crucial to study the intrinsic capability of the network to support data communication. It is worth pointing out that it is possible that, within a given time, a message can be routed from node $i$ to node $j$, but not *vice versa*. Hence, we must always consider ordered pairs of nodes. A preliminary question is to establish the maximum theoretical communications capabilities of the network, regardless of efficiency concerns. Thus, we define a quantity called *deliverability*: given a maximum delivery time, the ratio between all the ordered pairs of nodes for which message delivery is possible (i.e. it exists at least one time-dependent path connecting the source with the destination), divided by the total number of ordered pairs. Message delivery is established by unlimited flooding, which is the most effective way (but not the most efficient, of course). As we consider protocols where messages are not duplicated, but there is only a single copy in the network at any time, a useful quantity is the *delivery ratio*, defined as the fraction of distinct delivered messages over all the possible pairs source/destination within a given time. By definition, then, deliverability represents an upper bound on the delivery ratio of any routing protocol for a given network in a given time.

In Fig. 2 we show the comparison of the deliverability ratios between two different 3-week periods. It is interesting to note that week-ends constitute a real barrier for relatively fast deliveries: if the maximum delivery time is less than 4 days, there are days with very poor performance. On the whole, the deliverability is also quite low for higher waiting times, being around 0.5, but it stays constant over different days of the week. The comparison between the two periods also gives an interesting insight into the relationship between deliverability and short-time degree. The two periods only differ for the amount of activity (number of encounters, number of people involved in encounters, etc.), which is noticeably higher in the first set. We observe that this difference mostly affects the short delay deliveries (less than 3 days), whereas the deliverability after one week is more similar (the difference is about 0.1). This means that over time

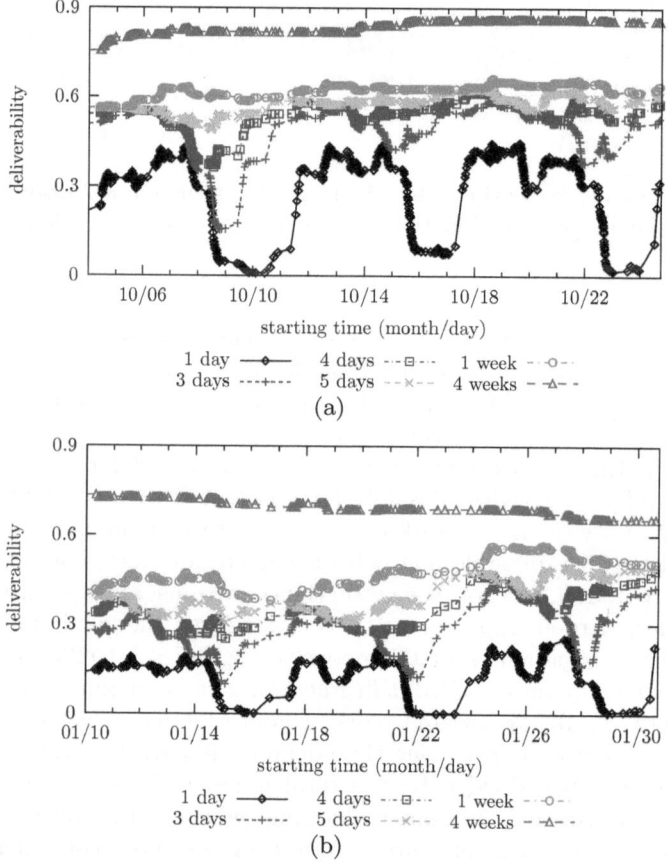

**Fig. 2.** Deliverability ratio based on flooding: comparison between two 3-week periods. It is necessary to allow at least 3 days for overcoming the drop of performance due to the small activity during week-ends. This comparison shows the similarity of the qualitative behaviour in two different periods of the data set.

scales of about a week individuals get close to people from other communities and thus improve the deliverability even in periods of low activity. From the perspective of self-management, the important point is that the network possesses characteristic and recurrent dynamical features.

## 2.3 Network Correlations

We examine the time dependence of some local properties of the nodes. To begin with, we divide the time into 6-hour slots, because, as observed by some authors [4], human daily routine can be divided into periods of activity which can be treated as roughly 6 hours long. We consider the aggregate graph of all the sightings happening in each time slot and calculate the betweenness centrality of each node, according to the Freeman algorithm [9]. In Fig. 3(a) we plot the time dependence of the Pearson correlation coefficient (pcc) of betweenness and node degree. We observe that there is a weak positive linear correlation between these two quantities: $0.653 \pm 0.18$. This means that, as an approximation, we can associate a high 6-hour degree to a high centrality of the node in the network, and *vice versa*. However, the value of the pcc has large oscillations with time with a standard deviation of 0.18.

This highly changing behaviour is not related to the incidence of holidays or singular events which may add noise to the measure. In fact, looking in Fig. 3(b) at the same curve restricted to the more regular 7-week period mentioned above, it emerges that the standard deviation of the pcc does not shrink at all, meaning that the oscillations are entirely due to the alternation of day/night and working days/week-ends. However, we observe a drop of the mean value of the correlation to 0.58, probably due to the occasional presence of starry structures during periods of low activity. (An explanation could be that, especially at the beginning of the experiment, some people were still not using the equipment properly, perhaps forgetting to enable the detection of other devices. If only one in a group is correctly recording neighbouring nodes, but all the others are not, the result would exactly be a star.)

The time dependence of the pcc between betweenness and number of connections in a time slot, shows a very similar behaviour, meaning that in a 6-hour

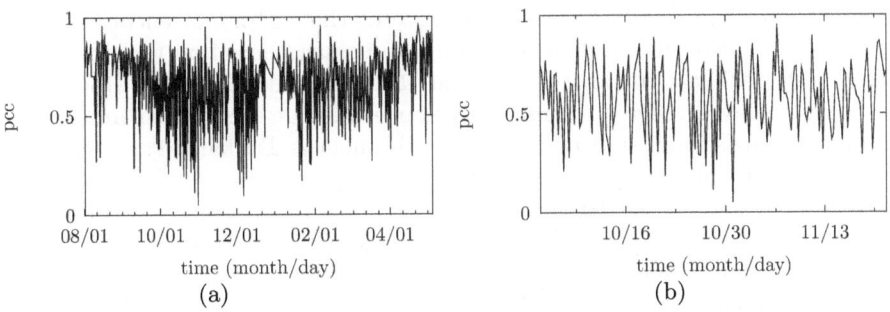

**Fig. 3.** (a) 6-hour pcc between betweenness centrality and node degree vs time. (b) 6-hour pcc between betweenness and degree vs time in a 7 week interval.

interval most encounters are distributed among the encountered nodes, so that there is not so much difference between number of sightings and degree.

As in Hui et al. [4], we calculate the *centrality* of a node as the number of times the node is on the shortest path between every ordered pair of nodes. If there are multiple shortest paths between two nodes, we divide the contribution of this value by the number of the paths. We compare the correlation between this centrality and other significant quantities over the 3 weeks starting with October 4th, 2004. We find that the correlation with betweenness, degree, and number of encounters is 0.41, 0.57, and 0.65, respectively. This implies that if we assume this notion of centrality as the most expressive quantity of the importance of a node in data dissemination, either the degree or the number of encounters seem to be the best candidates as local approximations. This means that a network protocol may, by making *local* observations, adapt its behaviour in a way that correlates strongly with an important *global* dynamic property of the network topology.

## 3   Dissemination Algorithms

A number of algorithms have been formulated to efficiently disseminate information on a dynamic network. The specific problem we consider is to send a message from node $i$ to node $j$, for arbitrary $i$ and $j$. The goal is to achieve a high fraction of successful deliveries, as well as low overhead, within a given time.

The most effective method consists of flooding the network with an arbitrarily large number of message copies. Flooding is characterized by the following policy: when a node receives a message it immediately broadcasts copies to all current neighbours. It also forwards copies of the message to all nodes it comes into contact with for the remaining duration of the simulation. This has a huge cost, due to the high number of redundant transmissions. Multiple-Copy-Multiple-Hop (MCP) consists of a type of limited flooding, where it has been established a maximum number of message copies and number of hops a message is allowed. Choosing a suitable number of copies and hops allows us to tune the efficiency of the algorithm [11].

Many protocols deal with dissemination in ad hoc networks. We have, for example, PROPHET, which uses knowledge of the history of encounters of a node and the clustering coefficient to route a message based on the probability that a node will lead to the destination [12]. Directed Diffusion [13] tackles dissemination by setting up communication gradients over which information is routed towards interested nodes. Protocols such as Trickle [14] use epidemic based routing to provide practical dissemination protocols.

Finally, we focus on BUBBLE, which uses both a measure of centrality, and the community of the destination as a rationale which routing is based on [4]. Centrality is defined as the number of times a node appears on a shortest (in terms of number of hops) path between two other nodes, normalised to the highest score. As an approximation of this notion of centrality, the node degree calculated over a suitable time interval is more practical.

## 4 Metric-based Routing and Community Structure

We now examine some important characteristics of the BUBBLE protocol, which we take as a benchmark. We use the *delivery ratio* as a measure of the performance of a dissemination process. We consider all the possible ordered pairs $(i, j)$ of nodes and run for each of them a dissemination protocol designed to send a message from node $i$ to node $j$. The delivery ratio at time $t$ is defined as the ratio between the number of delivered messages after time $t$ divided by the total number of messages (which equals the number of ordered pairs in the network). In Fig. 4(a) we plot the delivery ratio versus time in a 3-week period. The performance of the BUBBLE algorithm (with routing based on the pre-calculated centrality) is compared with routing without community knowledge. Quite remarkably, the behaviour of BUBBLE is very similar to an algorithm of routing based only on node degree (calculated over the previous 6 hour slot). This means that the community structure does not play a major role in improving the delivery ratio. It is interesting to note that applying the locally based degree routing to the algorithm, BUBBLE performs even better. Indeed, it emerges that the 6-hour degree, at least for this dataset, is the node property which best captures the importance of a node in data propagation.

In Fig. 4(b) we show the cost of the dissemination. Cost is defined as the total number of message hops from one node to another per message. We observe that centrality based BUBBLE is very efficient, with very low cost. However, centrality computation implies global knowledge of the network. Degree based BUBBLE is characterised by both a high delivery ratio and a significant cost, probably meaning that a portion of high degree nodes do not improve dissemination. The

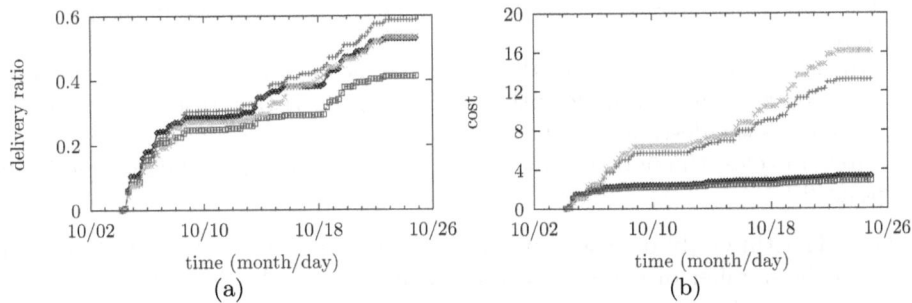

**Fig. 4.** Comparison of different implementations/modifications of the BUBBLE algorithm: delivery ratio (a) and cost (b) is plotted versus time over a 3-week period. Cost is defined as the total number of hops divided by the total number of messages sent between ordered pairs of nodes. Symbols refer to the same simulations in both figures. BUBBLE based on 3-day communities and centrality routing ($+$), is compared with the same algorithm based on 6-hour degree routing ($\times$). The performance of a simple protocol based on centrality routing regardless community awareness ($*$) leads to quite low delivery ratio, both with respect to 6-hour degree routing ($\square$) – basically BUBBLE without community awareness – and the original BUBBLE. Cost is higher for degree than centrality routing.

two plots show that centrality based routing leads to a large improvement in the cost, quite independently from the community structure. The cost advantage in introducing community routing into degree based BUBBLE is significant, but less important.

# 5   Self-management of Routing

## 5.1   Definition of the Self-management Algorithm

Now we want to develop a self-managing mechanism able to choose the best strategy for routing optimization. In order to do that, we have to provide a way to automatically detect the time scale which allows the best routing performance. In fact, we have seen that routing based on the 6-hour degree achieves good performance both in terms of delivery ratio and cost. However, the choice of the 6-hour time scale seems quite arbitrary and relies on external considerations. We want instead to be able to find an intrinsic mechanism to detect the most appropriate time scale. As human mobility follows periodic patterns, we expect this to affect also the related proximity network. We also expect that the degree calculated over a characteristic periodicity of network dynamics can achieve better results in the dissemination protocol, because it captures better the repetitive behaviour of a node.

Therefore, we can state a self-managing policy of locally adapting the routing rule to achieve good efficiency. So, each node acts according to the following algorithm:

1. During a preliminary time interval $\Delta t_0$, each node calculates the main periodicity $T$ in the number of contacts with other nodes.
2. The node calculates its degree over the time scale $T$ obtained in the previous step.
3. Whenever a message reaches the node, the message is forwarded as soon as the node encounters a node with a higher degree metric. If instead the node encounters the destination, the message is delivered and the algorithm ends.
4. After $n$ cycles, the node calculates again its periodicity $T$ and goes on from point 2.

This algorithm only depends on two external parameters: the initial interval $\Delta t_0$ and the number of cycles $n$ before calculating the dominant period again. $\Delta t_0$ can be based on some external considerations, including the requested maximum delivery time (e.g. a week may be a good choice for human activities). It is important to underline, though, that the characteristic period is re-calculated every $n$ cycles, so that the value can converge to an optimal period, hence the assigned value to $\Delta t_0$ is not particularly relevant, and could be set equal to the requested maximum delivery time. The number of cycles $n$ should be at least about 10, so that the period can be calculated on a time scale an order of magnitude larger than the previous period $T$. The period can be practically determined by calculating the highest peak of the Fourier transform of the number of connections over time. In this way, each node may calculate a different period, and therefore base its routing according to a different time scale degree, the one which best captures the periodic activities of the node.

## 5.2   Algorithm Evaluation

In order to evaluate the algorithm, we first have to investigate the importance of the degree criterion at different time scales. As a measure of efficiency, we calculate the ratio between delivery ratio and cost. This quantity can summarize the merit of a given protocol. In Fig. 5 and 6 we show the efficiency of routing based on node degree aggregated on different time scales for the Reality and the Cabspotting data sets. The plots show that the general behaviour is that higher efficiency corresponds to higher time scale. This can be explained by the fact that longer times allow to average over a larger number of events, and then to give a better estimate of the future importance of the considered node. However, Fig. 5 also shows that for the Reality data set this behaviour is not monotonic and there are time scales better than others. In particular, it appears that one day degree routing is more efficient than routing on a time scale of 2 or 3 days. 7 day routing obtains an even better performance than routing based on centrality (a property which implies global knowledge of network evolution). This behaviour is less important in the Cabspotting data set, but we can still notice that 1-day degree routing is sometimes more efficient than the 2-day degree routing.

In order to investigate the origin of this effect we look at the periodicities in node contacts. In fact, both the Reality and the Cabspotting datasets have periodicities in the hourly number of contacts. In Fig. 7 we show that by calculating

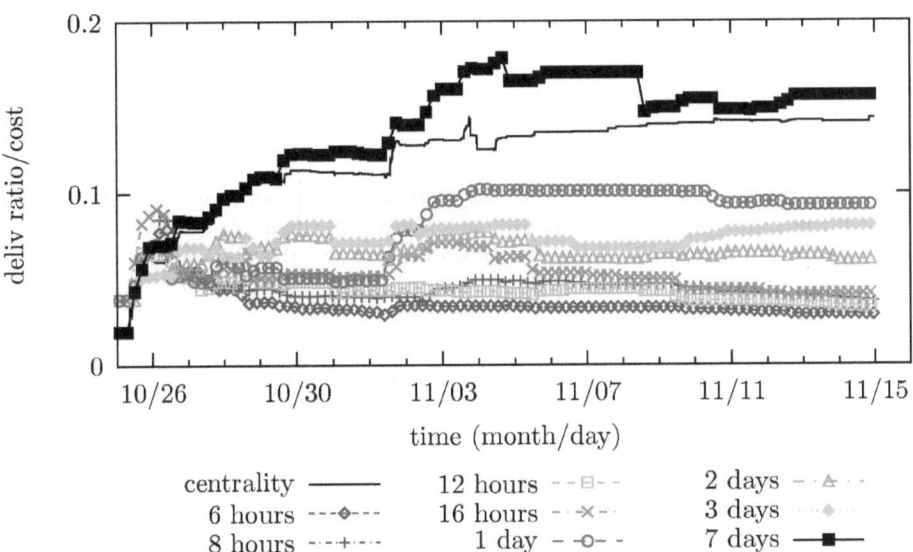

**Fig. 5.** Efficiency (defined as delivery ratio divided by cost) of routing based on the degree calculated over different time intervals for the Reality data set

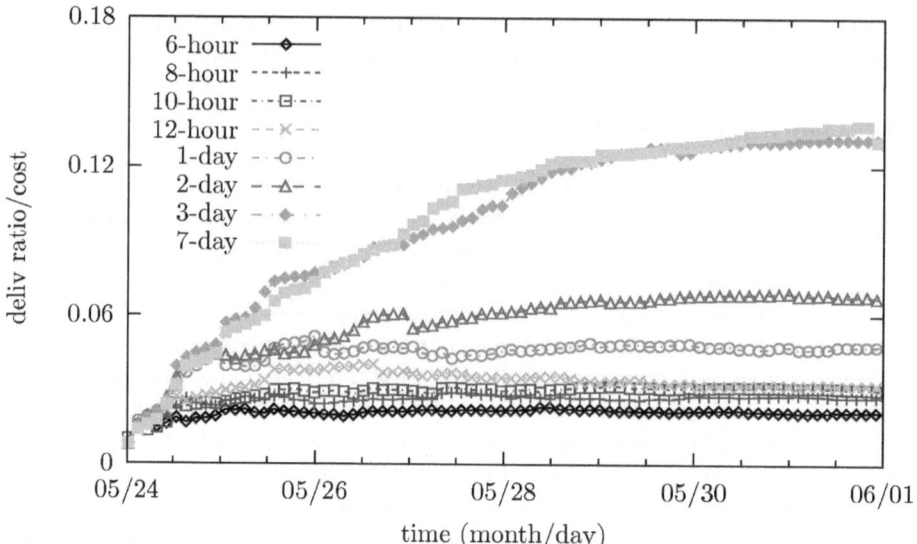

**Fig. 6.** Efficiency of routing based on the degree calculated over different time intervals for the the Cabspotting data set

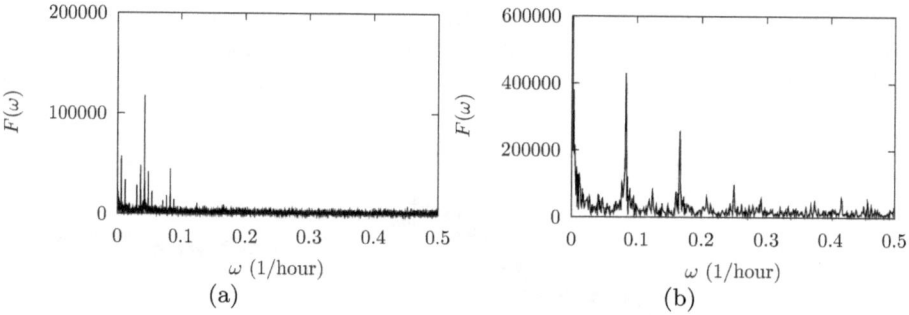

**Fig. 7.** Fourier transform of the time evolution of the number of connections, defined on time slots of 1 hour. The frequency $\omega$ has been normalized, so that a peak in $\bar{\omega}$ corresponds to a period $T = 1/\bar{\omega}$. The used data sets are Reality (a) and Cabspotting (b), respectively.

the Fourier transform of the number of node encounters per hour. The observed peaks are due to the periodicity of human activities, and in fact the largest ones occur at 6 hours, 1 and 7 days for the Reality data, and 12 hours and 1 day for the cabspotting data set. Thus, our algorithm detects from the highest peak of the Fourier transform that the most important periodicity in the Reality data set is 1 day, and routes messages based on the 1-day degree, which we have seen being a particularly efficient metric. Similarly, in the Cabspotting case, the algorithm chooses 12-hour degree routing. It can also be shown that most nodes detect the same main periodicity.

Therefore, our interpretation is that the presence of periodicities in the patterns of activity of the nodes has an effect in improving corresponding time scales in degree routing. The difference between the two data sets shown in Figures 5 and 6 can be explained by the fact that this periodicity effect is more important in networks which are sparse most of the time (as the Reality network), where the choice of a smart routing policy is more critical, than in networks where the connectivity is generally good most of the time, as in the Cabspotting data set.

As we have seen, the proposed algorithm generates a routing protocol which is highly efficient and does not rely on external assumptions. The only parameter to fix is the duration of the preliminary interval, but it will change after $n$ cycles if there is a better one in the system, or if the period of the node itself changes with time.

# 6  Conclusions

In this paper, we have investigated different approaches to data dissemination in dynamic human proximity networks. We have found that node degree is the best local property on which to base routing, and that there are time scales at which the protocol performs better. We have then formulated a self-management scheme where nodes automatically detect the best time scale and forward messages in an efficient way.

The significance of this approach is that it provides a mechanism by which to adapt data dissemination to the properties of the external processes affecting network dynamics, without having an explicit model of those dynamics embedded within the system. This makes the scheme purely topological and able to adapt autonomously to changing dynamics. Further work is needed to validate the approach against other kinds of dynamic networks (for example in environmental sensing), and to explore further local topological metrics that may be indicative of global properties useful for self-management.

# Acknowledgements

This work is supported by Science Foundation Ireland under grant 07/CE/I1147, and 03/CE2/I303-1, "Lero: the Irish Software Engineering Research Centre". The authors thank the anonymous reviewers for insightful comments.

# References

1. Fall, K.: A delay-tolerant network architecture for challenged internets. In: SIGCOMM 2003: Proceedings of the 2003 conference on Applications, technologies, architectures, and protocols for computer communications, pp. 27–34. ACM, New York (2003)
2. Zhao, W., Ammar, M., Zegura, E.: A message ferrying approach for data delivery in sparse mobile ad hoc networks. In: MobiHoc 2004: Proceedings of the 5th ACM international symposium on Mobile ad hoc networking and computing, pp. 187–198. ACM Press, New York (2004)

3. Pelusi, L., Passarella, A., Conti, M.: Opportunistic networking: data forwarding in disconnected mobile ad hoc networks. IEEE Communications Magazine 44(11), 134–141 (2006)
4. Hui, P., Crowcroft, J., Yoneki, E.: Bubble rap: social-based forwarding in delay tolerant networks. In: Proceedings of the 9th ACM International Symposium on Mobile Ad Hoc Networking and Computing (MobiHoc), pp. 241–250. ACM, New York (2008)
5. Eagle, N., Pentland, A.S.: CRAWDAD data set mit/reality (v. 2005-07-01) (July 2005), http://crawdad.cs.dartmouth.edu/mit/reality
6. Miklas, A., Gollu, K., Chan, K., Saroiu, S., Gummadi, K., de Lara, E.: Exploiting social interactions in mobile systems. In: Krumm, J., Abowd, G.D., Seneviratne, A., Strang, T. (eds.) UbiComp 2007. LNCS, vol. 4717, pp. 409–428. Springer, Heidelberg (2007)
7. Piorkowski, M., Sarafijanovic-Djukic, N., Grossglauser, M.: CRAWDAD data set epfl/mobility (v. 2009-02-24) (February 2009), http://crawdad.cs.dartmouth.edu/epfl/mobility
8. Piorkowski, M., Sarafijanovic-Djukic, N., Grossglauser, M.: A parsimonious model of mobile partitioned networks with clustering. In: COMSNETS 2009: Communication Systems and Networks and Workshops, pp. 1–10 (2009)
9. Freeman, L.C.: A set of measures of centrality based on betweenness. Sociometry 40(1), 35–41 (1977)
10. Onnela, J.P., Saramaki, J., Hyvonen, J., Szabo, G., Lazer, D., Kaski, K., Kertesz, J., Barabasi, A.L.: Structure and tie strengths in mobile communication networks. Proceedings of the National Academy of Sciences 104(18), 7332–7336 (2007)
11. Wang, Y., Jain, S., Martonosi, M., Fall, K.: Erasure-coding based routing for opportunistic networks. In: WDTN 2005: Proceeding of the, ACM SIGCOMM workshop on Delay-tolerant networking, pp. 229–236. ACM Press, New York (2005)
12. Lindgren, A., Doria, A., Schelén, O.: Probabilistic routing in intermittently connected networks. Service Assurance with Partial and Intermittent Resources, 239–254 (2004)
13. Intanagonwiwat, C., Govindan, R., Estrin, D.: Directed diffusion: a scalable and robust communication paradigm for sensor networks. In: MobiCom 2000: Proceedings of the 6th annual international conference on Mobile computing and networking, pp. 56–67. ACM, New York (2000)
14. Levis, P., Brewer, E., Culler, D., Gay, D., Madden, S., Patel, N., Polastre, J., Shenker, S., Szewczyk, R., Woo, A.: The emergence of a networking primitive in wireless sensor networks. Commun. ACM 51(7), 99–106 (2008)

# Revisiting P2P Content Sharing in Wireless Ad Hoc Networks*

Mohamed Karim Sbai and Chadi Barakat

EPI Planète, INRIA, France
{mksbai,cbarakat}@sophia.inria.fr

**Abstract.** Classical content sharing applications like BitTorrent are not designed to run over wireless networks. When adapting them to these constrained networks, two main problems arise. On one hand, exchanging content pieces with far nodes results in an important routing overhead. On the other hand, it is necessary to send some content pieces to far nodes to increase the diversity of information in the network, which fosters reciprocity and parallel exchanges. In this paper, we study both of these problems and propose a joint solution for them. Unlike uni-metric approaches, our solution considers relevant performance metrics together as throughput, sharing and routing overhead. We define a new neighbor selection strategy that balances sharing and diversification efforts and decides on the optimal neighboring scope of a node. We also consider the diversification incentives problem and evaluates the impact of nodes' mobility on the P2P strategy to be adopted. Through extensive simulations, we prove that our solution achieves both better download time and sharing ratio than uni-metric solutions.

## 1 Introduction

The proliferation of wireless devices (Laptops, PDAs, Smartphones, etc) motivates end users to connect to each other to form spontaneous communities. A wireless multi-hop network of devices, rendered possible by the use of ad hoc routing protocols, can be a good opportunity to share some contents (data, audio, video, etc) among the members of the same community without using any established infrastructure. As the resources of a wireless ad hoc network are scarce and shared among nodes, the application used for content sharing should not rely on any central service and should divide the replication effort fairly among the members of the community while reducing the overhead on the intermediate nodes serving as relays. Considering this, file sharing applications based on the peer-to-peer (P2P) paradigm are the best candidate solutions. First, in a few years, they have become the most popular applications on the Internet and users are familiar with their functionalities and features. Second, a P2P file sharing solution like BitTorrent [1] decentralizes the data transfer plane using the multi-sourcing concept and provides enough incentives to encourage fair

---

* This work was supported by the ITEA European project on experience sharing in mobile communities (ExpeShare).

T. Spyropoulos and K.A. Hummel (Eds.): IWSOS 2009, LNCS 5918, pp. 13–25, 2009.
© IFIP International Federation for Information Processing 2009

sharing. It is thus very beneficial to have the same principles applied in a wireless environment because nodes will tend to save capacity and energy. Multi-hop wireless communications consume resources in intermediate nodes and so there is a strong need for reducing the routing overhead.

Whereas efficient content localization in wireless ad hoc networks has attracted considerable research interest [4][5], the content replication problem is still in its first steps. BitTorrent [1] is the best known P2P content replication protocol that optimizes the data transfer plane. Previous studies focus on tuning BitTorrent algorithms to wireless networks to ameliorate a specific performance metric without considering all the metrics jointly. Some of them [2] aim to improve the global download time by reducing the routing overhead. In fact, the idea proposed is to make peers only concentrate on their nearby neighbors. We show in this paper that if this is done, the replication burden is unequally distributed among peers and that there is a poor transmission parallelism in the network. This is contradictory to the goals of BitTorrent and is not suitable for wireless ad hoc networks. In another previous work [3], we propose replicating the initial seed of the content at the edge of the network in order to increase the diversity and improve the parallelism. Although these policies recofd better download times and point to some new directions, they are limited to some specific cases that need to be generalized to clearly illustrate the relationship between content replication, user performance, fairness and overhead on the underlying network. On one side there is a need to diversify the content in the network to improve user perceived quality and enforce fairness, and on the other side, any diversification is costly because of the multi-hop routing. Optimal balance and how to achieve it are still not clear.

In this paper, we make an in-depth study of the routing overhead and content diversity problems in P2P applications run over wireless ad hoc networks. We observe the following dilemma: *How can the download time be reduced while maintaining sufficient parallel transmissions in the network and a fair distribution of replication load? How can fair sharing be boosted by diversifying pieces of the content without increasing the routing overhead?* Our objective is to come up with a joint solution for the routing overhead and content diversity problems. We mainly want to increase the sharing opportunities and have a minimum download time without overloading the network. In our ivestigation, we respect the natural tendency that peers, unless they have the entire content (called leechers in the former case and seeds in the latter case), have no incentives to participate in content diversification. Indeed, as in BitTorrent philosophy, leechers are only interested in sharing content with other peers who have new parts of the content to reciprocate. Only seeds are generous enough to participate in diversifying content for improving global performance.

We propose a new neighbor selection strategy that distinguishes between two main efforts of peer-to-peer file sharing application: the sharing effort and the diversification effort. First, we try to answer the following question: *What must be the importance of the sharing and diversification efforts?* We study in this work the best scope of sharing that minimizes the routing overhead. Our first finding is that the sharing effort must be made in a narrow area around each peer, otherwise this results in an important routing overhead. On the other side, the diversification effort aims to increase the entropy of information in the network to boost parallelism. That is why the diversification area should be taken wider than the sharing area. Clearly, diversifying content pieces in the network is very costly, thus this effort must be made less frequently than the sharing effort. In this paper, we study the impact of the diversification area around each seeding node and propose an efficient strategy for scheduling sharing and diversification connections of a peer. Another important question we address is the appropriate neighbor selection strategy while the diversification effort. By comparing different strategies, we prove that randomly choosing a peer in the diversification scope is the best approach. Our solution pays attention to balance the load equally among the different seeds in a diversification area. We also study the impact of node mobility on the diversity of pieces in the network and prove that the diversification effort must be slowed down in this case. The sharing area, however should be always limited to close physical neighbors, whatever the mobility pattern is.

Using our extension of the NS-2 network simulator [7] and realistic network realizations, we prove through extensive simulations that when using our neighbor selection strategy, BitTorrent achieves both better download time and better sharing ratio than its classical Internet version. It outperforms in all regards other solutions limiting the scope of the neighborhood without diversifying pieces. We can also achieve a better download time and better sharing ratio than when replicating the content at the edge of the network.

The remainder of this paper is organized as follows. Section 2 presents the background of our work and the methodology of our investigation. Section 3 studies the routing overhead problem and Section 4 mitigates the content diversity problem. Section 5 presents our solution in details. Section 7 summarizes our contribution and gives some future directions.

## 2    Background and Methodology

### 2.1    Background

BitTorrent [1] is a scalable and efficient P2P content replication protocol. Each peer shares some of its upload capacity with other peers in order to increase the global system capacity. Peers cooperating together to download a content form a sharing overlay called *Torrent*. To facilitate the replication of content in the network and to ensure multi-sourcing, each file is subdivided into a set of pieces. A peer who has all pieces of the file is called *a seed*. When the peer is downloading pieces, it is called *a leecher*. Among the members of the torrent, neighbors are those with whom a peer can open a TCP connection to exchange data and information. Only four simultaneous outgoing *active* TCP connections are allowed by the protocol. The corresponding neighbors are called effective neighbors. They are selected according to the *choking algorithm* of BitTorrent.

This algorithm is executed periodically and aims at identifying the best upload-ers. Once the choking period expires, a peer chooses to unchoke the 3 peers uploading to him at the highest rate. This strategy, called tit-for-tat, ensures reciprocity and enforces collaboration among peers. Now, to discover new up-load capacities, a peer randomly chooses a fourth peer to unchoke. All other neighbors are left choked. When unchoked, a peer selects a piece to download using a specific piece selection strategy. This strategy is called local rarest first. When selecting a piece, a peer chooses the piece with the least redundancy in its neighborhood. Rarest first is supposed to increase the diversity of pieces [6].

## 2.2   Scenario and Methodology

We start from the BitTorrent protocol and consider the interesting case where all nodes of a wireless ad hoc network are interested in sharing the same content. We want to understand the performance of BitTorrent in this challenging dense case before moving to more sparse torrents. Indeed, when the torrent is dense, the routing overhead on peers is at its maximum, since on one hand, the volume of exchanged data is large, and on the other hand any packet sent over multiple hops will steal bandwidth from all intermediate nodes, which are also peers. Note that a packet relayed by a node at the routing layer is not seen by the applications running in this node, in particular the BitTorrent application. We first consider that nodes are fixed and randomly distributed in the plane so that they form a big connected network. This means that content diversity cannot be obtained without sending data to far away nodes. Then, we extend the study in Section 6 to mobile scenarios and prove that if nodes are mobile, their movements help to increase the entropy of information in the network, hence making the scenario less challenging and more reciprocal.

In our investigation, we proceed with an experimental approach using the NS-2 network simulator [7]. To do this, we extend NS-2 by implementing a tunable BitTorrent-like module that allows content sharing in wireless ad hoc networks. With our module, the neighbor selection and piece selection strategies of the BitTorrent client can be changed and the resulting performance measured. In addition to the data transfer plane, our module implements a peer discovery mechanism on each peer. This mechanism emulates for the BitTorrent client the existence of a centralized tracker providing it with the list of torrent members. Since this work focuses on the data transfer plane, the optimization of the mem-bership management mechanism is out of the scope of this paper. Furthermore, our module profits from the existing NS-2 modules to ensure wireless communi-cation and multi-hop routing of packets. The wireless ad hoc network that we are simulating consists of 50 nodes randomly distributed in a $500m \times 80m$ square area. When nodes are taken fixed, we discard all realizations where the topology is not connected. Nodes connect to each other using the 802.11 MAC Layer with the RTS/CTS-Data/ACK mechanism enabled. The data rate is set to 11 Mb/s and the wireless range to 50m. The ad hoc routing service is ensured thanks to the DSDV proactive protocol [8]. At the beginning of each simulation, a ran-dom node is chosen as the seed of the content and the other nodes are leechers.

The content is a 100 Mbytes data file that is subdivided into 1000 pieces. All peers start downloading the file at the same time (a flash crowd scenario). The BitTorrent choking period is set to 40s.

## 3   The Routing Overhead Problem

The first question we address regards the optimal neighbor selection strategy for P2P content sharing in wireless ad hoc networks. In our investigation, we start from the classical version of BitTorrent and we vary the scope of the sharing area of the peers. This scope represents the maximum number of hops between peers authorized to exchange pieces. The version deployed in the Internet, called later classical version for short, corresponds to a sharing scope equal to the maximum number of hops in the network.

Figure 1 plots the average download time of the content per peer as a function of the number of hops of that peer to the initial seed for different values of the sharing scope. On one hand, the download time increases with the number of hops to the seed for all values of the sharing scope. This is mainly due to the fact that the achievable throughput of TCP decreases considerably with the number of hops in a wireless ad hoc network. Peers far from the seed get most, if not all, of the pieces of the content in multi-hop via other peers. On the other hand, the classical version of BitTorrent has the largest download time because of the routing overhead and the degradation of TCP performance in a multi-hop environment. For the best performances, the sharing scope needs to be limited to a small value, e.g. one or two. Figure 2 consolidates this observation by showing the average download time over all nodes as a function of the scope of sharing. This figure shows an amelioration in the download time when this sharing scope is reduced. In fact, the routing overhead is minimal when the scope is small, otherwise pieces of content are forwarded by intermediate nodes at the routing layer without profiting from them at the BitTorrent layer, which incurs a lot of overhead on these nodes. Moreover, additional transmissions are needed later to send the same pieces to these intermediate nodes. Another important factor is that the throughput of TCP is better over short distances and in this case, more pieces of data can be sent during the choking time slot. Figures show a

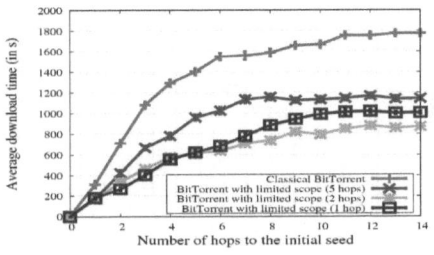
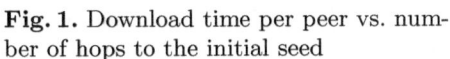
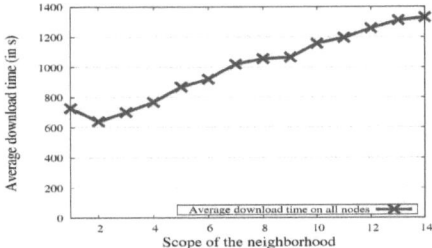

**Fig. 1.** Download time per peer vs. number of hops to the initial seed

**Fig. 2.** Download time per peer vs. scope of sharing

slightly better gain for a scope of two hops, compared to a scope of one hop, even though there is some routing overhead in the former case. In fact, the two-hop case allows more neighbors, which leads to a better sharing and a better forwarding of pieces. At the same time, the routing overhead is still small, so that we can notice an overall gain. Unfortunately, for scopes larger than two hops, the routing overhead becomes big enough to counteract any gain from having more neighbors.

## 4   The Piece Diversity Problem

In the previous section, we showed that decreasing the scope of sharing ameliorates the download time. In this paragraph, we try to answer the following questions: *By limiting the scope of sharing, are we limiting the sharing opportunities between nodes? Is there fair sharing among them in this case? Is there a piece diversity problem?* When sharing scope is very limited, the pieces of the content most likely propagate from the initial seed to the farthest nodes in a unique direction via other nodes in between. Far nodes do not have original pieces to provide to upstream nodes that are closer to the initial seed. That is why nodes fail to reciprocate data with each other, and hence, the load of sharing is not equally divided among them. In general, the farther the nodes are from the initial seed, the fewer packets they will have to send. Moreover, there will be no diversity of pieces in the network. The same pieces will propagate from one neighborhood to another, which cannot result in a fair exchange.

To strengthen this claim, we plot in Figure 3 the sharing ratio as a function of the number of hops to the initial seed for the same simulations used in the previous section. The sharing ratio between a couple $(i, j)$ of peers is defined as: $R_{ij} = \frac{min(D_{ij}, D_{ji})}{max(D_{ij}, D_{ji})}$, where $D_{ij}$ is the amount of data that node i has downloaded from node j during the torrent lifetime. This ratio measures the magnitude of the reciprocity between two nodes. A value nearing null means a one-way propagation of data. The ideal fair sharing case is obtained when the sharing ratio is equal to 1. From Figure 3, one can observe the following:

– The sharing ratio increases slightly with the number of hops to the initial seed. In fact, far peers can reciprocate some data with other far peers because they

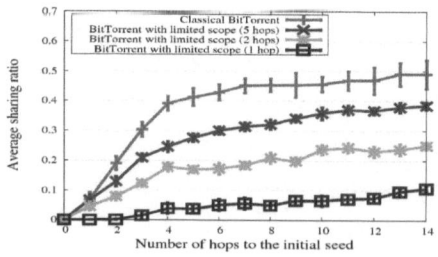
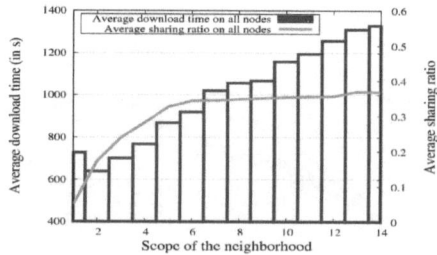

**Fig. 3.** Sharing ratio as a function of the number of hops to the initial seed

**Fig. 4.** Sharing ratio as a function of the scope of sharing

can both receive different pieces on different paths. The farther the nodes are from the initial seed, the more of these different paths exist. But the diversity of pieces is still low, as the dissemination is done in a unique direction.
– When the scope of sharing is decreased, the sharing ratio goes down dramatically. The cases where the scope is set to one or two hops yield the lowest sharing ratios, which can be explained by the fact that pieces propagate like a wave from the initial seed to the farthest nodes. The resources of the network are not fully used since nodes wait for pieces to arrive to their neighborhood and rarely have original pieces to reciprocate with their neighbors. The P2P file sharing application then behaves like a simple piece relaying protocol that ignores the parallel transmission capabilities of the network and the distribution of the load among peers.
– For large scopes of sharing, for instance the classical BitTorrent case, the sharing ratio is still lower than $\frac{1}{2}$ because the routing overhead is big and very few pieces can be downloaded during a choking slot, mainly when the number of hops between neighbors is high.

So, there is a trade-off between diminishing the routing overhead and increasing piece diversity in the network. Figure 4 plots both the sharing ratio and download time averaged over all nodes vs. the scope of sharing. From sharing perspectives, we can see that it is useless to increase the scope beyond 5 in our settings, because of the degradation of TCP throughput with the number of hops. From download time perspectives, the best solution is to limit the scope to a very low value, such as two hops to limit the routing overhead. *Can we do better?* In the following section, we prove that this is possible by decoupling the sharing effort from the diversification effort. Mainly, we introduce a new algorithm to increase the diversity of pieces in the network at a limited routing cost, and in parallel we limit the sharing scope to two hops in order not to overload the network. In this way, we can do better than the simple small scope case by having a better diversity of pieces, and hence more parallel transmissions and better reciprocity. At the same time, we are better than the simple large scope case in terms of sharing, because we can diversify pieces in the network to improve sharing, without suffering from the routing overhead problem.

## 5   Solving the Dilemma

In the two previous sections, we presented two problems related to content sharing in wireless ad hoc networks. We now face the following dilemma. On one hand, decreasing the scope of sharing ameliorates the download time but leads to very weak parallelism in the network due to the lack of piece diversity. On the other hand, increasing the scope of sharing increases the diversity of pieces in the network, but at the cost of more routing overhead and worse download time. In this paragraph, we present our solution to this dilemma. Our objective is to come up with a joint solution for the routing overhead and piece diversity problems. In designing our new neighbor selection strategy, we took into consideration the following points:

– Data transfers between distant peers suffer from very poor performances in wireless ad hoc networks. Hence, a leecher has no incentives to send pieces of the content to far nodes, as they will not be able to serve him back with a high throughput. Moreover, leechers that are far from the initial seed have less original pieces to reciprocate them with their nearer leechers. Considering this, we decided that in our neighbor selection strategy, only seeds send pieces to far peers. Indeed, a seed is a volunteer peer that serves others without expecting any return from them. The leechers have more incentives to concentrate on peers located in their close neighborhood, provided that there are original pieces to share with them.

– If all seeds send pieces to far nodes at the same time, the routing overhead will be large again and performance will decrease. In our solution, we subdivide the piece diversification effort among seeds both in space and time.

– If a seed cannot send a complete piece to the selected peer during the choking slot, the gain in diversity will be null since the smallest unit that a peer can share with others is the piece. In our solution, we limit the scope of diversification of seeds to the number of hops allowing the transfer of a complete piece. Pieces are spread in other parts of the network by other peers becoming seeds and deciding to stay in the torrent.


## 5.1   The Neighbor Selection Strategy

– In the leecher state, peers concentrate on their nearby neighborhood. The scope of sharing is fixed to 2 hops as it is proved to be the best regarding the routing overhead and transfer performance. A leecher maintains 4 simultaneous active outgoing connections. The first 3 connections are dedicated to best uploaders among peers in the sharing scope and the fourth connection consists in an optimistic unchoke allowing to discover new upload capacities and the bootstrap of the sharing. The fourth peer is chosen randomly among leechers within the sharing scope. The selection is done at the end of each choking time slot. Except the limitation of the scope to two hops, this is globally the classical BitTorrent algorithm for leechers.

– In the seed state, peers dedicate their first 3 connections to serve leechers within their sharing scopes (set to 2). These are the connections dedicated to injecting the content in the network by starting from the small sharing area. The fourth connection of a seed is mainly dedicated to diversify pieces in an area wider than the sharing area. This area is called the diversification area of the seed and contains all peers not belonging to its sharing area and that are located at a distance lower than the diversification scope. The scope of diversification is determined by observing the range of piece transmissions (see paragraph 5.4). Paragraph 5.3 studies the optimal way to choose a leecher in the diversification area and paragraph 5.2 shows how the fourth connection is used when there is more than one seed within the diversification area.

## 5.2   Dividing the Diversification Effort among Seeds

Sending pieces to far nodes engenders a big routing overhead. Hence, the diversification effort must be divided between all seeds, both in space and in time. In our solution, each seed is responsible for its own diversification area and does not have to serve the whole network. Moreover, when there are many seeds within the diversification scope of each other (for example when other peers finish the download and decide to stay), our solution reduces the routing overhead of the fourth connection of each of them, which is dedicated to diversification, by the number of seeds in its diversification area. This is done as follows. The seed pauses for a number of slots equal to the number of seeds in its diversification area between every two diversification time slots. During the pause, the seed can serve leechers in its sharing area. This scheduling is repeated periodically and follows the evolution of the number of seeds. In this way, the total diversification overhead is kept constant as there are more and more seeds in the network.

## 5.3   Optimal Diversification-Neighbor Selection Strategy

In this paragraph, we look for the best strategy used by seeds to select leechers in their diversification areas at time slots. The goal is to maximize the sharing ratio while minimizing the average download time. Let's note the sharing scope of a seed as $S_s$ and its diversification scope as $S_d$. Searching the optimal strategy, we define a parametric general probability distribution to tune the leecher selection and we study, through simulations and by varying the parameter of the distribution, the impact of the different strategies on the torrent performances. We model the probability to select a peer located at $h$ hops from a seed in its diversification scope ( $S_s < h \leq S_d$) as follows:

$$p(h) = \frac{h^\alpha}{\displaystyle\sum_{l=S_s+1}^{S_d} N_l l^\alpha} \qquad (1)$$

where $N_l$ is the number of peers located at $l$ hops and $\alpha$ is a parameter of the probability distribution. The sum of this probability function over all peers in

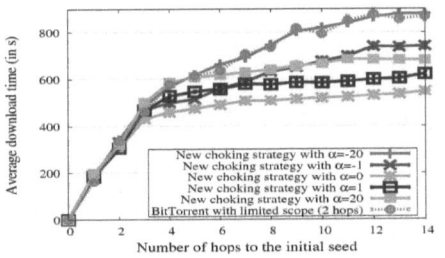

**Fig. 5.** Download time as a function of the number of hops to the initial seed

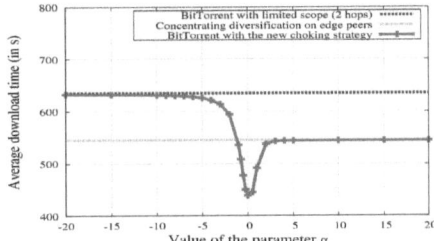

**Fig. 6.** Download time as a function of $\alpha$

a diversification area is clearly equal to 1. By setting $\alpha$ to 0, we can obtain the uniform probability distribution where peers are selected with the same probability independently of their location. For large positive values of $\alpha$, the probability to select the farthest peers becomes close to 1, and that to select peers near to the seed almost null. For large negative values of $\alpha$, the opposite occurs; the seed diversifies pieces over peers close to it. This parameter $\alpha$ then covers a large set of strategies, and its optimal value should point us to the optimal leecher selection strategy to use for diversification purposes. Next, we seek this optimal value using extensive simulations. Figure 5 plots the average download time as function of the number of hops to the initial seed for different values of the parameter $\alpha$. Figure 6 draws the average download time over all peers as a function of the parameter $\alpha$. For large negative values of $\alpha$, the download time is maximal and tends to the one obtained without diversifying the pieces (scope 2 in Figure 4). For large positive values of $\alpha$, one can obtain a better performance since there is the introduction of some diversity of pieces in the network but the concentration is only on leechers located at the edge of the diversification area. This is below the optimal because of the routing overhead and an inefficient spatial distribution of pieces. Our main observation is that a value of $\alpha$ equal to 0 gives the best performance. Indeed, for this optimal value, seeds distribute pieces uniformly in the network and then boosts fair sharing among peers and transmission parallelism while having a reasonable average routing overhead. Figure 7 plots the sharing ratio as a function of the number of hops to the initial seed for different values of the parameter $\alpha$. Figure 8 presents both the average sharing ratio and the average download time over all nodes as function of the parameter $\alpha$. These figures prove that the best choice of $\alpha$ is 0 as it also results in the best sharing ratio. Indeed, large negative values of $\alpha$ means no diversification of pieces and then the lowest sharing ratio is recorded, whereas for large positive values, one can obtain a better sharing ratio but the load is not equally divided among leechers. For the uniform probability case, the sharing opportunities are the best because original pieces are distributed over all peers and sharing areas and the load is not concentrated on any local neighborhood.

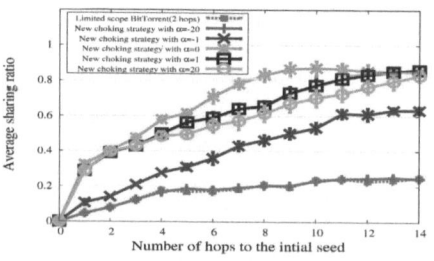

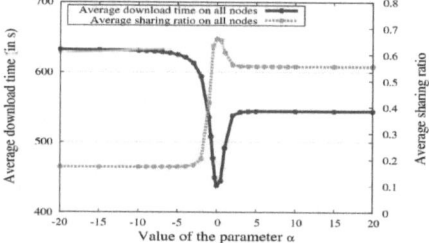

**Fig. 7.** Sharing ratio as a function of the number of hops to the initial seed

**Fig. 8.** Sharing ratio and download time as a function of $\alpha$

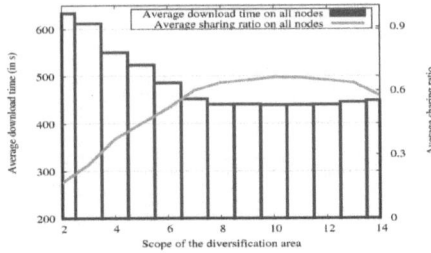

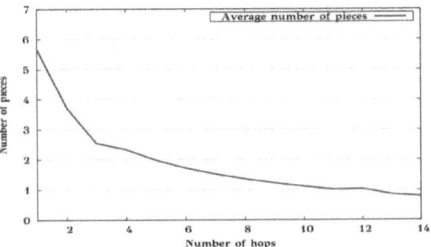

**Fig. 9.** Download time as a function of the diversification scope

**Fig. 10.** Average number of pieces sent during a choking slot

### 5.4   Choice of the Diversification Scope

The results shown up to now have been obtained for a diversification scope set to 10. In this paragraph, we study the impact of this scope both on download time and sharing ratio. Figure 9 plots the average download time and the average sharing ratio over all nodes vs. the scope of diversification. It shows that for small values of this scope, there is not enough diversity introduced into the network. Hence, the sharing ratio decreases considerably and the download time worsens. For large diversification scopes, again the sharing ratio and download time worsen for the simple reason that TCP becomes unable to send entire reusable pieces far away from the seed. It is clear that in our settings a diversification scope around 10 hops away from each seed leads to best performances both from download time and sharing ratio perspectives. This should be the largest scope where entire content piece could be sent. Figure 10 confirms this claim by showing the average number of pieces a seed can send as a function of the number of hops to the leecher. As a result, the diversification scope must be fixed so that it does not exceed the range of pieces. To support this in practice, the seed can adapt its diversification scope up or down by measuring the number of pieces that it can send during a choking slot to leechers located at the edge of the diversification area. We leave this online adaptation for a future research.

## 6   Mobility Helps Diversification

In the previous sections, we supposed that the network is fixed. In the case of mobility of nodes, two main factors must be considered. On one hand, the mobility naturally increases the diversity of pieces, since the neighborhood of a peer is changing while moving. In this case, one can hope that there will be enough sharing opportunities and hence there will be no need for sending pieces to far away nodes to boost diversity. On the other hand, as long paths suffer from bad performances in mobile ad hoc networks, it would be better to have a short range of sharing and diversification. Preliminary simulation results (see [9]) show indeed that the mobility has a beast and a beauty. The beast is that it increases packet losses over long multi-hop paths, thus it makes it

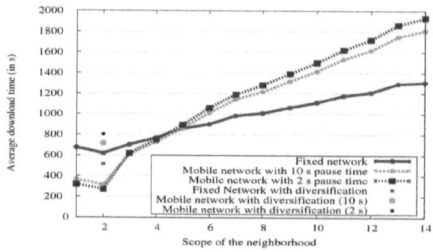
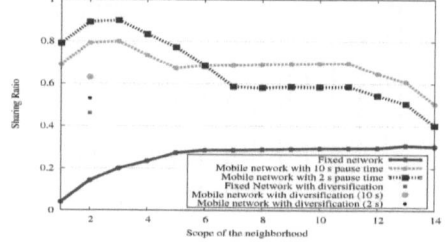

**Fig. 11.** Average download time Vs. Neighborhood scope

**Fig. 12.** Average sharing ratio Vs. Neighborhood scope

almost inefficient to send pieces to faraway peers. The beauty is that the mobility efficiently improves piece diversity since peers are continuously exchanging new pieces with the new peers they meet while moving across the network. We give a flavor of the results in Figures 11 and 12, where we compare respectively the download time and the average sharing ratio for both the fixed and the mobile scenarios when the diversification effort is activated or disabled. In these simulations, the RandomWay point mobility model is used. The speed of nodes is set to 2m/s and two pause times have been considered (2s and 10s). The curves in the figures correspond to no diversification and different sharing scopes. The dotes correspond to a fixed sharing scope of two hops and a diversification area of 10 hops. We can observe how the best performance is obtained for mobile networks limiting the sharing scope and disabling the diversification. Once nodes become fixed, the diversification by seeds becomes mandatory to replace the one inherent to mobility.

## 7   Conclusions and Perspectives

In this paper, we study the routing overhead and piece diversity dilemma and propose an efficient neighbor selection strategy that minimizes the download time while maximizing sharing opportunities. Our proposed choke/unchoke algorithm is practical and does not make any assumptions on the cooperation of leechers. We prove through extensive NS-2 simulations that when using our new neighbor selection strategy, the download time of BitTorrent is 30% of the one obtained with the classical version of BitTorrent and 65% of the one obtained with the version without diversification of pieces and a sharing scope equal to 2. As for the sharing, the upload is now better distributed among peers and more reciprocal. However, the solution and results presented in this paper relae to a dense challenging scenario. We will study sparse scenarios in our future work.

## References

1. BitTorrent protocol, http://wiki.theory.org/BitTorrentSpecification
2. Michiardi, P., Urvoy-Keller, G.: Performance analysis of cooperative content distribution for wireless ad hoc networks. In: WONS 2007, Obergurgl (2007)

3. Sbai, M.K., Barakat, C., Choi, J., Al Hamra, A., Turletti, T.: Adapting BitTorrent to wireless ad hoc networks. In: Ad-Hoc Now 2008, Sophia Antipolis, France (2008)
4. Klemm, A., Lindermann, C., Waldhorst, O.: A special-purpose peer-to-peer file sharing system for mobile ad hoc networks. In: VTC 2003 (2003)
5. Das, S.M., Pucha, H., Hu, Y.C.: Ekta: an efficient peer-to-peer substrate for distributed applications in mobile ad hoc networks. TR-ECE-04-04, Purdue University (2004)
6. Legout, A., Urvoy-Keller, G., Michiardi, P.: Rarest First and Choke Algorithms Are Enough. In: IMC 2006, Rio de Janeiro, Brazil (2006)
7. NS: The Network Simulator, http://www.isi.edu/nsnam/ns/
8. Perkins, C.E., Bhagwat, P.: Highly Dynamic Destination-Sequenced Distance-Vector routing (DSDV) for mobile computers. In: SIGCOMM 1994, London, UK (1994)
9. Salhi, E., Sbai, M.K., Barakat, C.: Neighborhood selection in mobile P2P networks. In: Algotel conference, Carry-Le-Rouet, France (2009)

# Event Detection Using Unmanned Aerial Vehicles: Ordered versus Self-organized Search

Evşen Yanmaz

Institute of Networked and Embedded Systems, Mobile Systems Group,
University of Klagenfurt, Austria
evsen.yanmaz@uni-klu.ac.at

**Abstract.** Event coverage problem in wireless sensor networks has drawn the interest of several researchers. While most of the previous work has been on static or ground mobile sensor networks, airborne sensor networks have also found its way into several civil and military applications such as environmental monitoring or battlefield assistance. In this work, we study the mobility pattern of an Unmanned Aerial Vehicle (UAV) network and explore the benefits of ordered and self-organized random mobility in terms of event detection performance, when the event is stationary and event duration is finite. Specifically, we compare the performance of a UAV network flying in parallel formation to a simple, distributed, locally-interactive coverage-based mobility model as well as legacy mobility models such as random walk and random direction. We study the event detection probability of the UAV network with both perfect and imperfect sensing capabilities. Our results show that when the timing constraints are highly stringent or when the UAV sensors have a high miss probability, flying in formation cannot achieve a high detection probability and a self-organized distributed mobility model is more effective.

## 1   Introduction

Wireless sensor networks have found various application venues in environmental monitoring, health monitoring, target tracking in hostile situations, etc. [1]-[2]. Especially, in the case of monitoring physically inaccessible or dangerous areas for humans to enter, such as wildfire tracking, glacier or volcano monitoring, liveliness detection in emergencies or hazardous material tracking, use of wireless sensor networks is expected to increase tremendously. Due to the inaccessibility of the geographical areas in these applications, the sensor nodes either need to be dropped forming a random static network or mobile ground or airborne robots equipped with sensors are needed to be deployed. Moreover, if the event (target) to be detected by the sensor network is of time-critical nature, the coverage of the network should be sufficiently high to be able to respond to the detected event in a timely manner; such as wildfire monitoring or liveliness detection under rubble in case of an earthquake, where the emergency personnel work against the clock. In such cases, using a mobile sensor network would be highly beneficial both in terms of event detection and utilization of the available system resources [3].

In this paper, we study the event detection performance of an unmanned aerial vehicle (UAV) network for different mobility models. Specifically, we aim to determine

T. Spyropoulos and K.A. Hummel (Eds.): IWSOS 2009, LNCS 5918, pp. 26–36, 2009.

when it would be beneficial for the UAV network to fly in a centralized, deterministic, parallel-formation. To this end, first, we derive the probability of detection of the UAV network flying in formation, when a finite-duration stationary event is assumed to occur at a random location in the geographical area to be monitored. We assume that sensing capabilities of the sensors on-board are imperfect (i.e., there is a non-zero probability that a UAV will miss an event in its sensing coverage). We compare the performance of parallel-formation with a distributed, coverage-based, cooperative mobility model that operates in a self-organizing manner and uses only local topology information to detect events without prior knowledge of the physical topology, by reducing the over-lapping covered areas. While determining the mobility path, no assumption is made on the application the sensor network is deployed for. Numerical studies are conducted to test the performance of the parallel-formation and coverage-based mobility models as well as legacy mobility models such as random walk and random direction. It is shown that while a centralized, deterministic, parallel-formation mobility model might be easier to implement, it does not always provide acceptable performance in terms of event detection probability. More specifically, when the event needs to be detected within a strict time interval or when the sensing capabilities on the UAVs are highly imperfect (unreliable) a more intelligent, adaptive, and preferably self-organizing mobility model is required to achieve a high probability of event detection. In such cases, our results show that the simple, distributed, random mobility models investigated in this paper can overcome the limitations of the parallel-formation model.

The remainder of the paper is organized as follows. In Section 2 background on mobility models and coverage problem in wireless sensor networks is summarized. A brief event detection analysis for the parallel-formation is provided in Section 3. Results are given in Section 4 and the paper is concluded in Section 5.

## 2   Background

### 2.1   Coverage in Wireless Sensor Networks

Coverage problems in wireless sensor networks are of great importance and have been investigated by several researchers. In static wireless sensor networks, in general, coverage problem is treated as a node-activation and scheduling problem [7]-[9]. More specifically, algorithms are proposed to determine which sensor nodes should be active such that an optimization criterion is satisfied. The criterion can for instance be minimizing the coverage time, achieving a certain event detection probability, or covering each point in the area by at least $k$ sensors, etc. In addition, there are also studies that take into account not only the event (or network) coverage, but the connectivity of the wireless sensor network as well [7]. While deciding which sensor nodes should be active at a given point in time, coverage and connectivity requirements are met.

Recently, mobile sensor networks have been under investigation and it has been shown that mobility, while complicating the design of higher layer algorithms, also can improve the network, for instance, in terms of capacity, coverage, etc. [10]-[11]. Optimum mobility patterns for certain applications are proposed, such as mobile target tracking, chemical detection, etc. using both ground and aerial vehicles. Mobile robots with swarming capability operate cooperatively and aim to achieve a global goal have also been considered [12]-[16].

## 2.2 Mobility Models

There are several mobility models that take into account the dependence on the mobility pattern of other nodes in the network [4], social relationships of the mobile nodes [5], or topographical information [6], etc. In this paper, the following mobility models are considered:

• *Random Walk*: A mobile node picks a random speed and direction from pre-defined uniform distributions either at fixed time intervals or after a certain fixed distance is traveled. The current speed and direction of the mobile node do not depend on the previous speeds and directions.

• *Random Direction*: A random direction drawn from a uniform distribution is assigned to a mobile node and the mobile node travels in that direction till it reaches the boundary of the simulation area. Once it reaches the boundary, it pauses there for a fixed amount of time, then moves along the new randomly selected direction. In this paper, for fair comparison, we assume that the pause time is zero.

• *Parallel-formation*: Mobile nodes sweep the geographical area from border to border following a direction parallel to the boundary line.

## 2.3 Coverage-Based Mobility

In this newly proposed mobility model, the objective is maximizing coverage in a given time duration. To this end, we aim to minimize the overlap between the coverage areas of different mobile nodes and as shown in Fig. 1, we model forces between mobile nodes that cause them to *repel* each other. The magnitude of the force that each node applies to others is inversely proportional to the distance between the nodes, i.e., the closer the nodes get the stronger they *push* each other. We also assume that the mobile node knows its current direction and a force with a magnitude inversely proportional to the node's transmission range (i.e., $r$) is applied to it in the direction of movement to avoid retracing the already covered areas by the mobile node. At the time of direction change, each mobile node computes the *resultant* force vector acting on them by themselves and their neighbors (i.e., the mobile nodes within their transmission range) and move in the direction of the resultant vector. The forces on mobile node 1 at the time of decision are illustrated in Fig. 1, where mobile node 1 is moving toward right in the previous step.

Observe from Fig. 1 that the resultant force on node $i$, $R_i = \sum_j F_{ji}$, where $F_{ii} \parallel V_i$ with $|F_{ii}| = \frac{1}{r}$ and $|F_{ji}| = \frac{1}{d_{ji}}$ when $j \neq i$, where $V_i$ is the velocity vector of mobile node $i$, $r$ is the transmission range of each mobile node, and $d_{ij}$ is the distance between nodes $i$ and $j$. The direction of $F_{ji}$ is parallel to the line drawn from node $j$ to node $i$. Mobile node $i$ will move in the direction of $R_i$ with a speed chosen from the range $[0, V_m]$ for a fixed time duration (i.e., a step length). Same algorithm is run for all the mobile nodes and the directions are updated accordingly. If, at the time of direction change, a mobile mode does not have any neighbors, the direction is not changed. Note that the step length is a design parameter and depends on the system parameters such as $N_m$ and $r$ among others.

Since the mobile airborne network is highly dynamic and the neighborhood of the mobile nodes constantly change, the mobile nodes need to decide based on only local

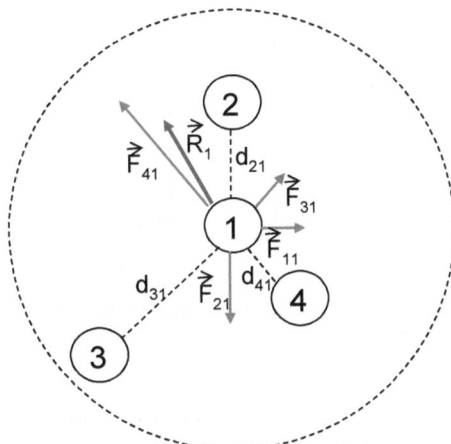

**Fig. 1.** Illustration of forces on mobile node 1, where the dashed circle is the transmission range of the node and mobile node 1 is moving toward right

interaction with other nodes and adapt to the changes in the environment in a distributed, self-organizing manner. Note that while the ultimate goal of maximizing coverage is not incorporated in the mobility model, as our results show a better coverage (i.e., event detection capability) emerges. Further work is necessary to design an analytical model that studies the performance of this mobility model and possibly to provide improvements.

## 3  Event Detection Analysis of Parallel-Formation

The system under investigation is a wireless sensor network that consists of airborne mobile nodes with the same transmission range. We assume that the UAVs fly at the same altitude and the directions are considered in a two-dimensional plane. The system parameters are summarized in Table 1.

**Table 1.** System Parameters

| Parameter | Definition |
|---|---|
| $N_m$ | Number of mobile nodes |
| $r$ | Transmission range |
| $a$ | Square simulation area length |
| $P_d$ | Total event (target) detection probability |
| $P_m$ | Event (target) miss probability for each UAV |
| $t_d$ | Event (target) duration |

In this section, we provide the event coverage (detection) probability by the mobile nodes flying in parallel formation (See Fig. 2 for a simple illustration) within a given time duration $t_d$. Since complete coordination between the mobile nodes is required, the mobile network needs to be connected at all times.

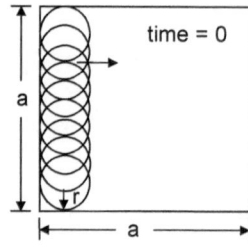 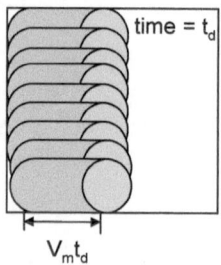

**Fig. 2.** Coverage illustration of mobile sensor nodes during time duration $t_d$, when the speed of the mobile nodes is $V_m$ for parallel-formation

Note that the event detection probability can be determined from the percentage area that is covered over time $t_d$. Assume that the transmission range of the mobile nodes is $r$ and their coverage area is of disc shape, i.e., area covered by each node at a given time is $\pi r^2$ and the nodes fly with a constant speed $V_m$. To better evaluate the limitations of the parallel-formation mobility, we assume that the number of mobile nodes, $N_m$ is such that one side of the square coverage area is fully-covered when the mobile nodes are aligned. Depending on the application of interest some overlap between the coverages of the UAVs might be assumed. In this work, we assume that the UAVs are placed such that the overlap is minimized given that airborne network is connected; i.e., the displacement between the UAVs is equal to $r$.

First, let's assume that the UAV sensors are accurate (i.e., probability of miss ($P_m$) = 0). The total covered area by the UAVs in time $t_d$, where $t_d \leq \frac{a}{V_m}$ is the sum of areas of the overlapping truncated cylinders shown in Fig. 2 and is given by:

$$A_c = aV_m t_d + A_{oc} \tag{1}$$

where $A_{oc}$ is the overlapping coverage between the discs and is given by:

$$A_{oc} = \begin{cases} \frac{2ar-(4-\pi)r^2}{2} - \left(\frac{a-2r}{r}\right)\frac{12r^2-2\pi r^2-3\sqrt{3}r^2}{12}, \\ \qquad\qquad 0 \leq t_d \leq \frac{a-r}{V_m} \\ a(r-x) - 2A_1 - \left(\frac{a-2r}{r}\right)A_2, \\ \qquad \frac{a-r}{V_m} \leq t_d \leq \frac{a-r+0.5r\tan(\pi/12)}{V_m} \\ a(r-x) - 2A_1, \quad \frac{a-r+0.5r\tan(\pi/12)}{V_m} \leq t_d \leq \frac{a}{V_m} \end{cases} \tag{2}$$

where

$$A_1 = (r-x)(r - \sqrt{r^2 - (r-x)^2}/2) \\ - 0.5r^2(\frac{\pi}{2} - \arccos(\frac{r-x}{r})) \tag{3}$$

$$A_2 = (r-x)(r - \sqrt{r^2 - (r-x)^2}) - 0.25r^2\sqrt{3} \\ - r^2(\frac{\pi}{6} - \arccos(\frac{r-x}{r})) \tag{4}$$

$$x = V_m t_d - (a-r) \tag{5}$$

Then, the probability of detection ($P_d$) for a single event, when $P_m = 0$ is given by:

$$P_d = \begin{cases} \frac{A_c}{a^2}, & 0 \leq t_d \leq \frac{a}{V_m} \\ 1, & t_d \geq \frac{a}{V_m} \end{cases} \quad (6)$$

The probability of detection for a known number of $N_{tar}$ targeted events, $P_{d_m}$, can be calculated by substituting $P_d$ from Eq. (6) into the following:

$$P_{d_m} = (P_d)^{N_{tar}} = \begin{cases} \left(\frac{A_c}{a^2}\right)^{N_{tar}}, & 0 \leq t_d \leq \frac{a}{V_m} \\ 1, & t_d \geq \frac{a}{V_m} \end{cases} \quad (7)$$

If on the other hand $P_m \neq 0$, i.e., there is a non-zero probability that an event may not be detected even if the whole area of interest is covered by the UAVs, then $P_d$ at a given time $t_d$ can be calculated to be:

$$P_d = (1 - P_m^{N_c}) \sum_{i=1}^{n-1} P_m^{N_c(n-1-i)}$$
$$+ P_m^{N_c(n-1)}(1 - P_m^{N_c})A_c/a^2 \quad (8)$$

where $n = \lceil \frac{V_m t_d}{a} \rceil$ is the total number of passes within $t_d$, $N_c$ is the number of checks a UAV does to detect an event, $A_c$ is the covered area within time $t_d - (n-1)a/V_m$ (i.e., the covered area during the n-th pass) and can be calculated using Eq. (1). Note that $N_c$ is a design parameter and depends on how often sensing is desired. As $N_c$ is increased, the effect of miss probability will be reduced, as expected.

## 4    Results and Discussion

In this section, performance comparison of several mobility models in terms of event detection probability is provided via Monte Carlo simulations, where each data point is computed over 2000 different runs. It is assumed that the range of the mobile nodes, $r$, is 500m. The simulation area is square-shaped with a length of 4000m. For parallel-formation, these values correspond to 8 mobile nodes and the mobile nodes are initially aligned along one side of the observation area. For the other mobility models under study in this paper, initially, mobile nodes are randomly distributed in the simulation area. When a mobile node approaches the boundary of the simulation area, a random direction toward the simulation area is assigned for random walk and coverage-based mobility models. The speed of the mobile nodes is assumed to be 5 m/s. The directions of the mobile nodes are updated every 50 m. Similarly, the step size for sensing is also assumed to be 50 m. We assume that a single event occurs at a random location within the simulation area and lasts for a duration of $t_d$ seconds.

### 4.1    Probability of Detection with Perfect Sensing

In this section, we study the probability of detection performance of several mobility models when the sensors on-board the UAVs are accurate with $P_m = 0$. Fig. 3 shows the

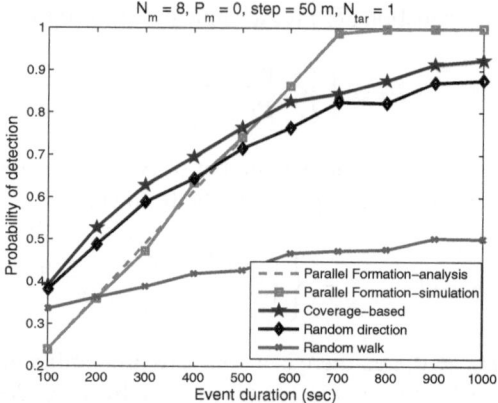

**Fig. 3.** Probability of detection versus event duration, when $N_m = 8$, $P_m = 0$, and number of targeted events is 1

probability of detection versus $t_d$ for random walk, random direction, coverage-based and parallel-formation models when $N_m = 8$. Observe that the detection performance of the mobility models under investigation strictly depends on the timing constraints of the application. If $t_d \geq a/V_m = 800$ sec, parallel-formation outperforms the rest of the models since the whole geographic area can be swept by then. However, if the timing constraints are strict and do not allow a network flying in formation to cover the area, then a more efficient mobility model is required. The simple coverage-based mobility model that inherently reduces the overlapped coverage areas between different mobile nodes can perform better than the rest, only using local information. More sophisticated mobility models can be designed that take into account the history of flight of the UAVs; however, this is beyond the scope of this paper.

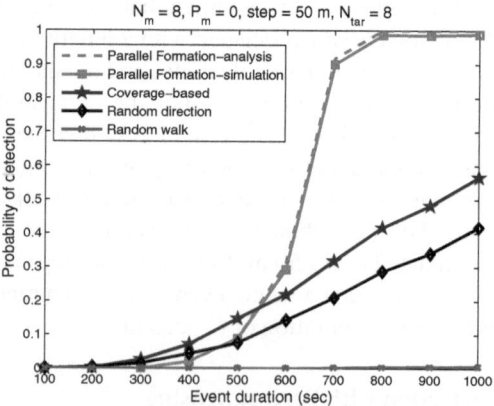

**Fig. 4.** Probability of detection versus event duration, when $N_m = 8$, $P_m = 0$, and number of targeted events is 8

As an illustration, we also studied the case with multiple stationary events. Fig. 4 shows the probability of detection versus event duration when the number of targeted events is 8. Note that while for short time values the probability of detection performance of all models suffer, the trends are still the same as the single event scenario. The random mobility models fail to detect all events as fast as parallel-formation. Also, observe from Figures 3 and 4 that the analytical and simulation results for parallel-formation are in excellent agreement verifying the findings, where the analytical results are obtained using Eq.'s (6) and (7), respectively.

## 4.2 Probability of Detection with Imperfect Sensing

Next, we study the case with imperfect sensing capabilities. We investigate the impact of event duration as well as the probability of miss of the sensors on the detection

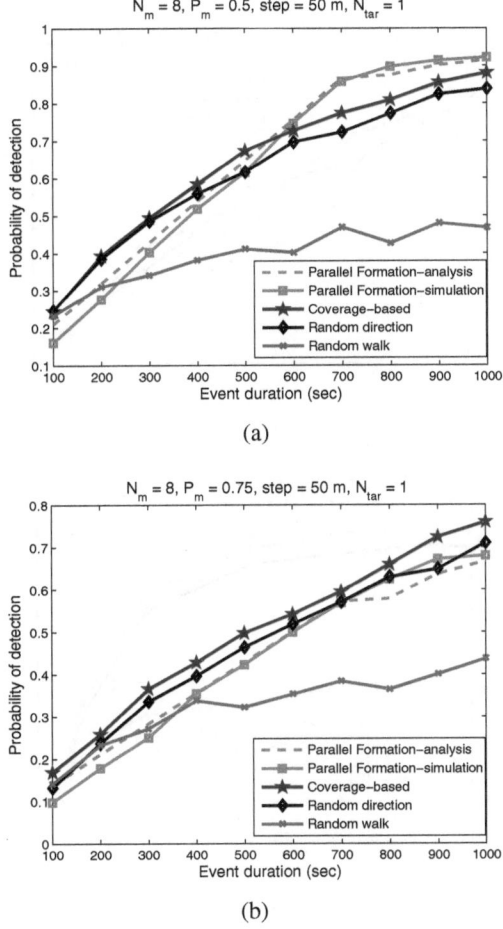

(a)

(b)

**Fig. 5.** Probability of detection versus event duration, when $N_m = 8$ and (a) $P_m = 0.5$ and (b) $P_m = 0.75$

performance of the chosen deterministic and random mobility models. Simulation and analytical results are provided, where the analytical results are obtained using Eq. (8).

Fig. 5 shows the probability of detection versus event duration when $P_m = \{0.5, 0.75\}$. In both cases, the probability of detection decreases with respect to the case with $P_m = 0$, as expected. Observe that parallel-formation significantly suffers from the imperfections of the sensing capabilities and although the whole observation area can be fully-swept, probability of detection stays at 0.9 and 0.6 for $P_m = 0.5$ and $P_m = 0.75$, respectively. While for $P_m = 0.5$, parallel-formation can still perform better than the other models for certain time durations, when $P_m$ is increased 0.75, coverage-based mobility model performs consistently better than parallel-formation.

To better illustrate the impact of $P_m$ on the performance, we studied the detection performance when $t_d = 100, 1000$ sec for several $P_m$ values. Results are shown in Fig. 6. Observe that when $t_d = 100$ sec, all random models outperform parallel-formation,

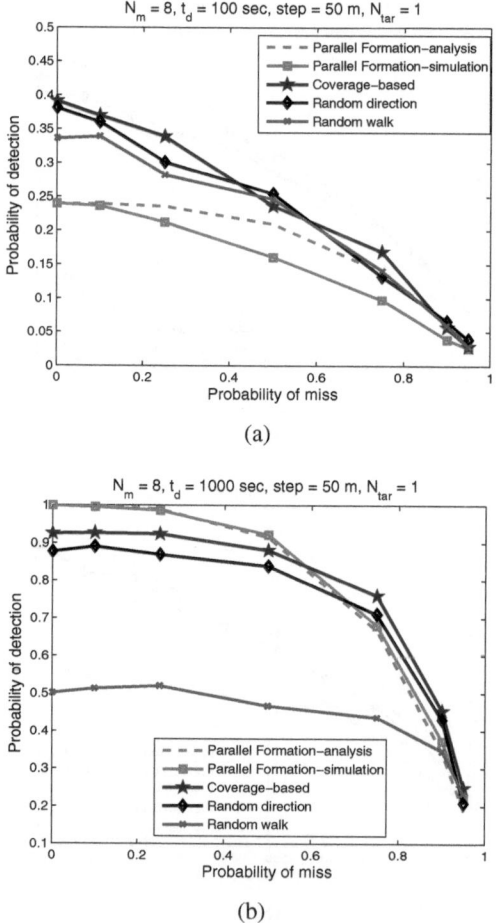

**Fig. 6.** Probability of detection versus probability of miss, when $N_m = 8$ and (a) $t_d = 100$ sec and (b) $t_d = 1000$ sec

for all $P_m$ values under investigation. On the other hand, when the timing constraints are loosened parallel-formation, coverage-based and random direction models all perform similarly. Random models exceed the performance of parallel-formation when the sensing capabilities are highly imperfect.

Finally, while not analyzed in this paper, a drawback of the centralized scheme is the requirement to be fully-connected at all times. In the configuration under study in this paper if a node in the middle of the formation breaks down the network itself becomes disconnected and a percentage of the area cannot be covered anymore, unless the remaining nodes regroup into a new formation. On the other hand, intuitively, the distributed mobility models studied in the paper are expected to be more robust to node failures, since system-wide connectivity is not required and the nodes communicate with each other only when they are within each other's coverage (for the coverage-based mobility) or do not communicate at all (random walk, random direction). The impact of *malfunctioned* nodes on the detection performance needs to be further investigated.

## 5 Conclusions

In this work, event detection performance of an airborne UAV sensor network that employs deterministic and random mobility models is investigated. Specifically, the limitations of a UAV network flying in parallel-formation is explored and its performance is compared with some legacy mobility models as well as a cooperative, coverage-based mobility model that uses local topology information. The results show that if timing-constraints are highly-stringent or the sensing capabilities on-board the UAVs are highly imperfect, parallel-formation might not be sufficient to detect the events in the observation area. While for such cases random mobility models are shown to improve the detection performance, further study is necessary to design an optimum mobility pattern that minimizes the event detection time and/or maximizes the probability of detection.

## Acknowledgements

The author would like to thank Prof. Christian Bettstetter of University of Klagenfurt and the attendees of Lakeside Research Days 2009 for valuable discussions and feedback on the paper. This work was supported by Lakeside Labs GmbH, Klagenfurt, Austria and funding from the European Regional Development Fund and the Carinthian Economic Promotion Fund (KWF) under grant 20214/17095/24772.

## References

1. Akyildiz, I.F., Su, W., Sankarasubramaniam, Y., Cayirci, E.: A Survey on Sensor Networks. IEEE Comm. Mag. 40(8), 102–114 (2002)
2. Sohrabi, K., Gao, J., Ailawadhi, V., Pottie, G.J.: Protocols for Self-organization of a Wireless sensor network. IEEE Pers. Comm. 7(5), 16–27 (2000)
3. Quaritsch, M., et al.: Collaborative Microdrones: Applications and Research Challenge. In: Proc. Intl. Conf. Autonomic Computing and Comm. Sys., September 2008, p. 7 (2008)

4. Camp, T., Boleng, J., Davies, V.: A Survey of Mobility Models for Ad Hoc Network Research. Proc. Wireless Commun. and Mobile Comp (WCMC): Special issue on Mobile Ad Hoc Networking: Research, Trends and Applications 2(5), 483–502 (2002)
5. Musolesi, M., Mascolo, C.: A Community Based Mobility Model for Ad Hoc Network Research. In: Proc. ACM/SIGMOBILE Intl. Workshop on Multi-hop Ad Hoc Networks: from theory to reality (REALMAN 2006), May 2006, pp. 31–38 (2006)
6. Jardosh, A., Belding-Royer, E.M., Almeroth, K.C., Suri, S.: Towards Realistic Mobility Models for Mobile Ad hoc Networks. In: Proc. Intl. Conf. Mobile Computing and Networking (MobiCom 2003), September 2003, pp. 217–229 (2003)
7. Wang, X., Xing, G., Zhang, Y., Lu, C., Pless, R., Gill, C.: Integrated Coverage and Connectivity Configuration in Wireless Sensor Networks. In: Proc. Int's. Conf. Emb. Net. Sens. Sys (SenSys 2003), pp. 28–39 (2003)
8. Megerian, S., Koushanfar, F., Potkonjak, M., Srivastava, M.B.: Worst and Best-Case Coverage in Sensor Networks. IEEE Trans. Mob. Comp. 4(1), 84–92 (2005)
9. Liu, B., Towsley, D.: A Study of the Coverage of Large-scale Sensor Networks. In: Proc. IEEE Int'l. Conf. Mob. Ad hoc Sens. Sys (IEEE MASS 2004), October 2004, pp. 475–483 (2004)
10. Grossglauser, M., Tse, D.N.C.: Mobility Increases the Capacity of Ad Hoc Wireless Networks. IEEE/ACM Trans. Networking 10(4), 477–486 (2002)
11. Liu, B., Brass, P., Dousse, O., Nain, P., Towsley, D.: Mobility Improves Coverage of Sensor Networks. In: Proc. ACM Int'l. Symp. Mob. Ad hoc Net. Comp (ACM MobiHoc 2005), pp. 300–308 (2005)
12. Poduri, S., Sukhatme, G.S.: Constrained Coverage for Mobile Sensor Networks. In: IEEE Int'l. Conf. Robotics and Autom (IEEE ICRA 2004), April 2004, pp. 165–172 (2004)
13. Vincent, P., Rubin, I.: A Framework and Analysis for Cooperative Search Using UAV Swarms. In: Proc. ACM Symp. on Applied Computing (2004)
14. Jin, Y., Liao, Y., Minai, A.A., Polycarpou, M.: Balancing Search and Target Response in Cooperative Unmanned Aerial Vehicle (UAV) Teams. IEEE Trans. On Systems, Man, and Cybernetics 36(3) (2006)
15. Kovacina, M.A., Palmer, D., Yang, G., Vaidyanathan, R.: Multi-Agent Algorithms for Chemical Cloud Detection and Mapping Using Unmanned Air Vehicles. In: IEE/RSJ Intl. Conf. on Intelligent Robots and Systems (2002)
16. Palat, R.C., Annamalai, A., Reed, J.H.: Cooperative Relaying for Ad hoc Ground Networks Using Swarms. In: Proc. IEEE Milit. Comm. Conf (MILCOM 2005), October 2005, vol. 3, pp. 1588–1594 (2005)

# A Survey of Models and Design Methods for Self-organizing Networked Systems

Wilfried Elmenreich[1], Raissa D'Souza[2], Christian Bettstetter[1],
and Hermann de Meer[3]

[1] University of Klagenfurt and Lakeside Labs, Austria
`firstname.lastname@uni-klu.ac.at`
[2] University of California at Davis and Santa Fe Institute, USA
`raissa@cse.ucdavis.edu`
[3] University of Passau, Germany
`demeer@uni-passau.de`

**Abstract.** Self-organization, whereby through purely local interactions, global order and structure emerge, is studied broadly across many fields of science, economics, and engineering. We review several existing methods and modeling techniques used to understand self-organization in a general manner. We then present implementation concepts and case studies for applying these principles for the design and deployment of robust self-organizing networked systems.

## 1 Introduction

The term self-organization was introduced by Ashby in the 1940s [1]. He referred to pattern formation occurring by the cooperative behavior of individual entities. Such formation can be described by entities achieving their structure without any external influence. The ideas behind self-organization were subsequently further studied and developed by a number of cyberneticians (e.g., von Foerster, Pask, Beer, and Wiener), chemists (e.g., Prigogine), and physicists (e.g., Haken). In the 1980s and 90s, the field was further fertilized by some applied mathematics disciplines, such as non-linear dynamics, chaos theory, and complex networks. Although there is still no commonly accepted *exact* definition of a self-organizing system that holds across several scientific disciplines, we refer to it as a set of entities that achieves a global system behavior via local interactions between its entities without centralized control [2].

Phenomena of self-organization can be found in many disciplines. A well-known example from nature is the flocking behavior in a school of fish. It is likely that there is no "leader fish," but each individual fish has knowledge only about its neighbors [3]. Despite (or probably because of) this localized and decentralized operation, the difficult task of forming and maintaining a scalable and highly adaptive shoal can be achieved.

With the increasing complexity in technology and its applications (more and more entities are interconnected to form a networked system), the notion of self-organization has also become an interesting paradigm for solving technical

T. Spyropoulos and K.A. Hummel (Eds.): IWSOS 2009, LNCS 5918, pp. 37–49, 2009.

problems (see, e.g., references in [4]). In order to design a self-organizing technical system, a set of modeling approaches and design methods are needed. The
goal of this paper is to give a survey of building blocks that can be used for
modeling and design of self-organization, with a special focus on information
and communications and traffic systems.

The paper is structured as follows: In Section 2, we take a closer look at specific modeling issues for self-organizing systems. Section 3 presents an overview
on different established models for self-organizing systems. Section 4 reviews different methods for designing a self-organizing system with an intended technical
effect. Section 5 give references to some case studies with respect to the concepts
described in Sections 3 and 4. Finally, Section 6 concludes the paper.

## 2    Two Perspectives on Self-Organizing Systems

A self-organizing system (SOS) consists of a set of entities that interact with each
other locally to obtain a global system behavior. The global system behavior
arises out of many simple interactions. It is an emergent property, i.e., it cannot
be explained by summation of the local interactions. A self-organizing system can
thus be viewed from two perspectives: the microscopic perspective describes the
entities and their behavior; the macroscopic perspective describes the (emergent)
behavior of the overall system.

An example for these two perspectives and their interrelation can be found
in the "predator-prey system" as described by Lotka and Volterra [5]. This
biologically-inspired model contains two types of animals, typically named "rabbits" (the prey) and "foxes" (the predators). It obeys simple microscopic rules:

- Rabbits reproduce at a given birth rate $\epsilon_1$.
- Rabbits die if caught by foxes, with rate for such a death event given by $\gamma_1$.
- Foxes reproduce at a rate $\gamma_2$ proportional to the number of caught rabbits.
- Foxes die according to a given death rate $\epsilon_2$.

These rules can be described in an aggregated form by the differential equations

$$\frac{dN_1}{dt} = N_1(\epsilon_1 - \gamma_1 N_2), \qquad \frac{dN_2}{dt} = -N_2(\epsilon_2 - \gamma_2 N_1),$$

where $N_1$ and $N_2$ is the number of rabbits or foxes, respectively. The rate of
interaction between both populations is a function of the product $N_1 N_2$.

When examining the behavior of this system over time, we observe emergent
phenomena on the macroscopic level, such as periodic oscillations of $N_1$ and
$N_2$. Unlike many oscillations in nature which can be described by trigonometric
functions, the solution of the Lotka-Volterra equations do not have a simple
expression in terms of trigonometric functions (see Fig. 1).

To describe the overall system behavior from the macroscopic perspective,
Lotka and Volterra found three laws:

1. The population sizes of rabbits and foxes oscillate with a particular phase
   offset. Period lengths depend on the initial population and model parameters.

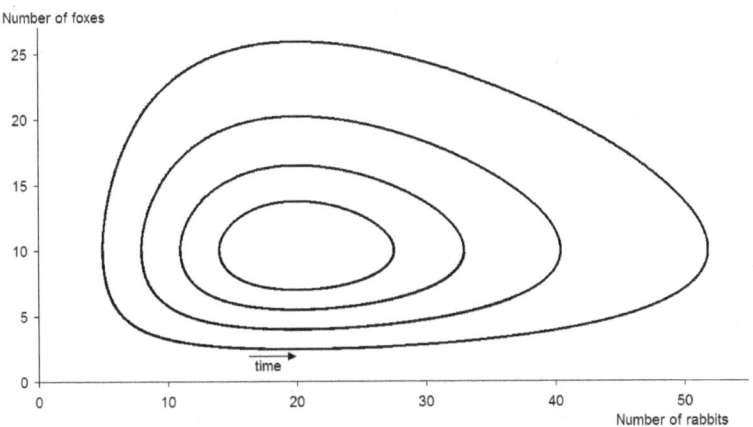

**Fig. 1.** Simulation with 4 different initial conditions of Lotka-Volterra system with parameters $\epsilon_1 = 1, \gamma_1 = 0.1, \epsilon_2 = 1, \gamma_2 = 0.05$

2. The averages of the two population sizes converge to a value that is determined by the model parameters and is independent of initial conditions.
3. If both populations are decimated proportionally to their size, the rabbit population will recover fast and will exceed their previous population.

These laws describe properties of the overall system, which cannot be immediately seen from the four microscopic rules.

The Lotka-Volterra model enables us to derive both perspectives from the differential equations. For more complex (more realistic) systems, however, the direct transformation between microscopic and macroscopic perspectives might not be possible. In this case, the macroscopic model could be built by (statistical) analysis of the overall behavior. This requires appropriate statistical tools to analyze emergent structures and behavior [6]. Such a model, although being potentially very useful in some cases, is unlikely to cover all possible aspects of the system behavior. An example is the global economy — although tremendous research efforts are put into understanding both its microscopic and macroscopic rules, the system is far from being exactly predictable.

To gain a deep understanding of a self-organizing system or to design a new system, it is beneficial to consider both the microscopic and macroscopic perspective. A microscopic perspective confers the advantage of providing an exact model, which directly supports implementation of the entities. However, it provides no mechanism for measuring or understanding emergent properties. A macroscopic perspective, in contrast, covers emergent behavior and a goal-oriented view. It is less exact, as it treats the system as a "black box." This obscures underlying dynamics and makes it difficult to verify the applicability of the model to a given system.

# 3    Models for Self-Organizing Systems

Before building and deploying an engineered SOS, it can be beneficial to develop a model of the system which can be analyzed mathematically or explored through computer simulation. This is especially important given that SOS typically display emergent behaviors which, by definition, can not be anticipated in advance. Models provide a virtual laboratory for exploring system design and function and give insights into the types of emergent behaviors that might be expected and the anticipated workings of real systems.

## 3.1    Differential Equations

Perhaps the first examples of modeling SOS reported in the literature are reaction-diffusion equations which generate patterns similar to those observed in Rayleigh-Bénard convection, viscous fingering, spiral waves in chemical reactions and a range of additional structures such as stripes and tilings [7,8]. Many of these patterns are also observed in biological systems, with differential equations providing models of organism growth and differentiation [7,9]. Ref [8] provides a comprehensive review of pattern formation as modeled by reaction-diffusion equations. Most recently, systems of differential equations have been used to model the emergence of synchronization in systems of coupled oscillators ranging from fireflies, to neurons, to Josephson junctions [8,10,11]. The Lotka-Volterra model discussed above is another prominent example of this approach.

Despite the broad range of applicability, this approach is thus far limited to describing pattern formation, oscillations and other simple collective phenomena. It does not allow much flexibility for diversity of components and assumes a uniform spatial interaction, neglecting more realistic types of connectivity in real-world systems which are often much better described by networks than uniform spatial fields.

## 3.2    Cellular Automata

Cellular automata models (CAs) are a simulation method for studying SOS which assume the world is a discrete grid (a lattice) and each site on the grid can be found in one of a set of possible discrete states. CAs have been successfully used to model a range of real physical systems from predator-prey systems, to chemical spiral waves, to hydrodynamics [12,13,14,15].

A highly desirable aspect of CAs, absent from differential equations, is that they allow for locality. In other words, sites on the grid interact directly only with neighboring sites, thus, consistent with physical law, there is no "action at a distance" and all signals must propagate along a path of connected neighbors. (Yet, similar to differential equations, this requirement of uniform spatial connectivity does not permit irregular networked patterns of interaction.) Another highly desirable quality of CAs is that they readily exhibit interesting behaviors such as forming a range of patterns similar to those observed with differential equations and also far more complex dynamics such as moving patterns ("gliders") in Conway's Game of Life [16]. Many CAs are computationally universal,

i. e., in theory CAs can be used to implement any computation. A fundamental requirement necessary for many (if not most) CAs to exhibit interesting behaviors is synchronous updating (all grid sites update simultaneously). Though not always realistic, as discussed in Sec. 5.1, there are simple algorithms which do allow collections of objects to synchronize, so it may be possible to engineer a system to fit the CA paradigm.

## 3.3   Agent-Based Models

Agent-based models are another modeling approach and a natural starting point if components in a system can vary dramatically from one another and display a range of behaviors and strategies (especially decision making). In contrast, CAs assume every site in the grid obeys the same rules. Furthermore, agent based models do not need to assume an underlying topology, specifying or restricting which components can interact; agents can come together and interact with one another in complex and dynamic ways. Such approaches are clearly necessary if, as in economics, we wish to model interactions amongst humans with complex decision making abilities. Agent based approaches can also be quite useful even if the agents are simple, such as capturing swarm behaviors [3].

Agent based approaches are extremely flexible, modeling the interactions of a collection of autonomous agents. Thus, unlike CAs and differential equations, they can model phenomena in social systems and agent based approaches lend themselves readily to incorporating game theoretic policies [17]. Many software platforms and techniques exist for building agent based models and Refs. [18] and [19] provide useful reviews.

A major drawback to agent based modeling is the lack of rigor. Due to the complexity of the model specifications (behaviors of agents, patterns of interaction) it is difficult to assess the robustness of observed phenomena to changes in the specifications and the accuracy by which the model describes real systems.

# 4   Design Methods for Self-Organizing Systems

The design of self-organizing systems differs from typical engineering approaches in the way that the system is rather built bottom-up than top down. At an early stage of development, it is necessary to tinker with the interactions between the individual system entities. In contrast, traditional systems are typically built starting with the overall system service and then approach the micro level only after several more and more fine-grained system models. Therefore, most standard design approaches do not fit well for the design of self-organizing systems. In the following, we discuss several design approaches that have been used or could be used for this purpose.

## 4.1   Analytical Approach

If the chosen model is abstract and simple enough, the settings for the desired global properties could be derived by an analytical solution. For example, if a system $S$ fed with a configuration $C$ gives the emergent behavior $B$, the task is to find the inverted function, i.e.,

$$S(C) = B \quad \Rightarrow \quad C = S^{-1}(B)$$

Unfortunately, even for moderately complex systems this is usually not feasible or would require a high effort for solving a mathematical problem which might already be represented by an less realistic abstraction from the actual problem. E. g., for the Lotka-Volterra example there exists is no complete analytical solution for the differential equation system, i.e., the equations have to be reduced or solved numerically.

An analytical approach may, however, help to discover certain aspects of the system. This might be achieved e.g. by assuming certain conditions or parameters. Thus it can help in the initial system design phase to predict certain aspects and in the final phase by verifying system aspects.

### 4.2    Applying a Reference Design

There exist many examples of self-organization in different domains, such as biology, physics, mathematics, economics. In order to find a working approach for achieving a particular behavior, a reference design from one of these disciplines can provide a major step towards a successful solution. There are two main paradigms for adopting a reference design: top-down and bottom-up.

In the top-down approach, a technical problem is tackled by looking for examples solving an equivalent problem. The found solution and its principles are then analyzed and re-built in a technical application. Examples of top-down approaches are (a) the design of aeroplane wings by observing the gliding flight of birds and (b) the design of turbulence-reducing winglets by analyzing the wingtips of birds [20].

In the bottom-up (or indirect) approach, the working principle of the system is first abstracted from its natural context. This step is done in a basic research effort that is not yet targeted at the specific application. Afterward, the results are used in particular technical applications. Thus, the indirect approach could also be called a "literature-inspired approach". Examples include (a) the concept of artificial neural networks and (b) the concept of ant foraging behavior being applied to mesh network packet routing [21].

### 4.3    Trial and Error

Another approach is to explore the effect of different interaction rules at the microscopic level on the global system behavior using a trial-and-error method.

The simplest trial-and-error method would be a Monte-Carlo method, where random configurations are created and tested until the global system shows the intended behavior. Due to the typically high-dimensional search space, however, randomized trial-and-error approaches are very unlikely to succeed within an acceptable time frame.

Alternatively, the trials can be used to learn about the causality of particular configurations and the global system behavior. Thus, after a reasonable number of test configurations, the tester might be able to apply his/her understanding of the emergent processes to find a local rule set with a desirable configuration.

An auxiliary concept to understand the causality between local interactions and global system behavior is introduced by Gershenson [22] as the the notion of *friction*. Friction is a property of interaction between two entities as well as a property of the overall system. This latter friction is to be minimized. By identifying and analyzing points of friction, an engineer can change the local rules towards better system performance. However, this is not straightforward: in several cases a higher friction for a particular entity is beneficial for the overall system.

Additionally, emergent behavior is often counterintuitive to what is expected by most people. Resnick [23] describes a simple simulation of a self-organizing slime mold. Several experts were asked to predict the influence of a specific parameter change on the system. The answer was binary, i. e., there was a 50% random change of guessing the correct answer. Nevertheless, a significant majority of people, including experts on complex and self-organizing systems, guessed the wrong answer.

### 4.4   Evolutionary Algorithms

Conventional search algorithms can be applied to search for an optimal or sufficiently good set of local rules. However, the search space is typically too large for an exhaustive search. For these cases, evolutionary algorithms and heuristic search algorithms can be a choice. Examples include genetic algorithms, simulated annealing, swarm-based optimization, and the Sintflut algorithm.

Using evolutionary algorithms requires a "testbed" that allows extensive and safe testing at low cost. Usually, such a testbed consists of a simulation of the target system with a model of the environment and the system itself. However, a simulation always implements an abstraction of the real environment, so after the experiments, a real-world validation is required to create trust in the derived solution.

The most prominent example for evolutionary algorithms are genetic algorithms. A genetic algorithm starts with an initial population of candidate solutions for a multidimensional optimization problem. It wishes to quickly find a near-optimal solution. At each generation, the candidates are randomly mutated or combined. The candidates with the best "fitness" establish the population of the next generation. An example where a genetic algorithm is used to design a self-organizing technical system is given in [24]. It was used to find the interaction behavior for a distributed robot soccer team. The behavior was modeled as an artificial neural network to support an implementation of mutation and combination.

### 4.5   Markov Models and Finite State Machines

Auer, Wüchner and De Meer propose a method to derive local interaction rules by learning from a reference solution [6]. The reference solution can be any algorithm that performs well for the problem. For example, the reference solution might be built as an omniscient system in a simulation. In many cases, it might not be possible to use this solution for a real application because the perfect information cannot be provided to the algorithm or the algorithm might be too complex to be implemented with reasonable response times. However, the omniscient algorithm can be used as an example for teaching its behavior to distributed entities that use only local information and interactions.

The behavior of the reference agent is analyzed using Markovian analysis and then rebuilt in a Finite State Machine (FSM). Thus, the state machine mimics the statistical behavior of the reference agent. The approach relies thus on the possibility that a suitable reference solution is available and that the behavior can be successfully used by an agent with local perception.

In [6], the application of this method is shown by designing an agent for the game theoretic problem "repeated prisoner's dilemma" [25]. In the design process, an agent having perfect knowledge (including the opponent's decision) is created. Then the behavior of this "perfect" agent is analyzed using Causal State Splitting Reconstruction [26], i. e., a method for building recursive hidden Markov models from discrete-valued time series.

The results are then implemented as FSM controling the behavior of a normal (non-omniscient) agent. The resulting behavior was similar to the well-known

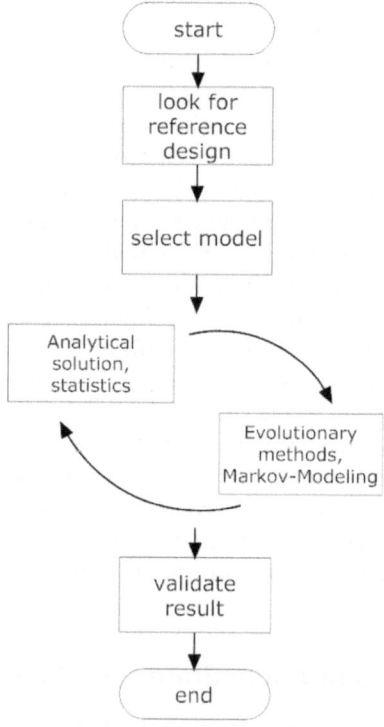

**Fig. 2.** Combined design approach

*tit-for-tat* strategy including *forgiveness*. Tit-for-tat with forgiveness is known to be a highly effective strategy in the repeated prisoner's dilemma.

## 4.6 Combining the Approaches

The presented design approaches can also be combined. For example, the reference design method may serve as a starting point, where the system designer applies one of the other methods subsequently after choosing a reference model. Another variant of a design process could involve a "bootstrapping" method, where efforts to understand the effect of local interactions are combined with an analysis of the global emergent behavior. Some effects of local rules could be predicted by mathematical analysis of the interaction. For example, in physics, only the gravitational system of two bodies is analytically solved. Still, the results can be applied to understand the movement of more bodies in our solar system, as long as some influences can be neglected. Statistical approaches such as Markov models or evolutionary algorithms can be a further step in the system design. In the combined approach as depicted in Figure 2, insights from the microscopic level are influencing and improving the design at the macroscopic level and vice versa.

## 5   Case Studies from Engineering

A nice feature of modeling approaches in self-organization is their simplicity at the microscopic level. Simple local rules lead to global structure and function. From an engineering perspective, we would like to apply these models to technical systems. This chapter will give some examples where models from self-organization and complex systems have been successfully applied to information and communications technology research. It will also show that a direct applicability of the theory of self-organization and complex systems is often not possible, but the design of self-organizing functions in technology often requires us to modify and extend the original schemes, taking into account some technological constraints and requirements.

### 5.1   Wireless Communications: Application of Coupled Oscillators

The pulse-coupled oscillator (PCO) model for firefly synchronization [10] has been employed in many fields of science and engineering. Prominent examples include self-organizing algorithms for time synchronization in wireless systems [27], resource scheduling [28], reducing energy consumption in sensor networks [29], and traffic light control systems [30].

The application of firefly synchronization to mobile and wireless systems requires us to perform some modifications and extensions to the original scheme. These changes are required, since the assumptions in [10] do not match with the system constraints of radio communications. In other words, the modeling assumptions in the original scheme are too simplistic compared to the modeling assumptions typically used in wireless and mobile systems.

First, in general, we cannot neglect inherent delays of the system, including propagation delays, decoding delays, and delays caused by signal processing. These delays make the synchronization scheme unstable, as nodes might receive "echos" of their own firing pulse. To regain stability, a refractory period can be introduced during which nodes do not increase their phase function [31,27]. Second, wireless communication technologies do typically not allow us to send infinitely short pulses over the air. This fact forces us to replace the infinitely short "firing pulses" by finitely long "synchronization signals" (see e.g. [32,33]). Third, the wireless medium suffers from noise and attenuation, which must be taken into account for the design of a synchronization scheme as well [34]. Last but not least, to minimize the use of radio resources, it would be beneficial to minimize the signaling overhead, such that nodes only send synchronization words when needed and not periodically as in the original scheme [35].

An approach for self-organized synchronization in wireless systems has been recently developed by Tyrrell, Auer, and Bettstetter [35]. The scheme, called *Meshed Emergent Firefly Synchronization (MEMFIS)*, applies a synchronization word that is common to all nodes in the network and is embedded into each payload data packet. This word is then detected at the receiver using a correlation unit. Starting from an unsynchronized network, synchronization emerges as nodes transmit data packets randomly according to some arrival process. In this way, the throughput of the network can increase gradually, e.g., from ALOHA to Slotted ALOHA.

Another example, where firefly synchronization has been applied to information and security technology is intrusion detection using sensor networks. One approach has been developed in [36].

## 5.2    Vehicular Traffic: Application of Cellular Automata and Agent-Based Approaches

An interesting application of cellular automata, studied extensively in both the physics and the engineering communities, is the modeling and analysis of urban vehicular traffic [37,38,39,40]. Gershenson and Rosenblueth [41] apply a two-dimensional model based on simple interaction roles. A street is modeled by a line of connected cells. Each cell can have two states, being empty (0) or occupied by a car (1). The interaction model defines that a car moves on to the next cell in its direction of motion, if this cell is empty; otherwise the car waits. At an intersection, two streets share a common cell. Depending on the state of the traffic lights, this cell operates as a forwarding cell either to the right or downwards, while blocking the other direction, respectively. In their work, Gershenson and Rosenblueth compare a traffic light control algorithm based on a green-wave method and a self-organizing approach. While the green-wave method requires the cars to match a predefined progression speed to show good throughput, the self-organizing approach shows to be more flexible in adapting to different load situations.

Resnick [23] describes an agent-based traffic model that explains the formation of traffic jams without a centralized cause (such as accidents). Each agent represents a car following a simple set of rules: it slows down if it detects a car close ahead; it speeds up if it does not see a car ahead. In this model, a traffic jam appears as a pattern moving in the opposite direction of the traffic flow. In contrast to cellular automata, the agent-based approach enables a more fine-grained model of the driver's action and decision process. For example, the model could be extended by drivers that have a bad reaction time due to distractions (e.g., phone calls).

## 6 Conclusions

Several problems in technology and society can be better understood and solved by modeling them as a self-organizing system. An engineer with the task of developing such a self-organizing system faces the problem of modeling and designing the local interactions which will achieve a desired global system behavior. In this paper, we have reviewed several modeling and design approaches suitable in this domain.

Differential equations can model a range of simple collective behaviors, such as pattern formation and the synchronization of coupled oscillators. E. g., the latter are an extremely useful paradigm for modeling self-organizing phenomena; they are often used to describe phenomena related to oscillation and synchronization within a system. Cellular automata are time-discrete and space-discrete models which are often used to display pattern formation phenomena or other phenomena related to the location of the entities. Agent-based models can be applied to both space-continuous and space-discrete phenomena. The are advantageous if entities display a range of behaviors and strategies such as decision making.

The design of a self-organizing system is difficult due to its emergent properties. This paper made an attempt to propose some design approaches, namely the analytic approach, working from a reference design, trial and error, evolutionary algorithms, and a statistic approach based on Markov models. Each of these has certain advantages and disadvantages, thus a combination of them can be useful in the system design process.

Finally, the paper gave examples from communications and traffic engineering where some of the presented models and design approaches have been successfully employed.

## Acknowledgments

This paper was supported in part by the KWF (contract KWF 20214/18124/26663 and KWF 20214—18128—26673), the ResumeNet project (EU Framework Programme 7, ICT-2007-2, Grant No. 224619), and the Forschungsrat at the University of Klagenfurt.

This work is an outcome of the Lakeside Research Days 2009 which took place at Lakeside Labs GmbH, Klagenfurt, Austria, from July 13, 2009 to July 17, 2009. The authors would like to thank all participants for the fruitful discussions.

# References

1. von Foerster, H.: Principles of the self-organizing system. In: von Foerster, H., Zopf Jr., G.W. (eds.) Principles of Self-organization, pp. 255–278. Pergamon Press, Oxford (1962)
2. Elmenreich, W., de Meer, H.: Self-organizing networked systems for technical applications: A discussion on open issues. In: Sterbenz, J.P.G., Hummel, K.A. (eds.) IWSOS 2008. LNCS, vol. 5343, pp. 1–9. Springer, Heidelberg (2008)
3. Reynolds, C.W.: Flocks, herds and schools: A distributed behavioral model. In: Proc. Annual Conf. on Computer Graphics and Interactive Techniques SIGGRAPH 1987, pp. 25–34 (1987)
4. Prehofer, C., Bettstetter, C.: Self-organization in communication networks: Principles and design paradigms. IEEE Communications Magazine, 78–85 (July 2005)
5. Volterra, V.: Leçons sur la théorie mathématique de la lutte pour la Vie, Gauthier-Villars, Paris (1931)
6. Auer, C., Wüchner, P., de Meer, H.: A method to derive local interaction strategies for improving cooperation in self-organizing systems. In: Hummel, K.A., Sterbenz, J.P.G. (eds.) IWSOS 2008. LNCS, vol. 5343, pp. 170–181. Springer, Heidelberg (2008)
7. Turing, A.M.: The chemical basis of morphogenisis. Philos. Trans. R. Soc. London Ser. B 237, 37–72 (1952)
8. Cross, M.C., Hohenberg, P.C.: Pattern formation outside of equilibrium. Reviews of Modern Physics 65(3), 851–1112 (1993)
9. Murray, J.D.: Mathematical Biology: I. An Introduction. Springer-Verlag, Berlin (1989)
10. Mirollo, R.E., Strogatz, S.H.: Synchronization of pulse-coupled biological oscillators. SIAM Journal on Applied Mathematics 50(6), 1645–1662 (1990)
11. Pikovsky, A., Rosenblum, M., Kurths, J.: Synchronization: A Universal Concept in Nonlinear Sciences. Cambridge University Press, Cambridge (2003)
12. Toffoli, T., Margolus, N.: Cellular automata machines. The MIT Press, Cambridge (1986)
13. Wolfram, S.: Theory and applications of cellular automata. World Scientific, Singapore (1986)
14. Chopard, B., Droz, M.: Cellular automata modeling of physical systems. Cambridge University Press, Cambridge (1998)
15. Kier, L.B., Seybold, P.G., Cheng, C.-K.: Modeling Chemical Systems Using Cellular Automata. Springer, Heidelberg (2005)
16. Gardner, M.: Mathematical games: The fantastic combinations of John Conway's new solitaire game "life". Scientific American 223, 120–123 (1970)
17. Axelrod, R.: The Complexity of Cooperation: Agent-Based Models of Competition and Collaboration. Princeton University Press, Princeton (1997)
18. Gilbert, N., Bankes, S.: Platforms and methods for agent-based modeling. Proc. Natl. Acad. Sci. 99(3), 7197–7198 (2002)
19. Railsback, S.F., Lytinen, S.L., Jackson, S.K.: Agent-based simulation platforms: Review and development recommendations. Simulation 82(9), 609–623 (2006)
20. Faye, R., Laprete, R., Winter, M.: Blended winglets. Aero, Boeing (17) (2002)
21. Di Caro, G., Ducatelle, F., Gambardella, L.M.: Anthocnet: An adaptive nature-inspired algorithm for routing in mobile ad hoc networks. In: Yao, X., Burke, E.K., Lozano, J.A., Smith, J., Merelo-Guervós, J.J., Bullinaria, J.A., Rowe, J.E., Tiňo, P., Kabán, A., Schwefel, H.-P. (eds.) PPSN 2004. LNCS, vol. 3242, pp. 461–470. Springer, Heidelberg (2004)

22. Gershenson, C.: Design and Control of Self-organizing Systems. PhD thesis, Vrije Universiteit Brussel (2007)
23. Resnick, M.: Turtles, Termites, and Traffic Jams: Explorations in Massively Parallel Microworlds (Complex Adaptive Systems). MIT Press, Cambridge (1997)
24. Fehérvári, I., Elmenreich, W.: Evolutionary methods in self-organizing system design. In: Proc. Intern. Conf. on Genetic and Evolutionary Methods (2009)
25. Tucker, A.: A two-person dilemma. Stanford University Press, Stanford (1950)
26. Shalizi, C.R., Shalizi, K.L.: Blind construction of optimal nonlinear recursive predictors for discrete sequences. In: Chickering, M., Halpern, J. (eds.) Proc. Conf. on Uncertainty in Artificial Intelligence, pp. 504–511 (2004)
27. Mathar, R., Mattfeldt, J.: Pulse-coupled decentral synchronization. SIAM Journal on Applied Mathematics 56(4), 1094–1106 (1996)
28. Patel, A., Degesys, J., Nagpal, R.: Desynchronization: Self-organizing algorithms for periodic resource scheduling. In: Proc. Intern. Conf. on Self-Adaptive and Self-Organizing Systems (July 2007)
29. Leidenfrost, R., Elmenreich, W.: Firefly clock synchronization in an 802.15.4 wireless network. EURASIP Journal on Embedded Systems, 17 (2009)
30. Bayraktaroglu, B.: Traffic light control system and method. United States Patent 4908615 (1990)
31. Ernst, U., Pawelzik, K., Geisel, T.: Synchronization induced by temporal delays in pulse-coupled oscillators. Phys. Rev. Lett. 74(9), 1570–1573 (1995)
32. Werner-Allen, G., Tewari, G., Patel, A., Welsh, M., Nagpal, R.: Firefly-inspired sensor network synchronicity with realistic radio effects. In: Proc. ACM Conf. Embedded Networked Sensor Systems (SenSys), San Diego, CA, USA (November 2005)
33. Tyrrell, A., Auer, G., Bettstetter, C.: Fireflies as role models for synchronization in ad hoc networks. In: Proc. Intern. Conf. on Bio-Inspired Models of Network, Information, and Computing Systems (BIONETICS), Cavalese, Italy (December 2006)
34. Hong, Y.-W., Scaglione, A.: A scalable synchronization protocol for large scale sensor networks and its applications. IEEE J. Select. Areas Commun. 23(5), 1085–1099 (2005)
35. Tyrrell, A., Auer, G., Bettstetter, C.: Emergent slot synchronization in wireless networks. IEEE Transactions on Mobile Computing (2010) (Under review )
36. Hong, Y.W., Scaglione, A.: Distributed change detection in large scale sensor networks through the synchronization of the pulse-coupled oscillators. In: Proc. IEEE Intern. Conf. Acoustics, Speech, and Signal Processing (ICASSP), Montreal, Canada (2004)
37. Biham, O., Middleton, A.A., Levine, D.: Self organization and a dynamical transition in traffic flow models. Phys. Rev. A 46, R6124 (1992)
38. Nagatani, T.: The physics of traffic jams. Rep. Prog. Phys. 65(9), 1331–1386 (2002)
39. Helbing, D., Nagel, K.: The physics of traffic and regional development. Contemporary Physics 45, 405–426 (2004)
40. Tonguz, O.K., Viriyasitavat, W., Bai, F.: Modeling urban traffic: A cellular automata approach. IEEE Communications Magazine (May 2009)
41. Gershenson, C., Rosenblueth, D.A.: Modeling self-organizing traffic lights with elementary cellular automata. C3 report 2009.06, Universidad Nacional Autónoma de México (2009)

# Laying Pheromone Trails for Balanced and Dependable Component Mappings

Máté J. Csorba[1], Hein Meling[2], and Poul E. Heegaard[1]

[1] Department of Telematics,
Norwegian University of Science and Technology, N-7491 Trondheim, Norway
{Mate.Csorba,Poul.Heegaard}@item.ntnu.no
[2] Department of Electrical Engineering and Computer Science,
University of Stavanger, N-4036 Stavanger, Norway
Hein.Meling@uis.no

**Abstract.** This paper presents an optimization framework for finding efficient deployment mappings of replicated service components (to nodes), while accounting for multiple services simultaneously and adhering to non-functional requirements. Currently, we consider load-balancing and dependability requirements. Our approach is based on a variant of Ant Colony Optimization and is completely decentralized, where ants communicate indirectly through pheromone tables in nodes. In this paper, we target scalability; however, existing encoding schemes for the pheromone tables did not scale. Hence, we propose and evaluate three different pheromone encodings. Using the most scalable encoding, we evaluate our approach in a significantly larger system than our previous work. We also evaluate the approach in terms of robustness to network partition failures.

## 1 Introduction

Data centers are increasingly used to host services over a virtualized infrastructure that permits on-demand resource scaling. Such systems are often comprised of multiple geographically dispersed data center sites to accommodate local demand with appropriate resources, and to ensure availability in case of outages. Major service providers, e.g. Amazon [1], Google, Yahoo! and others all use such infrastructures to power their world wide web offerings, popularly called cloud computing infrastructures. These systems are typically built using large numbers of cheap and less reliable blade servers, racks, hard disks, routers, etc., thus leading to higher failure rate [2]. To cope with increased failure rates, replication and repair mechanisms are absolutely crucial.

Another related and important concern in such data center infrastructures is the problem of *finding optimal deployment mappings for a multitude of services*, while ensuring proper balance between load characteristics and service availability in every infrastructure site. During execution a plethora of parameters can impact the deployment mapping, e.g due to the influence of concurrent services. Another set of parameters in the mix is the dependability requirements of services. Upholding such requirements not only demands replication protocols to ensure consistency, but also adds additional complexity to the optimization problem. Ideally, the deployment mappings should minimize resource consumption, yet provide enough resources to satisfy the dependability

T. Spyropoulos and K.A. Hummel (Eds.): IWSOS 2009, LNCS 5918, pp. 50–64, 2009.

requirements of services. However, Fernandez-Baca [3] showed that the general module allocation problem is NP-complete except for certain communication configurations, thus heuristics are required to obtain solutions efficiently.

This paper extends our previous work to find optimal deployment mappings [4], [5] based on a heuristic optimization method called the Cross-Entropy Ant System (CEAS). The strengths of the CEAS method is its capability to account for multiple parameters during the search for optimal deployment mappings [6]. The approach also enables us to perform optimizations in a decentralized manner, where replicated services can be deployed from anywhere within the system, avoiding the need for a centralized control for maintaining information about services and their deployment.

The main goal of this paper is twofold; to provide additional simulation results (i) involving scaling up the problem size, both in terms of number of nodes and replicas deployed, and (ii) evaluating its ability to tolerate network partition failures (split/merge). Scaling up the problem size turned out to be more challenging than first anticipated, and thus certain enhancements were necessary in the algorithm and the data representation. To tackle the challenges we met, we have introduced a new cost function, run-time binding of replicas, a new method for selecting next-hops and new pheromone encodings. In addition, we have used more simple service models in the current study. There are generally two branches of works where finding optimal replica deployment mappings are necessary and useful. On the one hand, virtual machine technology is increasingly being used in data centers for providing high availability and thus needs to consider the placement of replicas in the data centers to ensure efficient utilization of the system resources. The advantage of this approach is that (server) applications running on virtual machines can be repaired simply by regenerating them in another physical machine. This is the approach taken by the Amazon EC2 system [1] and in VMware, among others. The general drawback with virtualization for fault tolerance and high availability is that the storage system used to maintain application state must be independently replicated as it would otherwise constitute a single point of failure. On the other hand, server applications written specifically for fault tolerance typically replicate their application state to all replica processes, avoiding any single points of failure. These systems are typically built using a middleware based on a group communication system with support for repair mechanisms, e.g. DARM [7].

The importance and utility of deployment decision making and optimization has been identified previously, e.g. in [8]. Recently, Joshi et al [9] proposed a centralized approach in which an optimizer and model solver component is used to find optimal mappings specifically in the field of virtual machine technology. We however, intend to pursue a fully distributed solution that is based on optimization techniques and can support context-awareness and adaptation.

The paper is organized as follows. The next section presents our view on component replicas, their deployment, corresponding costs and requirements. In Sec. 3 the fundamentals of the CEAS are described. Sec. 4 proposes our algorithm for solving the deployment mapping problem and subsequently we demonstrate its operation in Sec. 5. Finally, we conclude and touch upon future work.

## 2   System Model, Assumptions and Notation

We consider a large-scale distributed system consisting of a collection of *nodes*, $\mathcal{N}$, connected through a network. Nodes are organized into a set of *domains*, $\mathcal{D}$, as illustrated by $d_1$ and $d_2$ in Fig. 1. All nodes within a domain are located at the same geographic site, whereas different domains are in separate sites. The objective of the distributed system is to provide an environment for hosting a set of *services*, $\mathcal{S} = \{S_1, S_2, \ldots\}$, to be provided to external clients. Let $C_i^k$ be the $i^{th}$ *component* of service $k$, and let $S_k = \{C_1^k, \ldots, C_q^k\}$ denote the set of components for service $k$, where $q = |S_k|$. Each component may be replicated for fault tolerance and/or load-balancing purposes. Thus, let $R_{ij}^k$ denote the $j^{th}$ replica of $C_i^k$. Hence, $C_i^k = \{R_{i1}^k, \ldots, R_{ip_i}^k\}$, where $p_i \geq 1$ is the redundancy level of $C_i^k$. Moreover, $S_k = \{R_{11}^k, \ldots, R_{1p_1}^k, \ldots, R_{i1}^k, \ldots, R_{ip_i}^k\}$ is the expansion of the component sets into replicas for service $k$.

The objective of the algorithms herein is to discover suitable *deployment mappings* between *component replicas* (*replicas* for brevity) and nodes in the network, such that the dependability requirements of all services are preserved with minimal resource consumption. To accomplish this, the CEAS optimization method is used, which works by evaluating a *cost function*, $F()$, for different deployment mappings. The CEAS method is implemented in the form of *ants* moving around in the network to identify potential locations where replicas might be placed. An ant is simply an agent with associated state; as such it is simply a message on which the ant algorithm is executed at each visited node. We say that different *ant species* are responsible for different services, e.g. the green and blue ants in Fig. 1 represent the green and blue services, respectively.

As Fig. 1 shows, each node contains an *execution runtime* whose tasks are to install, run and migrate replicas. A node also has a *pheromone table* which is manipulated by ants visiting the node to reflect their knowledge of existing mappings. Moreover,

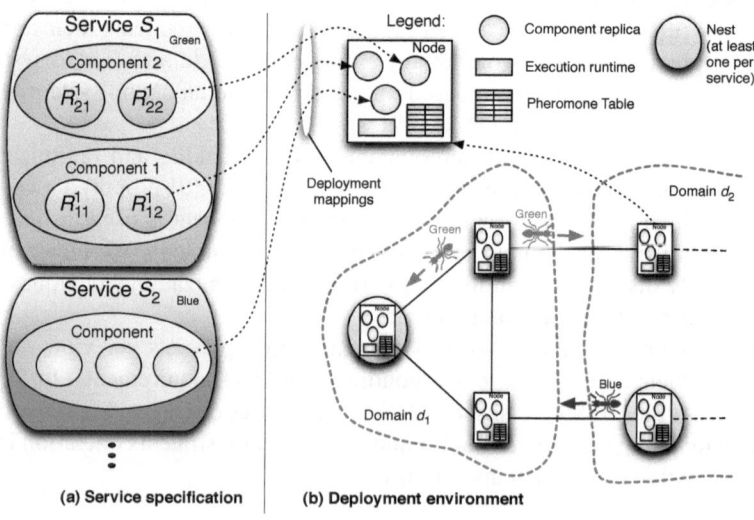

**Fig. 1.** Overview of the deployment environment and service specification

the pheromone table is used by ants for selecting suitable deployment mappings; it is not used for ant routing as in the Ant Colony Optimization (ACO) approach [10]. See Sec. 4.2 for details.

To deploy a service, at least one node must be running a *nest* for that service. The tasks of a nest are twofold: (i) to emit ants for its associated service, and (ii) trigger installation of replicas at nodes, once a predefined *convergence criteria* is satisfied, e.g. after a certain number of iterations of the algorithm. An iteration is defined as one round-trip trajectory of the ant, during which it builds a *hop list*, $H_r$, of visited nodes. A nest may be replicated for fault tolerance, and emit ants independently for the same service. During execution of the CEAS method, synchronization between nests is not necessary, but only a primary nest will execute deployment decisions. Fig. 1 shows a two-way replicated nest for the green service; nests for the blue service are not shown.

Initially, the composition of services to be deployed is specified as UML collaborations embellished with non-functional requirements that are used as input to the cost function of our algorithm to evaluate deployment mappings (cf. [11]). Our aim is not to find the globally optimal solution. The rationale for this is simple; by the time the optimal deployment mapping could be applied, it is likely to be suboptimal due to dynamics of the system. Rather, we aim to find a *feasible mapping*, meaning that it satisfies the requirements for the deployment of the service, e.g. in terms of redundancy and load-balancing. These requirements are specified as a set of rules, denoted $\Phi$. Thus, our objective function becomes $min \ F()$ subject to $\Phi$. Moreover, our algorithm can continue to optimize even though an appropriate mapping has been found and deployed into the network. Once a (significantly) better mapping is found, reconfiguration can take place.

Next we define the dependability rules, $\Phi$, that constrain the minimization problem, but first we define two mapping functions. These rules and functions apply to service $k$.

**Definition 1.** *Let* $f_{j,d} : R_{ij}^k \to d$ *be the mapping of replica* $R_{ij}^k$ *to domain* $d \in \mathcal{D}$.

**Definition 2.** *Let* $g_j : R_{ij}^k \to n$ *be the mapping of replica* $R_{ij}^k$ *to node* $n \in \mathcal{N}$.

Rule $\phi_1$ requires replicas to be dispersed over as many domains as possible, aimed to improve service availability despite potential network partitions[1]. Specifically, replicas of component $C_i^k$ shall be placed in different domains, until all domains are used. If there are more replicas than domains, i.e. $|C_i^k| > |\mathcal{D}|$, at least one replica shall be placed in each domain. The second rule, $\phi_2$, prohibits two replicas of $C_i^k$ to be placed on the same node, $n$.

**Rule 1.** $\phi_1 : \forall d \in \mathcal{D}, \forall R_{ij}^k \in C_i^k : f_{j,d} \neq f_{u,d} \Leftrightarrow (j \neq u) \wedge |C_i^k| < |\mathcal{D}|$

**Rule 2.** $\phi_2 : \forall R_{ij}^k \in C_i^k : g_j \neq g_u \Leftrightarrow (j \neq u)$

Combining these rules gives us the desired set of dependability rules, $\Phi = \phi_1 \wedge \phi_2$. In order to adhere to $\phi_1$, the ant gathers data about domains utilized for mapping replicas; hence, let $D_r$ denote the set of domains used in iteration $r$. The ant also collects information about replicas mapped to various nodes. Thus, we introduce $m_{n,r} \subseteq S_k$

---

[1] We assume network partitions are more likely to occur between domain boundaries.

as the set of service $k$ replicas mapped to node $n$ in iteration $r$. Moreover, let $M_r = \{m_{n,r}\}_{\forall n \in H_r}$ be the deployment mapping set at iteration $r$ for all visited nodes. Finally, ants also collect load-level samples, $l_{n,r}$, from every node $n \in H_r$ visited in iteration $r$; these samples are added to the *load list*, $L_r$, indexed by the node identifier, $n$.

The load-levels observed by an ant are a result of many concurrently executing ant species reserving resources for their respective services. For simplicity, all replicas have the same node-local execution cost, $w$, whereas communication costs are ignored. An ant during its visit to node $n$ reserves processing resources for the replicas, if any, that it has chosen to map to $n$. Mappings made at $n$ during iteration $r$ are stored in $m_{n,r}$, thus, resources of size $|m_{n,r}| \cdot w$ are reserved during a visit, assuming identical cost for all replicas. With this notational framework in place, we are now ready to introduce the cost function used by the deployment logic.

First, we define a list, $NC_x$, that can carry an element for each node visited by the ant and which elements account for specific execution costs imposed on those nodes. Elements of the list are calculated two different ways ($x = 1$ or $2$), using the observations on the services executed in parallel ($L_r$), and the mappings of replicas made by the ant itself ($M_r$).

$$NC_x[n] = (\sum_{i=0}^{\vartheta_x} \frac{1}{\Theta_x + 1 - i})^2 \tag{1}$$

Parametrization of the list is done by changing the upper bound of the summation, $\vartheta_x$ and the constant in the denominator, $\Theta_x$. Accordingly, $\vartheta_x$ and $\Theta_x$ are defined as follows.

$$\vartheta_x = \begin{cases} |m_{n,r}| \cdot w, & x = 1 \\ |m_{n,r}| \cdot w + L_r(n), & x = 2 \end{cases} \tag{2}$$

The constant, $\Theta_x$ represents the overall execution load of one service or all services. In other words, $\Theta_1$ is the total processing resource demand of the service deployed by the ant, whereas $\Theta_2$ represents the overall joint load of the service being deployed and the load of replicas executed in parallel.

$$\Theta_x = \begin{cases} \sum_{\forall n \in H_r} |m_{n,r}| \cdot w, & x = 1 \\ \sum_{\forall n \in H_r} (|m_{n,r}| \cdot w + L_r(n)), & x = 2 \end{cases} \tag{3}$$

Importantly, the equations only have to be applied on the subset of nodes an ant has actually visited ($H_r$), which is beneficial for scalability as there is no need for exploring the total amount of available nodes. Finally, to build a cost function that satisfies our requirements with regard to $\Phi$, while maintaining load-balancing, we formulate $F()$ using a combination of terms, as shown in (4).

$$F(D_r, M_r, L_r) = \frac{1}{|D_r|} \cdot \sum_{\forall n \in H_r} NC_1(n) \cdot \sum_{\forall n \in H_r} NC_2(n) \tag{4}$$

Thus, we use (1) for load-balancing, i.e. to distribute replicas to the largest extent possible. The three terms correspond to our goals in the optimization process. The first reciprocal term caters for $\phi_1$. Applying (1) solely on the replicas of the service the ant species is responsible for ($x = 1$) penalizes violation of $\phi_2$, i.e. favors a mapping where replicas are not collocated, but distributed evenly. Lastly, the standard application of (1), $x = 2$, balances the load taking into account the presence of other services during the deployment mapping. A more detailed introduction to the application of the load-balancing term can be found in [5] and [11]. The next section describes how the cost function plays a role in driving the optimization using the CEAS.

## 3 The Cross-Entropy Ant System

We build our algorithm around the CEAS to obtain optimal deployment mappings with high confidence. CEAS can be considered as a subclass of ACO algorithms [10], which have been proven to be able to find the optimum at least once with probability close to one; once the optimum has been found, convergence is assured within a finite number of iterations. The key idea is to have many ants, search iteratively for a solution according to a cost function defined according to problem constraints. Each iteration is divided into two phases. Ants conduct *forward search* until all the replicas are mapped successfully. After that, the solution is evaluated using the cost function, the ants continue with *backtracking* leaving *pheromone* markings at nodes. This resembles real-world ants foraging for food. The pheromone values are proportional to the solution quality determined by the cost function. These pheromone markings are distributed to nodes in the network, and are used during forward search to select replica sets for deployment mapping, gradually approaching the lowest cost solution. In forward search, a certain proportion of ants do a random *exploration* of the state space, ignoring the pheromone trails. Exploration reduces the occurrence of premature convergence leading to sub-optimal solutions. The CEAS uses the *Cross-Entropy (CE) method* introduced by Rubinstein [12] to evaluate solutions and update the pheromones. The CE method is applied to gradually change a probability matrix $\mathbf{p}_r$ according to the cost of the mappings with the objective of minimizing the cross entropy between two consecutive probability matrices $\mathbf{p}_r$ and $\mathbf{p}_{r-1}$. The method itself has been successfully applied in different fields of network and path management, for examples and an intuitive introduction we refer to [6].

In our algorithm, the CEAS is applied to obtain an appropriate deployment mapping, $\mathcal{M} : C_i^k \rightarrow \mathcal{N}$, of the replicas ($C_i^k$) of service $S_k$ onto a set of nodes. A deployment mapping is evaluated by applying the cost function as $F(M_r)$. In the following, let $\tau_{mn,r}$ be the pheromone value corresponding to, $m_{n,r}$, the set of replicas mapped to node $n$ in iteration $r$. Various pheromone encoding schemes are discussed in Sec. 4.2.

To select a set of replicas to map to a given node, ants use the so-called *random proportional rule* matrix, $\mathbf{p}_r = \{p_{mn,r}\}$ presented below. Similarly, *explorer ants* select a set of replicas with uniform probability $1/|C_i^k|$, where $|C_i^k|$ is the number of replicas to be deployed.

$$p_{mn,r} = \frac{\tau_{mn,r}}{\sum_{l \in M_{n,r}} \tau_{ln,r}} \tag{5}$$

A parameter $\gamma_r$ denoted the *temperature*, controls the update of the pheromone values and is chosen to minimize the performance function, which has the following form

$$H(F(M_r), \gamma_r) = e^{-F(M_r)/\gamma_r} \tag{6}$$

and is applied to all $r$ samples. The expected overall performance satisfies the equation

$$h(p_{mn,r}, \gamma_r) = E_{\mathbf{p}_{r-1}}(H(F(M_r), \gamma_r)) \geq \rho \tag{7}$$

$E_{\mathbf{p}_{r-1}}(X)$ is the expected value of $X$ s.t. the rules in $\mathbf{p}_{r-1}$, and $\rho$ is a parameter (denoted *search focus*) close to 0 (in our examples 0.01). Finally, a new updated set of rules, $\mathbf{p}_r$, is determined by minimizing the cross entropy between $\mathbf{p}_{r-1}$ and $\mathbf{p}_r$ with respect to $\gamma_r$ and $H(F(M_r), \gamma_t)$. Minimized cross entropy is achieved by applying the random proportional rule in (5) for $\forall_{mn}$ with

$$\tau_{mn,r} = \sum_{k=1}^{r} I(l \in M_{n,r}) \beta^{\sum_{j=k+1}^{r} I(j \in M_k)} H(F(M_k), \gamma_r) \tag{8}$$

where $I(x) = 1$ if $x$ is true, 0 otherwise. See [12] for further details and proof.

As we target a distributed algorithm that does not rely on centralized tables or control, neither on batches of synchronized iterations, the cost values obtained by applying Eq. (4) are calculated *immediately* after each sample, i.e. in each iteration $r$. Then, an auto-regressive performance function, $h_r(\gamma_r) = \beta h_{r-1}(\gamma_r) + (1 - \beta) H(F(\mathbf{M}_r), \gamma_r)$ is applied, where $\beta \in< 0, 1 >$ is a *memory factor* that gives weights to the output of the performance function. The performance function smoothes variations in the cost function and helps avoiding undesirable rapid changes in the deployment mappings.

The *temperature* required for the CEAS, e.g. in Eq. 6, is determined by minimizing it subject to $h(\gamma) \geq \rho$ (cf. [13])

$$\gamma_r = \{\gamma \mid \frac{1 - \beta}{1 - \beta^r} \sum_{i=1}^{r} \beta^{r-i} H(F(M_i), \gamma) = \rho\} \tag{9}$$

However, (9) is a complicated function that is storage and processing intensive since all observations up to the current sample, i.e. the entire mapping cost history $F(M_r) = \{F(M_1), \ldots, F(M_r)\}$ must be stored, and weights for all observations have to be recalculated. This would be an impractical burden to on-line execution of the logic. Instead, given that $\beta$ is close to 1, it is assumed that changes in $\gamma_r$ are relatively small in subsequent iterations, which enables a first order Taylor expansion of (9), and a second order Taylor expansion of (8), see [13], thus saving memory and processing power.

## 4   Ant Species Mapping Replicas

In this section we present our deployment algorithm, how we apply the CEAS method, and three different ways of encoding replica mappings into pheromone values.

### 4.1   Swarm-Based Component Deployment

Our algorithm has successfully been applied for obtaining component mappings that satisfy non-functional requirements. In addition, the algorithm's capability to adapt to changing network conditions, for example caused by node-failures, has been investigated, cf. [5]. However, from a dependability point of view it is interesting to equip the logic with the capability to adapt to dynamicity of domains, i.e. splitting/merging of domains. To also cater for domain splits and merges we propose to initiate an ant-nest in multiple nodes belonging to separate domains. These ant-nests will emit ants corresponding to the same set of services, this however, will not result in flooding the network with ants as the rate of emission in a stable network can be divided equally between the nests. Besides, ants emitted from different nests but optimizing mappings for the same service will update the same pheromone tables in the nodes they visit during their search for a solution. The concept of multiple nests has been introduced in [11].

**Algorithm 1.** Code for $Nest_k$ corresponding to service $S_l$ at any node $n \in \mathcal{N}$

```
1: Initialization:
2:    r ← 0                                              {Number of iterations}
3:    γ_r ← 0                                                    {Temperature}

4: while r < R                                              {Stopping criteria}
5:    M_r ← antAlgo(r, k)                           {Emit new ant, obtain M_r}
6:    update(availableDomains)        {Check the number of available domains}
7:    if splitDetected() ∨ mergeDetected()
8:       release(S_l)                 {Delete existing bindings for all replicas c_i ∈ C_i^l}
9:    if φ_1(M_r, availableDomains) ∧ φ_2(M_r)
10:      bind1(M_r)               {Bind one of the still unbound replicas in C_i^l}
11:   r ← r + 1                                    {Increment iteration counter}
```

Algorithm 1 shows the code of a single ant-nest that sends out an ant in every iteration. The idea is that when a coherent network suffers a split, there shall be at least one nest in each region after the split event that will maintain a pheromone database in each region. (By a region we denote a set of nodes partitioned into one or more domains.)

To ease convergence of the mappings made by the ants the nests are allowed to bind one replica at a time if some condition applies. Here we check rules $\phi_1$ and $\phi_2$ against the mapping obtained in the current iteration, $M_r$. Replica bindings are indicated in the service specification that is derived from the model of the service. After a replica has been bound to a specific host ants in subsequent iterations will not try to find a new mapping for it, instead these bound mappings are maintained and the search is conducted for the remaining replicas only. Importantly however, bound replicas are also taken into account when the cost of the total mapping is evaluated by the ant. When a split or a merge event occurs these soft-bindings are flushed by the ant nest and, for example in case of a merge, two nests being in the same region can start to cooperate and share bindings and pheromone tables again.

Here it is important to clearly distinguish between the notions of replica mapping, binding and deployment. We use the term mapping during the optimization process, where our algorithm is constantly optimizing an ordering of replicas of a service to underlying execution hosts, but only internally to the algorithm itself. When a replica is bound to a host it means that from that point the algorithm does not change the mapping between that replica and a host. By deployment however, we refer to the actual physical placement of a software component replica to a node, which is triggered after the mappings obtained by our algorithm have converged to a satisfactory solution. The latter property ensures that there is no undesirable fluctuation in the migration of replicas using our method. In Algorithm 2 we present the steps executed by the ants emitted from a nest.

Each species of ants retrieves and updates the temperature used in the CEAS method from the nest where they are emitted from. First, an ant visits the nodes, if any, that already have a bound replica mapped to maintain these mappings, which will be taken into account when the cost of the total mapping is evaluated. The pheromones corresponding to these bound mappings will also be updated during *backtracking*. Besides, ants allocate processing power corresponding to the execution costs of the bound replicas,

**Algorithm 2.** Ant code for mapping of replicas $C_i^l \in S_l \subset S$ from $Nest_k$

```
 1: Initialization:
 2:     H_r ← ∅                                    {Hop-list; insertion-ordered set}
 3:     M_r ← ∅                                    {Deployment mapping set}
 4:     D_r ← ∅                                    {Set of utilized domains}
 5:     L_r ← ∅                                    {Set of load samples}

 6: function antAlgo(r, k)
 7:     γ_r ← Nest_k.getTemperature()              {Read the current temperature}
 8:     foreach c_i ∈ C_i^l                        {Maintain bound replica mappings}
 9:         if c_i.bound()
10:             n ← c_i.boundTo()                  {Jump to the node where this comp. is bound}
11:             n.reallocProcLoad(S_k, w)          {Allocate processing power needed by comp.}
12:             l_{n,r} ← n.getEstProcLoad()       {Get the estimated processing load at node n}
13:             L_r ← L_r ∪ {l_{n,r}}              {Add to the list of samples}

14:     while C_i^l ≠ ∅                            {More replicas to map}
15:         n ← selectNextNode()                   {Select next node to visit}
16:         if explorerAnt
17:             m_{n,r} ← random(⊆ C_i^l)          {Explorer ant; randomly select a set of replicas}
18:         else
19:             m_{n,r} ← rndProp(⊆ C_i^l)         {Normal ant; select replicas according to Eq. (5)}
20:         if {m_{n,r}} ≠ ∅, n ∈ d_k              {At least one replica mapped to this domain}
21:             D_r ← D_r ∪ d_k                    {Update the set of domains utilized}
22:         M_r ← M_r ∪ {m_{n,r}}                  {Update the ant's deployment mapping set}
23:         C_i^l ← C_i^l − {m_{n,r}}             {Update the set of replicas to be deployed}
24:         l_{n,r} ← n.getEstProcLoad()          {Get the estimated processing load at node n}
25:         L_r ← L_r ∪ {l_{n,r}}                 {Add to the list of samples}

26:     cost ← F(M_r, D_r, L_r)           {Calculate the cost of this given mapping, using Eq. (4)}
27:     γ_r ← updateTemp(cost)     {Given cost, recalculate temperature according to Eq. (9)}
28:     foreach n ∈ H_r.reverse()                  {Backtrack along the hop-list}
29:         n.updatePheromone(m_{n,r}, γ_r)        {Update pheromone table in n, Eq. (8)}
30:     Nest_k.setTemperature(γ_r)                 {Update the temperature at Nest_k}
```

derived from the service specification. After maintenance the ants jump over to nodes selected in a guided random manner and attempt to map some replicas to the node they reside in. This selection of the next node to visit, in contrast to e.g. ant-based routing algorithms, is independent from the pheromone markings laid by the ants. The selection of replica mappings in each node, however, is influenced by the pheromones.

Here, we distinguish between *explorer* and *normal* ants, where the former selects a set of replicas to map randomly and the latter uses the pheromone table at the current node. In case of a *normal* ant the selection process varies depending on the form of the pheromone tables (cf. Sec. 4.2). After some variables carried along by the ant ($M_r$, $D_r$, $C_i^k$) are updated a sample of the sum of execution load on the current node is taken by the ant. This replica load reservation mechanism is intended to function as an indirect way of communication between species executed in parallel. At the end of the *forward search* phase, when the ant has managed to map all the replicas of the service, the mapping is evaluated using the cost function and the temperature is recalculated using the obtained cost value. The last part in the lifetime of a single ant is the *backtracking* phase, during which the ant revisits the nodes that have been used for the mapping of the service and updates the pheromone database.

The gain in using a guided but random hop-selection instead of a pure random walk lies in that with the proper guidance the frequency of finding an efficient mapping is greater. The idea is that at first the next node is selected from a domain that has not yet been utilized until all visible domains are covered, leading to better satisfaction of $\phi_1$. Then the next hop selection continues with drawing destinations from the set of nodes not yet used in the mapping by checking with the variable $M_r$, before reverting to totally random drawing. The guided hopping strategy for the selection of a next node to visit is summarized in Algorithm 3.

**Algorithm 3.** Procedure to select the next hop for an ant

```
 1: function selectNextNode()                         {Guided random hop}
 2:    if H_r = N                                       {All nodes visited}
 3:       n ← random(N)                        {Select candidate node at random}
 4:    else
 5:       if D_r = D                            {All available domains utilized}
 6:          n ← random(N \ M_r)          {Select a node that has not been used yet}
 7:       else
 8:          d_i ← random(D \ D_r)               {Select a domain not yet used}
 9:          n ← random(d_i)                   {Select a node within this domain}
10:       H_r ← H_r ∪ {n}                         {Add node to the hop-list}
11:    return n
```

### 4.2  Encoding Sets of Replicas into Pheromone Entries

Generally, pheromone entries can be viewed as a distributed database located in the nodes available in the network considered for deployment. This distributed database has to be built so that it is able to describe arbitrary combinations of replicas of a given service component. At the same time the size of this database is crucial for obtaining better scalability for our approach. The reasons are twofold. The first reason is related to memory consumption as each participating node has to cater for a pheromone database for each service being deployed. Thus, memory consumption grows with the database size (depending on the encoding) and with the number of parallel services, where we can influence the former. Second, as we can see in the algorithm description in Sec. 4.1, an individual ant agent has to browse through the pheromone entries during its visit at a node, so clearly, a more compact encoding helps speeding up execution of the tasks an ant has to perform. The different encodings we proposed are shown in Table 1.

The *bitstring* encoding is the largest as it has a single value for all possible combinations of replica mappings in every node, which results in prohibitively large memory need. For example, in case of 20 replicas per service this encoding leads to $2^{20}$ pheromone values, which by using 4 byte long floating point numbers would require

**Table 1.** Three pheromone encodings for a service with $|C_i^k|$ replicas

| Encoding | DB size in a node | Encoding example w/ $|C_i^k| = 4$ |
|---|---|---|
| bitstring | $2^{|C_i^k|}$ | $[0000]b \dots [1111]b$ |
| per comp. | $2 \cdot |C_i^k|$ | $[0/1]; [0/1]; [0/1]; [0/1]$ |
| # replicas | $|C_i^k| + 1$ | $[0] \dots [4]$ |

4 MB of memory for each of such services at every node. To tackle this problem we might reduce the pheromone table size by applying more simple bookkeeping taking into account solely the number of replicas mapped to a given node (# *replicas*). This results in the most compact pheromone database, however it comes with a drawback that it can only be applied if there is no need to distinguish between replicas in the service specification (for example considering replication and dependability aspects only). As a trade-off we developed a third encoding (*per comp.*) that results in no information loss and still linear growth of the pheromone database. *per comp.* uses one distinct pheromone entry for every replica instance indicating whether or not to deploy a replica at a given node. The drawback is that an ant arriving at a node has to decide on the deployment mapping of each replica, one-by-one reading the multiple pheromone entries. Nevertheless, a reduction in the database structure size is necessary for scaling the algorithm up to larger amounts of nodes and replicas. How the various encodings perform will be demonstrated with an example in Sec. 5.

## 5   Simulation Results

To evaluate the deployment mapping logic proposed above we start with an example where 10 services ($S_1 \dots S_{10}$) are being deployed simultaneously, that means 10 independent species are released. Besides, we apply 20 ant nests to look at a simple split/merge scenario where 1 nest for every service remains in each region after the split. Each service has a redundancy level as shown in Table 2. The simulation of the logic's behavior is conducted in a custom built discrete event simulator.

Mapping of the services is conducted in a network of 11 interconnected hosts, where we assume full mesh connectivity and do not consider the underlying network layer. The 11 nodes are partitioned into 5 domains as depicted in Fig. 2.

In this setting we conducted simulations with all three pheromone encodings. To test our concept of tackling domain splitting we have used a basic setting where domain $d_1$ containing 4 nodes has been split from the rest of the domains and later the two regions merged again. We then compared the resulting deployment mappings with the mappings obtained by executing our logic with no splitting. To demonstrate how the cost evaluation works in the optimization process the evolution of the cost output is displayed in case of service $S_{10}$ in Fig. 3 with the three pheromone encodings introduced

**Table 2.** Service instances in the example

| Service | $S_1$ | $S_2$ | $S_3$ | $S_4$ | $S_5$ | $S_6$ | $S_7$ | $S_8$ | $S_9$ | $S_{10}$ |
|---|---|---|---|---|---|---|---|---|---|---|
| # replicas | 2 | 3 | 4 | 5 | 6 | 7 | 8 | 9 | 10 | 11 |

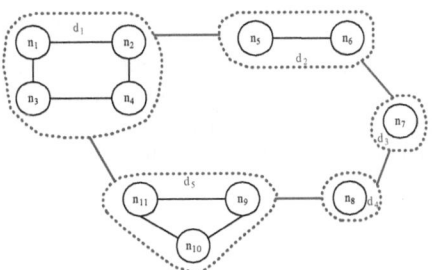

**Fig. 2.** Test network of hosts clustered into 5 domains

in Sec. 4.2. Fig. 3(a) shows how the optimal mappings are found and kept maintained iteration by iteration. The experiment is repeated with the introduction of the splitting of $d_1$ after 4000 iterations, the evolution of mapping costs is shown in Fig. 3(b).

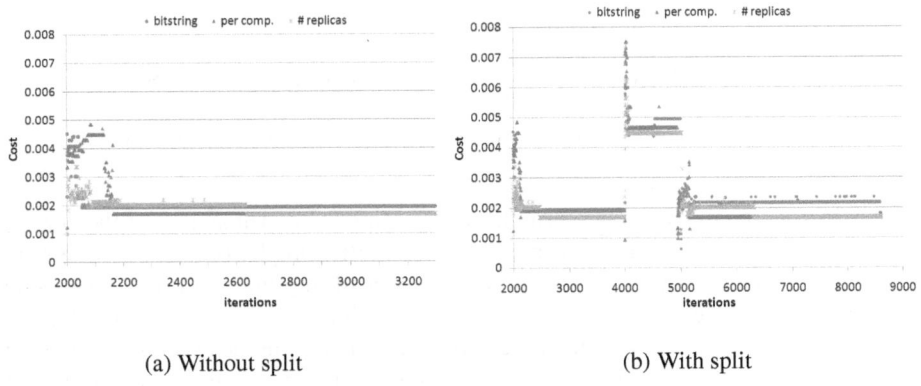

(a) Without split                                                                 (b) With split

**Fig. 3.** Mapping costs of $S_{10}$

An appropriate solution is found almost identically with the three different encodings. However, the *bitstring* encoding converges to a solution with slightly higher overall cost, whereas the lowest cost is obtained first by *per comp.* and somewhat later by *# replicas* too. In Fig. 3 the first 2000 iterations are not shown as the simulations start with 2000 explorer iterations for the sake of comparability. Initially, a random cost figure appears corresponding to exploration that is omitted here. The amount of initial exploration was constrained by the *bitstring* encoding. The more compact encodings would require significantly less iterations, e.g. one tenth of that. In Fig. 3(b), where a domain splits at iteration 4000 we can observe how the swarm adapts the mappings to a more expensive configuration after the event has happened. Similarly, as the domains merge the deployment mappings are adapted to utilize a more optimal configuration. The *bitstring* encoding in this test case is unable to find exactly the same mapping and converges to a somewhat more costly solution. *per comp.* is the fastest to obtain the lowest cost mapping followed by the third encoding about 1000 iterations later.

**Table 3.** Success rate of the three encodings

| wo/ split | $\phi_1$ | $\phi_2$ | w/ split | $\phi_1$ | $\phi_2$ |
|---|---|---|---|---|---|
| bitstring | 100% | 88% | bitstring | 100% | 87% |
| per comp. | 100% | 100% | per comp. | 100% | 100% |
| # replicas | 100% | 100% | # replicas | 100% | 99% |

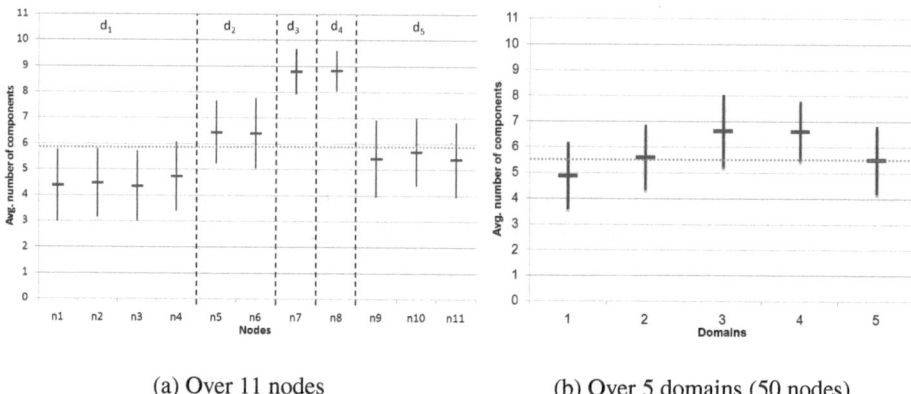

(a) Over 11 nodes                    (b) Over 5 domains (50 nodes)

**Fig. 4.** Load-balancing (average number of replicas and deviation per node)

Considering the rules that we formulated regarding the dependability of the deployment mapping (cf. Sec. 2) Table 3 shows the three different pheromone encodings and the percentage of test cases, which succeeded in satisfying the two rules. The results are obtained by executing the algorithm 100 times with different input seeds. Our first objective was load-balancing among the nodes participating in the execution of the services, while basic dependability rules are satisfied too. To investigate that aspect we can look at the average number of replicas placed onto the nodes after convergence. Here, we chose the best encoding (cf. Table 3), i.e. *per comp.*. In Fig. 4(a) the average load placed on the 11 nodes ($n_1 \ldots n_{11}$) partitioned into the 5 domains is depicted. A total of 65 replicas constituted the ten service instances giving an average of 5.91 replicas per node; shown as a dotted horizontal line. We observed that the smaller domains, e.g. $d_3$, $d_4$, were overloaded compared to the rest due to $\phi_1$, but generally replicas were placed quite evenly, showing that cooperation between the species worked.

As a next step towards developing our logic further for larger scales we repeated our experiment with a setting consisting of 50 nodes in 5 domains (containing 20-10-5-5-10 nodes respectively). Naturally, an increased amount of available resources for placement would make the deployment mapping problem actually easier, so we have used larger service specifications too to scale up the problem. Accordingly, the 10 services assigned to ant species were sized as $|C_i^k| = i \cdot 5$ replicas for $S_i$, where $i = 1 \ldots 10$, thus giving a total amount of 275 replicas.

**Table 4.** Collocation within the 10 large services

| service | $S_1$ | $S_2$ | $S_3$ | $S_4$ | $S_5$ | $S_6$ | $S_7$ | $S_8$ | $S_9$ | $S_{10}$ |
|---|---|---|---|---|---|---|---|---|---|---|
| collocation ($\phi_2$) | 0 | 1 | 0 | 3 | 1 | 1 | 0 | 3 | 1 | 13 |

We repeated the experiment 50 times using the selected encoding, *per comp.*, and allowing a maximum amount of 10000 iterations in each run. The resulting average execution load in the 5 available domains is depicted in Fig. 4(b), where the average number of replicas per node, using identical domains, would be 5.5 that is shown with a dotted horizontal line. Regarding load-balancing similar effects are observed as in the previous example. We can look at the dependability aspects of the solutions obtained in the 50 runs of the simulation too. As the problem size was significantly larger and the number of allowed iterations was constrained too, collocation is observed in some cases (in Table 4), while rule $\phi_1$ is never violated. Violations of $\phi_2$ are more frequent in case the number of replicas was close to the number of available nodes (e.g. $S_{10}$), which makes satisfying $\phi_2$ harder when load-balancing has to be performed simultaneously.

## 6 Closing Remarks

Our focus has been on applying swarm intelligence, in particular the CEAS method to manage the deployment of collaborating software components. While developing our distributed approach we targeted a logic that shall not be over-engineered and uses only a few parameters that do not depend on the problem at hand (e.g. to avoid having to adjust parameters and cost functions manually). It is also required to be able to handle certain degrees of dynamics and adaptation to changes in the execution context. These are the reasons that lead us to nature inspired methods and systems that do not have to be altered significantly for every new target system. We have tested the ability of the logic to handle domain splitting and dealing with dependability requirements as well as load-balancing using two example settings and a custom built simulator. We believe that applying CEAS will not only result in a tailored optimization method but, at least on the long run, it will allow the implementation of a prototype of a truly distributed system that will support run-time deployment within software architectures.

Our results are promising and are inline with our efforts to further develop the deployment logic and increase its scalability and adaptability. Furthermore, we plan to experiment with another dimension of dynamicity by introducing run-time component replication that means that the amount of replicas in a service might change at run-time. Moreover, extensive simulations will be conducted to test scalability and convergence of the algorithm and also to evaluate its behavior compared to other relevant optimization methods that support distributed execution. This is a possible direction for future work, however, we advocate that a thorough comparison could in fact be a separate paper in itself as it would require fine tuning of multiple parameters in case of many available methods to be able to look into the scenario at hand with confidence.

# References

1. Amazon Elastic Compute Cloud (2009), `http://aws.amazon.com/ec2` (Last checked: September 28, 2009)
2. Dean, J.: Software engineering advice from building large-scale distributed systems (2009), `http://research.google.com/people/jeff/stanford-295-talk.pdf` (Last checked: September 28, 2009)
3. Fernandez-Baca, D.: Allocating modules to processors in a distributed system. IEEE Tran. on Software Engineering 15(11) (1989)
4. Csorba, M.J., Meling, H., Heegaard, P.E., Herrmann, P.: Foraging for Better Deployment of Replicated Service Components. In: Senivongse, T., Oliveira, R. (eds.) 9th Int'l Conf. on Distributed Applications and Interoperable Systems (DAIS 2009), June 2009. LNCS, vol. 5523, pp. 87–101. Springer, Heidelberg (2009)
5. Csorba, M.J., Heegaard, P.E., Herrmann, P.: Adaptable model-based component deployment guided by artificial ants. In: 2nd Int'l Conf. on Autonomic Computing and Communication Systems (Autonomics), September 2008, ICST/ACM (2008)
6. Heegaard, P.E., Helvik, B.E., Wittner, O.J.: The Cross Entropy Ant System for Network Path Management. *Telektronikk* 104(01), 19–40 (2008)
7. Meling, H., Gilje, J.L.: A Distributed Approach to Autonomous Fault Treatment in Spread. In: 7th European Dependable Computing Conference, May 2008, IEEE Computer Society Press, Los Alamitos (2008)
8. Kusber, R., Haseloff, S., David, K.: An Approach to Autonomic Deployment Decision Making. In: Hummel, K.A., Sterbenz, J.P.G. (eds.) IWSOS 2008. LNCS, vol. 5343, pp. 121–132. Springer, Heidelberg (2008)
9. Joshi, K., Hiltunen, M., Jung, G.: Performance Aware Regeneration in Virtualized Multi-tier Applications. In: DSN 2009 Workshop on Proactive Failure Avoidance, Recovery and Maintenance (PFARM), June 2009, IEEE Computer Society Press, Los Alamitos (2009)
10. Dorigo, M., Maniezzo, V., Colorni, A.: The Ant System: Optimization by a colony of co-operating agents. IEEE Tran. on Systems, Man, and Cybernetics Part B: Cybernetics 26(1) (1996)
11. Csorba, M.J., Heegaard, P.E., Herrmann, P.: Component Deployment Using Parallel Ant-nests. Int'l Journal on Autonomous and Adaptive Communications Systems, IJAACS (to appear, 2009); ISSN (Online): 1754-8640. ISSN (Print): 1754-8632
12. Rubinstein, R.Y.: The Cross-Entropy Method for Combinatorial and Continuous Optimization. Methodology and Computing in Applied Probability 2, 127–190 (1999)
13. Helvik, B.E., Wittner, O.: Using the Cross Entropy Method to Guide/Govern Mobile Agent's Path Finding in Networks. In: Pierre, S., Glitho, R.H. (eds.) MATA 2001. LNCS, vol. 2164, p. 255. Springer, Heidelberg (2001)

# Self-organized Data Redundancy Management for Peer-to-Peer Storage Systems

Yaser Houri, Manfred Jobmann, and Thomas Fuhrmann

Computer Science Department
Technische Universität München
Munich, Germany
{houri,jobmann,fuhrmann}@in.tum.de

**Abstract.** In peer-to-peer storage systems, peers can freely join and leave the system at any time. Ensuring high data availability in such an environment is a challenging task. In this paper we analyze the costs of achieving data availability in fully decentralized peer-to-peer systems. We mainly address the problem of churn and what effect maintaining availability has on network bandwidth. We discuss two different redundancy techniques – replication and erasure coding – and consider their monitoring and repairing costs analytically. We calculate the bandwidth costs using basic costs equations and two different Markov reward models. One for centralized monitoring system and the other for distributed monitoring. We show a comparison of the numerical results accordingly. Depending on these results, we determine the best redundancy and maintenance strategy that corresponds to peer's failure probability.

## 1  Introduction

Since the advent of the peer-to-peer (P2P) paradigm, many P2P storage systems have been designed and implemented, for example DHash [1], OceanStore [2], TotalRecall [3], and pStore [4]. Their main objective is to reliably provide persistent storage on top of unreliable, yet collaborating peers.

Persistent storage implies the availability and durability of the stored data: *Availability* assures that data can be retrieved at any time. *Durability* means that once the data are stored, they are never lost. Durable data can be unavailable for a certain period of time. Thus availability implies durability, while durable data are not always available. Maintaining data durability is less expensive than maintaining availability in terms of bandwidth and storage overhead. In this paper, we will concentrate on availability and its costs in a fully decentralized self-organizing environment.

Maintaining availability of the stored data is a challenging task since in P2P systems, peers join and leave the network dynamically. Even worse, peers often leave the network ungracefully, e.g., due to link outage, disk failure, or unexpected user behavior. In any such case it is hard to determine whether the departure of a peer is temporary or permanent. If in case of a temporary outage the redundancy is replaced too early, the system unnecessarily consumes bandwidth. On the other hand, delaying the redundancy repair too long can put the resilience of the system at risk.

T. Spyropoulos and K.A. Hummel (Eds.): IWSOS 2009, LNCS 5918, pp. 65–76, 2009.
© IFIP International Federation for Information Processing 2009

In this paper, we analytically discuss the cost of maintaining data availability in a DHT based storage system in terms of network bandwidth usage. We compare two architectures using two Markov reward models. We discuss the bandwidth cost only, because this is the bottleneck in today's systems.

The remainder of this paper is organized as follows: In section 2 we summarize the relevant related work. In section 3 we describe how we model the various redundancy systems for our analysis. Section 4 contains our analysis. In section 5 we show the numerical results of our analysis. Section 6 concludes with an outlook to future work.

## 2   Related Work

In recent years, many studies and implemented prototypes discussed the issues of data redundancy strategies, fragments placement and redundancy maintenance, both analytically and through simulation. They all argue about which redundancy strategy is better: replication or erasure codes, but to the best of our knowledge none of them addresses the issue of under which conditions and with which parameters it is better to use replication or erasure codes. In this paper, we will address this issue.

As we have mentioned above, many storage systems prototypes were implemented with different redundancy and maintenance strategies. DHash [1] uses replication to ensure data availability. DHash places the data on selected peers and uses eager repairing policy to maintain redundancy. TotalRecall [3] and OceanStore [2] both use erasure codes to reduce storage overhead. They place the data randomly on the participating peers. Unlike OceanStore, TotalRecall uses a lazy maintenance policy which allows the reintegration of the temporarily unavailable fragments and therefore reduces the maintenance bandwidth. Only the temporarily unavailable fragments that return before starting the maintenance process are reintegrated. Carbonite [5] extends this policy to allow full reintegration of the temporarily unavailable fragments. Carbonite uses a specific multicast mechanism to monitor the fragments availability.

Besides the mentioned prototypes, many analytical studies explore what redundancy configuration achieves a desired level of reliability. Many of them compare replication strategies to erasure code strategies. We believe that all of them neglect important aspects that we intend to cover with this paper.

Tati et al. [6] gave a simple analysis on how the temporary and permanent failures affect the maintenance overhead. They also studied how the fragment placement schemes affect the system capacity. But they addressed the redundancy maintenance only from the storage overhead perspective. Lin et al. [7] also compared replication and erasure codes in terms of storage overhead. They showed that replication strategy is more suitable in a low peer availability environment.

In [1, 3, 8] the authors argued that erasure code reach the same level of availability as simple replication while using much less storage space. Rodrigues et al. [9] argued back that erasure coding has its limitations, when taking the network bandwidth into account. They show that sustaining a high data availability level using erasure coding generates a great bandwidth overhead due to churn: When a peer leaves the network and another one joins, ideally the new peer would take over the data fragments that have just been lost. But in order to regenerate a data fragment, the whole object that it belongs to needs to be reconstructed first. Therefore, a new peer has to download all the fragments that are needed to reconstruct the objects for which fragments have been lost. This consumes a large amount of bandwidth, which is a limiting factor for the scalability of P2P storage systems [10]. Their solution is to maintain a copy of the whole object at one of the peers, while replicating fragments of the stored objects to other peers. Such a hybrid solution creates a bottleneck, because a peer has to replace all its objects' fragments when they get lost in the system. Vice versa the other peers have to ensure that peer's availability. Such a hybrid strategy adds great complexity because the system needs to maintain two types of redundancy.

Motivated by network coding [11,12], Dimakis et al. developed a solution to overcome the complexity of the hybrid strategy [13]. They applied the random linear network coding approach of Ho et al. [12] to a *Maximum-Distance Separable* (MDS) erasure code for storing data. But unless the new peers keep all the data they download, the performance proved to be worse than erasure codes.

Acedaski et al. showed mathematically and through simulation [14] that random linear coding strategy achieves a higher availability level than erasure codes while consuming much less storage space. They suggest to cut a file into $m$ pieces and store $k$ random linear combinations of these pieces with their associated vectors in each peer. A similar strategy was developed by Gkantsidis et al. in [15] for content distribution networks. But in both papers, the authors neglected the cost for repairing the redundancy. Therefore their results come in favor of random linear coding.

Duminuco et al. [16] showed that using linear coding causes high costs for redundancy maintenance compared to simple replication. They introduced *hierarchical codes*, a new coding scheme which offers a trade-off between storage efficiency and repairing costs. But their scheme has higher repairing costs than replication.

Wu et al. [17] studied the availability of a data object analytically and through simulation. They developed their model based on stochastic differential equations. They analyzed the performance of the distributed storage system in terms of effective storage capacity and maintenance bandwidth considering proactive and reactive maintenance strategies. But in their analysis, the authors neglected the monitoring costs. Alouf et al. [18] did a similar analysis. But their concern was availability in terms of redundancy level, not costs.

Part of our analysis is similar to the one used by Ramabhadran et al. [19], but we calculate the bandwidth costs differently. Moreover, our concern, in contrast to Ramabhadran et al. [19], is the comparison between replication and erasure coding in terms of bandwidth costs.

Other studies, e.g., Datta et al. [20], did the comparison between replication and erasure coding not in terms of bandwidth usage, but in terms of durability using different repairing strategies and resilience to correlated failures. This is not in the scope of our paper.

The main problem is that increasing the amount of replication also increases the probability that one of the involved peers fails. The probability of losing a fragment increases proportionally to the number of peers that are involved in storing the fragments. For each peer that fails, we need to repair the redundancy and thereby consume network bandwidth. The more redundancy we have, the higher are the monitoring, maintenance, and storage costs.

In this paper, we investigate different redundancy and maintenance strategies and relate their bandwidth cost to the achieved availability. We consider this the single most important criterion for selecting a redundancy strategy, because today bandwidth costs dominate over storage costs: A typical DSL up-link provides up to 1 MBit/s, whereas a typical home user HD drive provides up to 1000 GB. To the best of our knowledge that relation has not been addressed analytically, so far.

## 3   System Model

We consider a P2P storage system where the peers randomly join and leave the system. When a peer disconnects, all the data that has been stored on that peer is no longer available. Therefore, a storage system must create redundancy in order to better ensure data availability.

In this paper we compare two redundancy mechanisms: replication and erasure coding. When applying the *replication* strategy, each block of data is replicated so that we altogether have $n$ copies of that data block. Using the *erasure coding* strategy means to create $n$ fragments of data, $m$ of which suffice to reconstruct the data block. For the purpose of the paper we assume that all replica or fragments are distributed so that their potential loss is independent and identically distributed (i.i.d.). Furthermore, we attribute all loss events to ungracefully disconnecting peers that may or may not return. Concerning our analysis all other potential reasons are indistinguishable from these two reasons.

We have two possibilities of how to respond to disconnecting peers: In the first case, we consider a replica or fragment permanently lost, when the respective peer disconnects. Therefore, we immediately invoke the repair process (*eager repair strategy*). When the peer rejoins the network later, there is no point in reintegrating the recovered data again, because it has already been replaced. In the second case, we wait some time before we consider the data as lost and accordingly delay the repair process (*lazy repair strategy*). Therefore, when the peer rejoins early enough, we can omit the repair process and save the according costs.

Even though both strategies are different from a system perspective, they map to the same analytical model: Both cases can be described by a probing rate, a repair rate, the probing costs, and the repair costs. The probing rate or a repair rate can be related by the rejoining probability: The entire cost are the sum of all probes and the repair cost, when the strategy decides that the data has been permanently lost. (See analysis below.)

In order to manage the appropriate level of redundancy the system has to check which peers have disconnected. We assume a heartbeat model, where the peers mutually probe the existence of the other peers. In practice these probes must ensure that the peers actually take the effort of storing the data. But for the purpose of this paper, it is out of scope to discuss strategies that cope with malicious peers, because regardless of their actual workings all probing mechanisms map to a certain bandwidth effort. We only consider that effort.

We study two heartbeat models: In *model 1* a peer monitors one other peer at a time. If that peer is considered lost, the monitoring peer recreates one replica or redundancy fragment. In *model 2* a peer simultaneously monitors all other peers that store a replica or fragment of the respective block, so that it can recreate the redundancy level in one go, even if multiple replica or fragments have been lost.

## 4  Analysis

In this section we present a mathematical analysis of the data redundancy management models that we have introduced in section 3. As described already, we study only the effect of churn in terms of bandwidth consumption, because all peer failures can be mapped to peer outages, and bandwidth cost are the dominant cost component today. The goal of our analysis is to find the redundancy strategies that fits best the different application scenarios.

### 4.1  Redundancy Repairing Costs

For our redundancy cost discussion we need to first define the peers' failure model. Ramabhadran et al. [19] found that peer failures can be modeled with an exponential distribution, i.e. we assume a failure model with a rate $\lambda$. Without loss of generality, we simplify and assume that each peer in our system stores only one unique block or its replica. From that we calculate the cost for monitoring and restoring the desired level of redundancy. In our equations we use the superscript "$r$" for replication and the superscript "$e$" for erasure codes.

**Replication Redundancy Repairing Costs**  As peers fail, replicas are lost and need to be repaired. The resulting costs for the entire process are the sum of the monitoring costs (for $t \gg \lambda^{-1}$) and the repairing costs for the failed replicas.

$$RC^r(t) = \lambda \cdot t \cdot size(b) + M(t) \tag{1}$$

$\lambda t$ is the number of failed peers at time $t$.

Even though the monitoring costs depend on the used monitoring protocol, they can be modeled by a simple *heartbeat protocol*.

$$M(t) = \mu \cdot t \cdot N \cdot size(heartbeat) \tag{2}$$

$N$ is the number of peers that store replica or fragments in the system. In the case of replication, $N$ is also the number of replica including the original data block. In the case of erasure codes, $N$ corresponds to the number of fragments. $size(heartbeat)$ denotes the sum of all messages required for one heartbeat, and $\mu$ is the average heartbeat rate.

The monitoring costs are proportional to the redundancy level $R_L$. For replication, the redundancy level equals the number of desired replicas in the system. For erasure codes, the redundancy level is calculated as follows:

$$R_L = \frac{n}{m} , \tag{3}$$

where $n$ is the total number of fragments including the redundant ones, and $m$ is the required fragments to reconstruct the data object.

**Erasure Codes Redundancy Repairing Costs.** For erasure codes, the repairing costs consists of the sum of all messages that are needed to reconstruct a lost fragment, and again the monitoring costs.

$$RC^e(t) = \lambda \cdot t \cdot m \cdot size(f) + M(t) \tag{4}$$

$f$ denotes a the fragment and $m$ denote the number of needed fragments to reconstruct the data block in order to regenerate the lost fragments.

It seems that both equation 1 and equation 4 deliver the same results, since $size(b) \simeq m \cdot size(f)$, but this is not the case. The costs for erasure codes are slightly higher due to the higher monitoring costs. This is because to reach the same replication redundancy level using erasure codes, we will need more peers than replication.

## 4.2  Markov Reward Model

In this section, we analyze the models from section 3 with the help of two different Markov chain models. From them we calculate the repairing costs using reward functions.

Our system consists of peers which store replicas or erasure code fragments. Each peer joins the system for some limited duration $t_{on}$. If a peer that recently left rejoins before the next probe discovers its absence, we treat the peer as if it never left the system. The same principle applies when the probing strategy is such that it waits for rejoining peers.

Measurements have shown [19] that $t_{on}$ can be modeled as an exponentially distributed random variable with mean $\frac{1}{\lambda}$, where $\lambda$ is the departure rate of the peers. We assume that $t_{on}$ is independent and identically distributed for all peers in our system.

A data block $b$ on a peer is available as long as the peer participates in the system. When a peer leaves, the data block $b$ is considered lost. The system's repair mechanism compensates for this attrition and replenishes the lost replicas. To this end, it must first detect the loss of a replica or fragment, and then create a new one. In the case of erasure coding the peer must also download enough fragments before it can create the new fragment.

The whole repair process thus takes the system some time. We assume that this time, $t_{rep}$, is also an exponentially distributed random variable with mean $\frac{1}{\mu}$, where $\mu$ is the repair rate. That leads us to describe our system as a Markov chain.

At some point of time, the system has $k$ available replicas ($0 \leq k \leq n$); the remaining $n - k$ are being repaired. In state $k$, any one of the $k$ available replicas can fail. In such a case, the system goes to the state $k - 1$. When one of the $n - k$ unavailable replicas is repaired, the system goes to the state $k + 1$. The system moves from the state $k$ to the state $k - 1$ with rate $k\lambda$ and for the state $k$ to state $k + 1$ with rate $(n - k)\mu$.

For replication, the state 0 is an absorbing state. In the case of erasure codes, the state $m - 1$ is an absorbing state. In the absorbing state, the system can no longer recover the original data block. The average time span between the system initialization and this event is the expected lifetime for that block.

Figure 1 shows Markov models for our two monitoring models.

In this paper we compare both models in terms of bandwidth costs and mean life time. To this end, we add reward functions to our model.

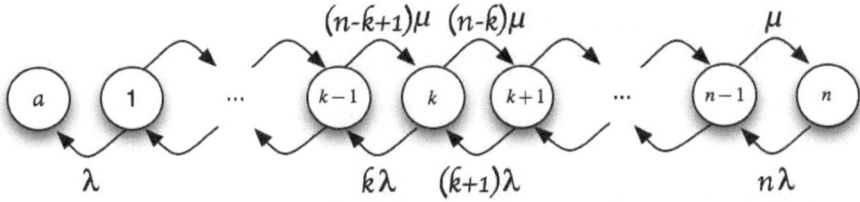

(a) Model 1: Monitor one replica or fragment only

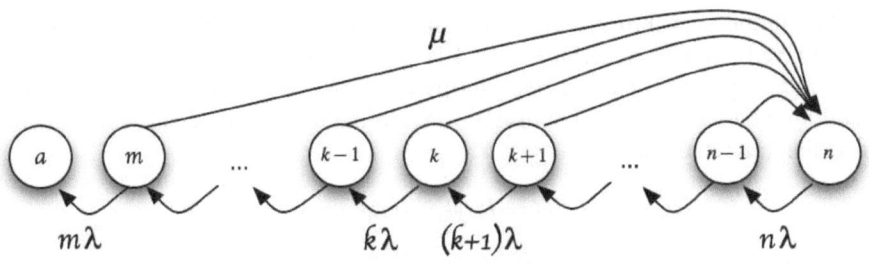

(b) Model 2: Monitor all replica or fragments

**Fig. 1.** Markov Chain Models

In state $i$ the system occupies an amount $RR_{ii}$ of storage space. This is called the *rate reward*. In our case we find

$$RR_{ii}^r = i \cdot size(b) \qquad (5)$$
$$RR_{ii}^e = i \cdot size(f) \qquad (6)$$

Each transition from state $i$ to state $j$ consumes an amount $IR_{ij}$ of bandwidth. This is called the *impulse reward*. In the case of replication the number of repaired replicas and the block size drive the impulse reward. In the erasure code case, it is driven by the number of fragments needed for reconstruction and the newly created fragments.

$$IR_{ij}^r = (j - i)size(b) \text{ , where } j > i \qquad (7)$$
$$IR_{ij}^e = (m - 1)size(f) + (j - i)size(f) \text{ , where } j > i \qquad (8)$$

Since our comparison is in terms of bandwidth costs, we only take the impulse rewards into account.

The impulse reward during the time interval $[t, t + \Delta t]$ is calculated as follows, see for example [21]:

$$IR_{ij}(t, t + \Delta t) := \overline{P}_i(t, t + \Delta t) \cdot \Delta t \cdot IR_{ij} \cdot \mu_{ij} \qquad (9)$$

$\mu_{ij}$ is the mean repair rate for a transition from state $i$ to state $j$, where $j > i$.

$$\overline{P}_i(t, t + \Delta t) := \frac{1}{\Delta t} \int_{\tau=t}^{t+\Delta t} P_i(\tau)\delta\tau \qquad (10)$$

$P_i(t)$ is the probability of being in state $i$ at time $t$. $\overline{P}_i(t, t + \Delta t)$ is the average probability of being in state $i$ during the time interval $[t, t + \Delta t]$.

The impulse reward for all states at time interval $[t, t + \Delta t]$ is simply the sum over all states:

$$IR(t, t + \Delta t) := \sum_{i \in \mathbb{S}} \sum_{j \in \mathbb{S}} IR_{ij}(t, t + \Delta t) \tag{11}$$

To calculate the average total reward (i.e. bandwidth costs), we need to calculate the mean life time, $T_l$:

$$E[T_l] = -P_0 \cdot (Q_c)^{-1} \cdot \mathbf{1} \tag{12}$$

$P_0$ is the initial probability vector excluding the absorbing state. $Q_c$ is the transition matrix $Q$ excluding the absorbing state. $\mathbf{1}$ is a vector of ones of appropriate size.

The total expected repairing costs during time interval $[t, t + \Delta t]$ is then calculated by adding the total expected impulse rewards to the monitoring costs from equation 2 during that time interval:

$$RC(t, t + \Delta t) = IR(t, t + \Delta t) + M(t, t + \Delta t) \tag{13}$$

In our numerical evaluation, $t = 0$ and $\Delta t = E[T_l]$.

## 5   Numerical Results

In this section we evaluate our developed models with the following parameters:

- Block size: 512 KB and 4 KB
- Heartbeat messages (total volume): 512 Byte
- Replication redundancy level: $[3, 6]$
- Erasure codes with $n = [3, 14]$ and $m = [2, 10]$

In the calculations we use $\mu$ and $\lambda$ values from company context measurement. This means that the presented life times are on the order of hundreds of days.

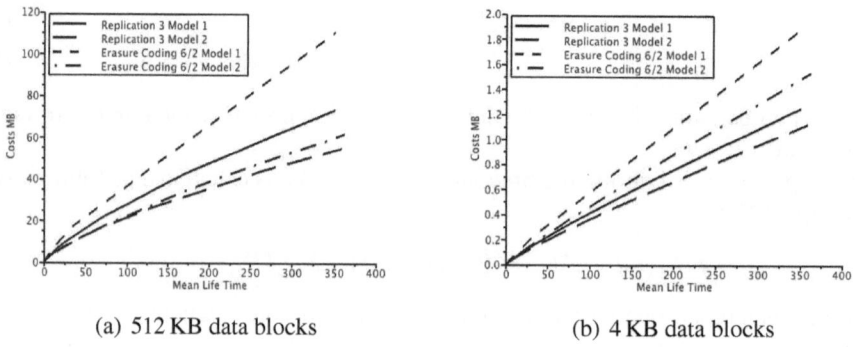

(a) 512 KB data blocks                (b) 4 KB data blocks

**Fig. 2.** Replication vs. Erasure codes

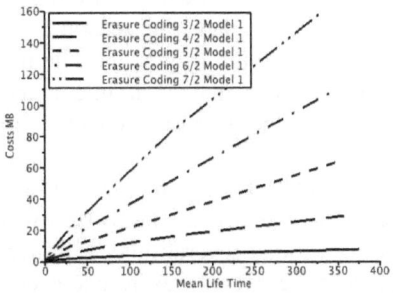
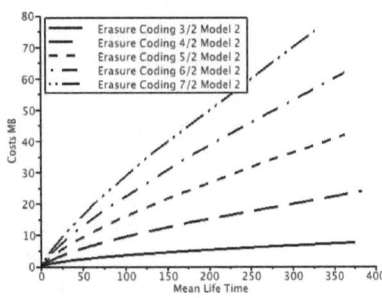

(a) Different erasure codes schemes with the (b) Different erasure codes schemes with the same $m$ using model 1 and 512 KB data blocks same $m$ using model 2 and 512 KB data blocks

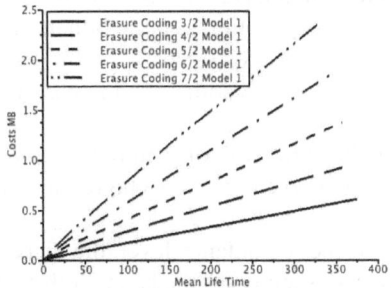
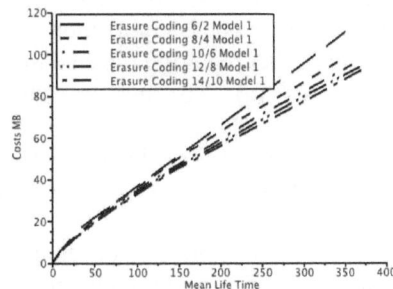

(c) Different erasure codes schemes with the (d) Different erasure codes schemes with con-
same $m$ using model 1 and 4 KB data blocks stant $n - m$ using model 1 and 512 KB data
blocks

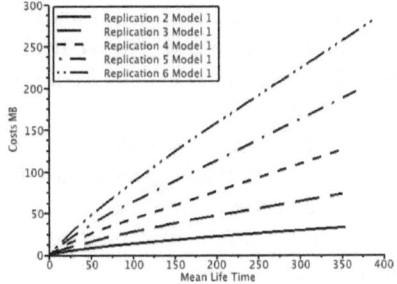
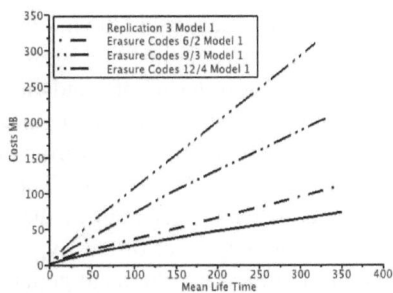

(e) Different replication levels using model 1 (f) Replication vs. different erasure codes
and 512 KB data blocks schemes with the same redundancy level, using
model 1 and 512 KB data blocks

**Fig. 3.** Effect of varying the parameters

We evaluated our models according with different $\rho = \frac{\mu}{\lambda}$. Figure 2(a) shows the bandwidth costs difference between replication and erasure codes for block size 512 KB. In *model 1*, both replication 3 and erasure codes 6/2 reach the same mean life time for the same $\rho$, but the bandwidth costs for erasure codes are higher. The reason behind this difference is that erasure codes use more peers to store the fragments. The more peers, the larger is the probability that one of them disconnects in a given time interval. Thus, erasure codes need more repairing than replication and therefore consume more bandwidth.

Erasure codes need a lower monitoring frequency than replication to reach the same lifetime. Nevertheless, in model 1 erasure codes consume slightly more bandwidth than replication, because the repairing peer has to download several other fragments. In model 2 we repair all lost blocks or fragments at once. Here, erasure codes come cheaper than replication (in model 1), because for each erasure code repair, the required fragments need to be downloaded only once, and the size of a fragment is smaller than the size of a block.

When using a block size of 4 KB, erasure codes lose their advantages (cf. figure 2(b)). Now they consume considerably more bandwidth than replication in both models. This is due to the higher repairing rate and the higher monitoring costs.

From figures 3(a), 3(b), and 3(e) we can see that the bandwidth costs increase with the redundancy level.

From figures 3(f) we can deduce that increasing the number $m$ of required fragments for erasure code scheme increases the bandwidth costs, even for the same redundancy level.

In figures 3(d) we notice that even if we decrease the redundancy level, the increase of the number $m$ of required fragments for erasure codes increases the bandwidth costs.

The above results are also valid when using smaller block sizes (not shown here).

## 6   Conclusions and Future Work

In this paper we have analyzed the influence of the various parameters on the bandwidth costs of the respective redundancy strategies. In *model 2*, erasure codes prove itself as good as replication in terms of bandwidth consumption. But erasure codes requires a higher repair rate to achieve the same mean lifetime as replication,.

In *model 1*, replication outperforms erasure codes, because erasure codes involve more peers to store the fragments. Therefore the probability that a peer fails and a fragment gets lost is larger than in the replication case. Thus erasure codes need to repair the redundancy more often. Even worse, for each lost fragment, a peer needs to download $m$ required fragments to generate the lost one.

In general, the repair costs increase with the redundancy level. The higher the redundancy level, the more peers are needed to store the blocks or fragments. The more peers involved, the higher the probability that one of them fails and the larger the effort to replace the lost replica or fragment. Thus, an increased monitoring frequency outperforms an increase level of redundancy.

The higher the ratio of required fragments to the total number of fragments, the smaller the repair costs, because the more required fragments we have, the smaller the fragment size and therefore the smaller the consumed bandwidth. The higher the repairing rate, the less replica or fragments we need to reach the desired average lifetime of the data.

Depending on the failure rate and the number of allowed simultaneous peer failures in our network, we can tune the redundancy level, the repairing rate and the redundancy scheme to achieve the required availability in the system.

There are some other techniques like *network coding* [13] and *hierarchical codes* [16]. These techniques trade bandwidth for storage space. As described above, we consider this a bad trade, because today the bandwidth costs dominate over the storage costs. Even more, the available bandwidth of the individual peers in the system determines the rate of the repair process. Therefore, the value of the repair rate is limited by the available bandwidth at the peers. It is among our future work to study the effect of the available bandwidth on the repair process in terms of delay. We also intended to study the effects of read requests and correlated failures on the repairing process.

Finally we note that this analytical work is part of the effort to improve IgorFS. In the course of these workings, we will implement a data redundancy management protocol in IgorFs [22] and evaluate it in PlanetLab.

# References

1. Dabek, F., Li, J., Sit, E., Robertson, J., Kaashoek, M.F., Morris, R.: Designing a DHT for low latency and high throughput. In: NSDI, pp. 85–98 (2004)
2. Kubiatowicz, J., Bindel, D., Chen, Y., Czerwinski, S., Eaton, P., Geels, D., Gummadi, R., Rhea, S., Weatherspoon, H., Weimer, W., Wells, C., Zhao, B.: OceanStore: An architecture for global-scale persistent storage. In: Proceeedings of the Ninth International Conference on Architectural Support for Programming Languages and Operating Systems (ASPLOS 2000), Boston, MA, November 2000, pp. 190–201 (2000)
3. Bhagwan, R., Tati, K., chung Cheng, Y., Savage, S., Voelker, G.M.: Total recall: System support for automated availability management. In: Proc. of NSDI, pp. 337–350 (2004)
4. Batten, C., Barr, K., Saraf, A., Trepetin, S.: pStore: A secure peer-to-peer backup system. Technical Memo MIT-LCS-TM-632, Massachusetts Institute of Technology Laboratory for Computer Science (2002)
5. Chun, B., Dabek, F., Haeberlen, A., Sit, E., Weatherspoon, H., Kaashoek, M., Kubiatowicz, J., Morris, R.: Efficient replica maintenance for distributed storage systems (2006)
6. Tati, K., Voelker, G.: On object maintenance in peer-to-peer systems (2006)
7. Lin, W.K., Chiu, D.M., Lee, Y.B.: Erasure code replication revisited. In: Fourth International Conference on Peer-to-Peer Computing (P2P 2004), pp. 90–97 (2004)
8. Weatherspoon, H., Kubiatowicz, J.: Erasure Coding vs. Replication: A Quantitative Comparison. In: Druschel, P., Kaashoek, M.F., Rowstron, A. (eds.) IPTPS 2002. LNCS, vol. 2429, p. 328. Springer, Heidelberg (2002)
9. Rodrigues, R., Liskov, B.: High availability in dhts: Erasure coding vs. replication. In: Castro, M., van Renesse, R. (eds.) IPTPS 2005. LNCS, vol. 3640, pp. 226–239. Springer, Heidelberg (2005)
10. Blake, C., Rodrigues, R.: High availability, scalable storage, dynamic peer networks: pick two. In: HOTOS 2003: Proceedings of the 9th conference on Hot Topics in Operating Systems, Berkeley, CA, USA, USENIX Association, pp. 1–1 (2003)
11. Ahlswede, R., Cai, N., Li, S.-y.R., Yeung, R.W., Member, S., Member, S.: Network information flow. IEEE Transactions on Information Theory 46, 1204–1216 (2000)

12. Ho, T., Mdard, M., Koetter, R., Karger, D.R., Member, A., Effros, M., Member, S., Member, S., Member, S., Shi, J., Leong, B.: A random linear network coding approach to multicast. IEEE Trans. Inform. Theory 52, 4413–4430 (2006)
13. Dimakis, R.G., Godfrey, P.B., Wainwright, M.J., Ramch, K.: Network coding for distributed storage systems. In: Proc. of IEEE INFOCOM (2007)
14. Acedański, S., Deb, S., Mdard, M., Koetter, R.: How good is random linear coding based distributed networked storage. In: NetCod (2005)
15. Gkantsidis, C., Rodriguez, P.: Network coding for large scale content distribution. In: IN-FOCOM 2005: Proceedings IEEE 24th Annual Joint Conference of the IEEE Computer and Communications Societies, vol. 4, pp. 2235–2245 (2005)
16. Duminuco, A., Biersack, E.: Hierarchical codes: How to make erasure codes attractive for peer-to-peer storage systems. In: Eighth International Conference on Peer-to-Peer Computing, P2P 2008, pp. 89–98 (2008)
17. Wu, D., Tian, Y., Ng, K.W., Datta, A.: Stochastic analysis of the interplay between object maintenance and churn. Comput. Commun. 31(2), 220–239 (2008)
18. Alouf, S., Dandoush, A., Nain, P.: Performance Analysis of Peer-to-Peer Storage Systems. Research Report RR-6044, INRIA (2006)
19. Ramabhadran, S., Pasquale, J.: Analysis of long-running replicated systems. In: INFOCOM 2006: Proceedings of 25th IEEE International Conference on Computer Communications, pp. 1–9 (2006)
20. Datta, A., Aberer, K.: Internet-scale storage systems under churn – a study of the steady-state using markov models. In: P2P 2006: Proceedings of the Sixth IEEE International Conference on Peer-to-Peer Computing, pp. 133–144. IEEE Computer Society, Los Alamitos (2006)
21. Lampka, K.: A symbolic approach to the state graph based analysis of high-level Markov Reward Models. PhD thesis, University Erlangen-Nuremberg (2007)
22. Kutzner, K., Fuhrmann, T.: The IGOR file system for efficient data distribution in the GRID. In: Proc. Cracow Grid Workshop CGW 2006 (2006)

# Self-organization of Internet Paths*

T. Kleiberg and P. Van Mieghem

Faculty of Electrical Engineering Mathematics and Computer Science
Delft University of Technology, P.O. Box 5031, 2600 GA Delft, The Netherlands
{T.J.Kleiberg,P.F.A.VanMieghem}@tudelft.nl

**Abstract.** The Internet consists of a constantly evolving complex hierarchical architecture where routers are grouped into autonomous systems (ASes) that interconnect to provide global connectivity. Routing is generally performed in a decentralized fashion, where each router determines the route to the destination based on the information gathered from neighboring routers. Consequently, the impact of a route update broadcasted by one router may affect many other routers, causing an avalanche of update messages broadcasted throughout the network. In this paper we analyze an extensive dataset with measurements on Internet routes between a set of highly stable testboxes for a period of five years. The measurements provide insight into the coherence between routing events in the Internet and we argue that the routing dynamics exhibit self-organized criticality (SOC). The SOC property provides an explanation for the power-law behavior that we observe in the operational times of routes.

## 1 Introduction

Interactive services in the Internet place strict bounds on the performance of end-to-end paths. Packet delay, delay variations and packet-loss have a severe impact on the quality of the Internet service and therefore it is important that end-to-end communication is reliable and predictable. The ability to control the end-to-end performance is seriously complicated by the connectionless nature of the Internet Protocol and the lack of any widely deployed Quality-of-Service implementation. As a consequence, the packets in the Internet are exposed to erratic network performance due to traffic fluctuations and routing dynamics. Traffic fluctuations can lead to temporary congestion of the router buffers, causing delay variations between the packets and packet-loss. Although congestion occurs very frequent in the Internet, measurements indicate that the traffic fluctuations are highly transient and the impact on the service performance often remains within bounds [25]. Routing dynamics correspond to the process where routing messages are propagated between sets of routers to advertise a route change. When a network event, for example a link or node failure, causes a route change, the network temporarily resides in a transient state while the routing tables of other

---

* The work is funded by the Next Generation Infrastructures foundation as part of the "Robustness and optimization of complex networks" project. The authors wish to acknowledge RIPE for providing the data and Fernando Kuipers, Steve Uhlig and Henk Uijterwaal for the valuable discussions.

T. Spyropoulos and K.A. Hummel (Eds.): IWSOS 2009, LNCS 5918, pp. 77–88, 2009.

routers are being updated. Routing dynamics contribute to most prolonged path disruptions and can last as long as 10 minutes, leading to serious degradation of Internet services [18, 20, 26, 25, 29].

In this work we study the dynamics of Internet paths with the use of an extensive dataset of traceroute measurements. In particular, we analyze how many routes are used between two end-hosts and how long a route remains operational. We regard the Internet as a "black box" and consider the path dynamics as the result of a collective behavior that organizes the thousands of autonomous nodes into a single complex system. Routing dynamics correspond to perturbations in the Internet and we argue that the statistical properties of the measured perturbations hint towards self-organized critical (SOC) behavior in the Internet. Self-organized criticality is often found in nature and other complex systems and arises as a collective result of the interaction between many autonomous sub-systems. When routers are considered as autonomous (sub-)systems that communicate via route-updates, the routing plane in the Internet can be considered as a SOC system. By comparing the characteristic features of SOC systems we argue that routing in the Internet also exhibits SOC. The presence of the SOC mechanism in the Internet implies that routing in the Internet is unpredictable in the sense that routes can change unexpectedly. The inter-event time between two routing events has no typical value and is widely varying. Packets associated with the same stream may follow entirely different routes, introducing a wide spread between the delivered packets and large delay variations. For the increasing number of real-time applications, such as interactive gaming, IP telephony, video and others, these variations can lead to dramatic degradation of the experienced quality. Furthermore, we find that permanent changes in the Internet cause the breakdown of existing routes and the discovery of new ones. On average, the number of discovered routes increases linearly in time at a fixed rate.

This paper is organized as follows: in Section 2 we briefly introduce self-organized criticality and present some features that are typical for SOC systems. Next, the measurements are described in Section 3. Section 4 contains the observations, followed by a discussion in Section 5. Section 6 presents an overview of related work. Finally, Section 7 presents the conclusions.

## 2  Self-Organized Criticality

Self-organized criticality was introduced by Bak *et al.* [3] as a property of a system that, through a self-organized process, always evolves to a "critical state", regardless of the initial state of the system. A common feature observed in SOC systems is the power-law temporal or spatial correlations that can extend over several decades. SOC systems organize into clusters, with a scale-free spatial distribution, with minimally stable, critical states. A perturbation in that system in a critical state can propagate through the system at any length scale from a local change to an avalanche by upsetting the minimally stable clusters. The magnitude of the perturbations is only limited by the size of the system. The lack of a characteristic length leads directly to the lack of a characteristic time for the resulting fluctuations. Hence, a power-law distribution arises for the lifetime distribution of the fluctuations in the system. From the inter-event time of the fluctuations a time signal can be constructed, where the perturbations are modeled as a series of Dirac pulses [12],

$$I(t) = \sum_k \delta(t - t_k) \tag{1}$$

where $t_k$ corresponds to the time of the $k$-th event. The time signal can now be transformed into the frequency domain and the power-spectrum of (1) is found as,

$$S(f) = \lim_{T \to \infty} \left\langle \frac{2}{T} \left| \sum_{k=k_{min}}^{k_{max}} e^{-j2\pi f t_k} \right|^2 \right\rangle \tag{2}$$

where $T$ denotes the whole observation time, $k_{min}$ and $k_{max}$ are the minimal and maximal values of the index $k$ in the interval of observation and the brackets $\langle \dots \rangle$ denote the averaging over realizations of the process. The averaging is necessary since we are only interested in the process that leads to the power-spectrum and not a realization of the process. From (2) it follows that the estimation of the spectrum improves as the observation time increases. It can be shown that a power-law distribution in the inter-event time leads to a power-law spectral density [11],

$$S(f) \propto 1/f^\beta \tag{3}$$

where $\beta$ is typically close to 1. The power-law spectrum in (3) is referred to as $1/f$ noise and is widely found in nature. The $1/f$ noise phenomenon is often observed in large systems that act together in some connected way and can be seen as a measure of the complexity of the system. The noise can arise as the result of the coherence between events in the system, the so-called long-range-dependence. It is also seen as a naturally emergent phenomenon of the SOC mechanism [3, 4, 28].

## 3   Measurements

### 3.1   Methodology

The measurement apparatus consists of a set of testboxes that are deployed by RIPE as part of the Test Traffic Measurements service (TTM) project[1]. A testbox measures the Internet paths towards a set of pre-determined destinations by repeatedly probing the router-level path from source to destination. The path is measured by the traceroute tool, which sends probes to the destination host and infers the forwarding path by analyzing the response from the intermediate hosts. The traceroute messages are transmitted at exponentially distributed random intervals, with an average of 10 messages per hour from one source to one destination. Besides the IP path, the AS path is obtained by inspection of BGP data and matching each address in the IP path with an AS prefix[2]. In

---

[1] A detailed technical description of the design and features of the TTM testboxes can be found in [8] and on the TTM website, http://www.ripe.net/projects/ttm/.

[2] The AS path information has not been recorded in the initial phase of the project and is available only from the beginning of 2003.

the translation from the IP route to the AS path, the duplicate AS entries are removed, which result from the multiple IP hops in one AS. Hence, the AS path consists of a list of unique AS numbers.

Each source maintains its own list of destinations, which is a subset of the other testboxes deployed by RIPE. The testboxes are placed at customers' sites, typically ISPs residing in various countries, just behind their border routers. Since the enrollment of the TTM project in 1999, the number of testboxes has increased from approximately 30 up to around 150 active boxes, today. Between the roughly 160 testboxes that exist, or have existed, around 10,000 source-destination pairs have been registered, where pair (A,B) is different from (B,A). Hence, the data that is available from 1999 does not include all the testboxes available today. In addition, testboxes can be temporarily offline for managerial or other purposes and several testboxes have disappeared completely, indicating the termination of the TTM service at the customer's site. The configuration and (geo)location of the testboxes is very stable. The IP address of a testbox seldom changes and only few testboxes are discontinued. The stability of the testboxes facilitates the measurement trustworthiness in the sense that the observations indeed reflect the network state and not so much the measurement setup. The high fidelity of the measurements and the extent of the observation time make the TTM measurements an excellent set to study long-term Internet path dynamics. In fact, it is the only publicly available dataset containing router-level information for this time span with such high accuracy.

### 3.2 Dataset

Section 2 emphasizes the importance that the observation time is long with respect to the interval times between subsequent events. On the other hand, increasing the observation time reduces the number of usable source–destination pairs, because less testboxes were available in the early stage of the project. Furthermore, measurements between source–destination pairs can fail: the list of destinations of each testbox can change over time and testboxes can be temporarily offline. To decrease the influence of these dynamics in the analysis, the set of usable source–destination pairs is restricted by the maximum time a source–destination pair was inactive. Increasing the stringency on the outage restrictions will reduce the number of usable source–destination pairs. Hence, a trade-off is made between the number of usable source–destination pairs versus the observation time and outage restrictions. The resulting dataset consists of the traceroutes between all the source-destination pairs that were active the entire period from January 1, 2003 until January 1, 2008, where any outage between a source–destination pair is restricted to maximally 28 days. Source–destinations pairs of which both source and destination belong to the same AS are excluded from the dataset. The dataset, that we will denote by $\mathcal{D}$, contains 64 source-destination pairs out of a set of 10 testboxes located within 8 different European countries.

### 3.3 Measurement Artifacts

The anonymous and dynamic nature of the Internet inherently adds noise to the measurements which consequently incurs errors in the analysis. These measurement artifacts include persistent forwarding loops, transient routing loops, infrastructure failures

**Table 1.** Statistical overview of the pathologic routes and probes in $\mathcal{D}$

| Route result | probes | % probes | routes | % routes |
|---|---|---|---|---|
| Successful delivery | 30986955 | 98.90 | 13496 | 43.19 |
| Persistent forwarding loop | 24732 | 0.07 | 1422 | 4.55 |
| Transient routing loop | 411 | 0.00 | 231 | 0.73 |
| Infrastructure failure, destination not reached | 36240 | 0.12 | 2561 | 8.20 |
| Packet delivered at non-listed IP address | 3995 | 0.01 | 619 | 1.98 |
| Anonymous reply, destination reached | 278066 | 0.88 | 12920 | 41.34 |

and anonymous replies by the intermediate routers. Persistent forwarding loops are generally related to mis-configured routers, while transient forwarding loops are often a manifestation of routing dynamics. Infrastructure failures lead to premature termination of the traceroute probe. Anonymous replies are due to unresponsive routers or rejected probes. For a detailed discussion on these pathologies we refer to [18, 19]. We have also found several cases where the packet was not delivered at the correct destination address. This may be the result of erroneous routing or a configuration problem in the measurement setup such that the packet is delivered at an unknown IP address. Finally, we would like to mention the presence of "third-party" addresses as a source of noise in the traceroute measurements [9]. But since such occurrences are rare [9] we disregard them in our analysis. The classic traceroute tool developed by Van Jacobson and used in the TTM project is unable to detect such pathologies and different modifications have been developed to address short-comings of the classic traceroute tool [7, 24, 1]. These recent changes were not available in 1999 and are therefore not included in the TTM project. Routes which exhibit the above mentioned artifacts have been filtered from the dataset before processing. Table 1 presents an overview of the frequency of the measurement artifacts. From Table 1, we can deduce that the pathologies contribute to slightly more than 1 percent of the measured probes. Hence, we argue that the measurements are barely affected by the pathologies.

### 3.4    Route Fluttering

Between the successful routes, there is also a significant fraction of aliasing routes. Route aliasing can be a manifestation of load-balancing, where a group of packets that are traveling between the same source and destination traverse different routes. The packets are separated based on their packet header or simply in a round-robin fashion. As a result, the samples of the IP path in the presence of load-balancing routers will consist of rapidly alternating routes, so called fluttering routes [18]. Load-balancing is the result of a decision process inside one router, it does not involve the advertisement of any routing updates to neighboring routers and does not affect the state of the routing tables[3]. Hence, load balancing does not contribute to the routing dynamics and we will handle fluttering routes as a single route, i.e. as if the packets were sent along one route.

To identify fluttering routes, we will adopt the heuristics presented by Paxson [18]: two routes are considered the same when the paths have an equal length and differ at

---

[3] Here we assume that the routing tables are not affected by the actual traffic due to some form of traffic engineering.

maximally one consecutive hop. When routes are considered the same route, the samples of all the routes are aggregated as if they were samples from one route. E.g., if route $R_1$ is observed 1000 times and route $R_2$ is observed 500 times, then the aggregated route has 1500 observations. Prior to filtering the fluttering routes, the dataset $\mathcal{D}$ contained 13496 routes. Afterwards 8612 routes remained.

## 3.5  Metrics

The routing dynamics in the Internet can be measured by means of the time intervals between subsequent route events and their coherence. A route event can affect one or more routers on the route between a source–destination pair, such that the route is changed. Hence, the time-interval between two route events corresponds to the time that a route is operational without being interrupted, which we will denote by the *route duration*. In the RIPE measurement setup, the path between a source–destination pair is sampled at independent, exponentially distributed random intervals with an average of 360 seconds. The exact time of the traceroute call is rounded to seconds and recorded in the database. When the same path is sampled multiple, consecutive, times, only the time of the first and last call are recorded. Hence, there is no accurate information of the exact time of each call, only the number of calls and the start and end time of the sequence of calls. The number of calls qualifies as an alternative measure for the duration of the path, hence the route duration is defined as the number of successive occurrences once it is selected. Due to the Poisson measurement times, the PASTA property applies and the sampled time averages indeed reflect the real time averages [18]. The sampling rate prevents us from detecting the typically highly transient failures at the data plane due to congestion, which are typically in the order of seconds. Yet, the average inter-arrival time is sufficiently small to detect the slow route dynamics, which can last many minutes.

## 4  Observations

Figure 1 shows the IP routes between a typical source–destination pair, considered over a long period of five years. Only the 40 most dominant routes are displayed. Figure 1 depicts when routes are operational and demonstrates the presence of several phenomena in the Internet, which we will discuss here. First, Figure 1 exemplifies that route fluttering is a common artifact in the Internet and it is important to consider these oscillations in the analysis. Figure 1a shows several cases of route fluttering, e.g. at the end of 2006 equivalent routes appear that overlap in time. After resolving the equivalent IP paths, the fluttering routes have been merged, as illustrated in Figure 1b. The routes presented in Figure 1b are considered unique routes and a route change indeed corresponds to a change in the routing table and not to load-balancing.

Second, Figure 1 illustrates that route events occur frequently and at all time scales. Most of the time a single route prevails between a source–destination pair. The dominant route is sometimes interrupted, where the interruption can be very brief or sometimes last for days. The inset in Figure 1b shows all the routes between this source–destination

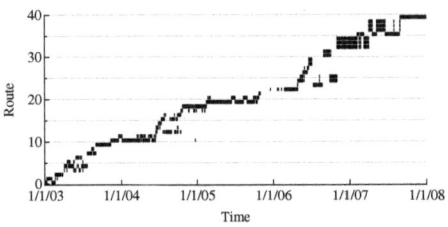

(a) Before filtering the route oscillations.

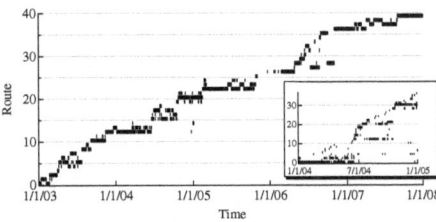

(b) After filtering the route oscillations.

**Fig. 1.** Example of a set of observed routes between one source–destination pair from the dataset. The horizontal axis represents the time axis, ranging from January 1, 2003 to January 1, 2008. The vertical axis indicates the different routes between the source–destination pair. The marked areas indicate when a particular route is operational. Only the 40 most prevalent routes have been shown out of a total of 213 routes for (a) and 158 for (b). The routes are sorted from bottom to top in order of their first appearance. The inset in (b) shows *all* the observed routes for the year 2004.

pair for the whole 2004. The inset reflects that many routes are observed only occasionally and briefly. These routes can be the result of temporal route failures or routing dynamics.

Finally, we can conclude from Figure 1 that the routes have a limited lifetime. Between a route's first and last appearance many other routes can be operational, however, in all cases the route eventually disappears and is never seen again. The restriction on these lifetimes is a consequence of the evolution taking place in the Internet, that is stimulated by changes in peering relations, the birth and death of ASes and reconfigurations at the intranet level.

Figure 2 presents the measurements on the number of *unique* routes learned since January 1, 2003 till January 1 2008, on both the AS- and IP-level, averaged over the source–destination pairs. We consider an IP route *unique* when there exists no other route with the same sequence of IP addresses. Similarly, we consider an AS path *unique* when there exists no other AS path with the same sequence of AS numbers.

The measurements in Figure 2 exhibit a remarkable linear behavior on both the AS and IP level, which is in agreement with the findings in [17]. If $R(t)$ represents the number of routes that are learned as a function of the time $t$, then according to Figure 2 we can write $R(t) = \alpha t - g(t)$, where $g(t) = o(t)$ for large $t$. Hence, $\lim_{t \to \infty} R(t)/t = \alpha - \lim_{t \to \infty} g(t)/t = \alpha$, which corresponds to the "rate" at which new routes are discovered. Figure 2 shows that the discovery rate remains fairly constant for the entire observation period. A new IP route is learned approximately every 14 to 15 days, on average. Note that this is not the same as the average duration: the route that is actually used still frequently changes back to a previous route. The first few months of 2003 the discovery rate is slightly higher due to a learning phase. At the AS-level, a new route is discovered every 38 to 39 days, on average.

Figure 3 shows the complementary cumulative distribution function (CCDF) of the duration of a route at the IP- and AS-level. The duration is measured as the number of successive occurrences of the same route until a new route becomes operational or the

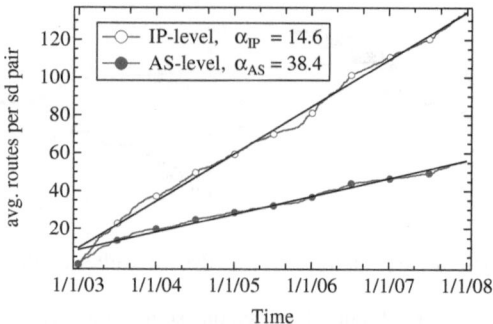

**Fig. 2.** The average number of discovered routes between the source–destination pairs at both the AS- and IP-level, counting from January 1, 2003. The measurements have been fitted with a line. The fit at the IP-level resulted in a discovery rate of 14.6 days per route, while at the AS-level the fit provided 38.4 days per route.

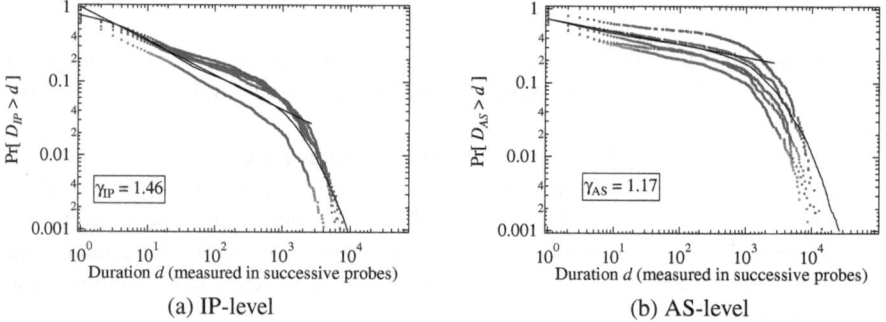

(a) IP-level                                      (b) AS-level

**Fig. 3.** The complementary cumulative distribution function of the route duration on log-log scale. The markers represent distributions for five individual source–destination pairs with typical behavior. The measurements are aggregated into a single distribution (solid line) and fitted with $F(d) = Cd^{1-\gamma}$. The fit results are presented as the straight lines; the power-law coefficients are found as $\gamma_{IP} = 1.46$ and $\gamma_{AS} = 1.17$.

routing fails and an error message is received. We have selected five source–destination pairs that do not have any end-points in common and for which the routes show typical behavior. For each pair we have computed the route duration and plotted the CCDFs in Figure 3. Figure 3 demonstrates that the shape of the distributions in both the IP-level and AS-level cases, are very similar. The requirement that the source–destination pairs do not have any node in common argues that the behavior that we observe in Figure 3 is a true feature of the Internet and not an artifact of the dataset. The aggregate result of the entire dataset, which is obtained by accumulating the results for all the source–destination pairs, is presented by the solid line. In both cases, IP and AS, the aggregate distributions strongly resemble that of the individual pairs, which indicates that the mean reflects a real property and can be considered as the average behavior of any source–destination pair in $\mathcal{D}$.

Figure 3 illustrates that the aggregate distributions of the duration relatively closely follow a power-law for the first three decades. At approximately $t = 10^3$ the distribution is cutoff followed by a steep decline. The CCDF of the aggregate result has been fitted with a power-law and is presented by the straight lines. The power-law exponent of the fits are $\gamma_{IP} \approx 1.46$ and $\gamma_{AS} \approx 1.17$. Power-law distributions with an exponent $\gamma < 2$ exhibit extreme behavior and have a divergent mean [15]. When sampling a power-law with extreme behavior, the mean is determined by the sample with the highest value, which will go to infinity when the number of samples becomes large. In real-world systems the mean is finite: the distribution is cut off in the tail because the system has a limited size. The measurements on the route durations are restricted by the limited sample space of the Internet (we cannot sample the entire Internet) and possibly the limitations of the measurement architecture (e.g. the limited uptime of testboxes due to managerial purposes, etc.) Compared to the IP routes, the power-law exponent at the AS-level is smaller, yielding a fatter tail. At the AS-level there are more long-lasting routes, which can be explained by considering that a route change at the IP level does not necessarily lead to a different AS path. This observation agrees with our finding from Figure 2. The discovery rate of AS paths is smaller than that of the IP routes, implying that multiple IP routes exists per AS path. The average duration of an AS path must therefore be greater than that of the IP routes.

Finally we will compute the spectral density of the routing dynamics and examine the existence of $1/f$ noise. The time signal (1) requires the time of occurrence of the routing events, which can be constructed from the route duration measurements by,

$$t_k = \sum_{i=0}^{k-1} D_i \qquad (4)$$

where $D_i$ is the duration of the route after $i$ route changes. We have computed $I(t)$ separately for all the source–destination pairs. The power-spectrum $S(f)$ is then computed by averaging the transformed signal between the source–destination pairs. The result is presented in Figure 4, which demonstrates the presence of $1/f$ noise.

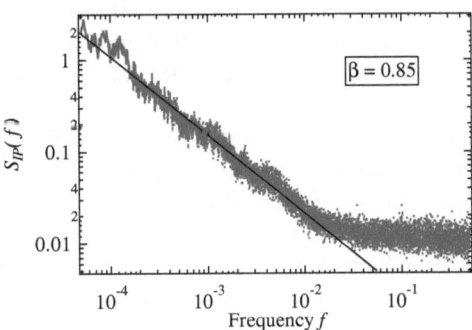

**Fig. 4.** Power specrum $S(f)$ of the inter-event time-signal $I(t)$ printed on log-log scale

# 5   Discussion

The extreme power-law behavior observed in Figures 3a and 3b is often seen in real world systems [15]. When a distribution possesses such a heavy tail, the expected value and the variance tend to infinity. In practice this implies that the observed measure is highly unpredictable. In our case it means that the route is unpredictable and packets associated with the same stream may follow (many) different routes. At first glance, the unpredictable nature of the routes seems remarkable. The packets in the Internet are routed by the connectionless Internet Protocol, where each packet is routed individually without establishing a connection prior to the transmission. However, most of the lower layers, the datalink and physical layer, use connection-oriented technologies, such that the route towards the destination is known in advance.

The changes of the topology are relatively sparse and slow, because it is often manually managed. Hence, one would expect more stable paths. Yet, the measurements indicate that route changes occur at all time scales. Such frequent changes can have negative impact on streaming services that rely on packets arriving on time and in the correct order [26,23]. At the same time, the high variability demonstrates that the Internet is resilient and fastly adapts to changes.

The power-law spectral density and power-law behavior of the inter-event time can be seen as a manifestation of self-organized criticality. Bak *et al.* [3,2] argue that SOC naturally arises in interactive dynamical systems with many degrees of freedom. The Internet is clearly a dynamical system with self-organizing behavior. The notion of SOC is a plausible explanation of the observed dynamics.

# 6   Related Work

There has been considerable work devoted to Internet path measurement in several projects, such as Skitter[4] and Rocketfuel[5]. The Skitter project, which also measures IP path information, was initiated around the same time as the RIPE TTM project, however its goal is Internet topology research which it does by querying roughly 400,000 destinations several times per day using only 20 source nodes. The Rocketfuel project combines BGP data with traceroute measurements to infer the ISP topologies [22]. To our knowledge, the database from RIPE is the only database that actually has IP-level measurements for a prolonged period (more than 9 years) at a relatively high sample rate between fixed and stable testboxes.

Several works have been published routing dynamics in the Internet, but none of them have examined the long-term correlations between routing events. Early work by Paxson [18] studies the stability of Internet paths through traceroute measurements performed during several months. Labovitz *et al.* [13] studies the path stability by combining IP and AS information. Iannaccone *et al.* [10] and Markopoulou *et al.* [14] study the path properties by monitoring the route update-messages within an AS and conclude that a small fraction of the links contributes to a large fraction of the route updates. Pucha *et al.* [20] and Wang *et al.* [26] study the impact of route changes on

---

[4] http://www.caida.org/tools/measurement/skitter/
[5] http://www.cs.washington.edu/research/networking/rocketfuel/

the path performance w.r.t. packet delay and jitter. Our work differs from the previous works in that the dataset that we use extends over a period of five years, which exposes long-term effects. Furthermore, we associate the routing dynamics with self-organized criticality and argue that routing is unpredictable leading to large variations in the path performance. Several works have related the SOC mechanism and $1/f$-noise to the Internet. The self-similarity and long-range dependence of traffic patterns often reported in the Internet are considered related to the $1/f$ noise phenomenon [6, 27]. Csabai [5] measured the round trip times of packets and showed that the correlation between the round-trip times produces $1/f$ noise in the power spectrum. Ohira *et al.* [16] and Solé *et al.* [21] demonstrated that computer networks with self-organizing behavior show the maximum information transfer and efficiency at the critical state. In addition, Solé *et al.* demonstrate that near criticality, the network performance shows the highest variability in terms of packet latency. In this work we study the inter-event times of route changes in the Internet and argue that unpredictability and instability of Internet routes may be related to SOC.

## 7 Conclusions

Through analysis of traceroute measurements we have studied the lifetime of routes and the route dynamics. Based on the observations from our dataset we find that routing in the Internet is highly dynamic and results in unpredictable route durations. The actual cause of the perturbations in the Internet is hard to retrieve, since it is often related to configuration changes within and between ASes. The extreme power-law behavior suggests that the Internet exhibits self-organized criticality. The power spectrum obtained from the inter-event time of route changes confirms our conjectures. The impact of SOC at performance of applications and services on the Internet remains difficult to evaluate and requires further investigation. The unpredictability introduced by SOC may lead to quality degradation of Internet services, in particular services that heavily depend on a timely delivery of the packets. At the same time, the adaptability of the Internet paths demonstrates the resilience against failures and attacks.

## References

1. Augustin, B., Cuvellier, X., Orgogozo, B., Viger, F., Friedman, T., Latapy, M., Magnien, C., Teixeira, R.: Avoiding traceroute anomalies with Paris traceroute. In: Internet Measurement Conference (IMC), October 2006, pp. 153–158. ACM Press, New York (2006)
2. Bak, P.: How Nature Works: Copernicus. Springer, New York (1996)
3. Bak, P., Tang, C., Wiesenfeld, K.: Self-organized criticality: An explanation of the 1/f noise. Physical Review Letters 59(4), 381–384 (1987)
4. Christensen, K., Olami, Z., Bak, P.: Deterministic 1/f noise in nonconserative models of self-organized criticality. Physical Review Letters 68(16), 2417–2420 (1992)
5. Csabai, I.: 1/f noise in computer network traffic. Journal of Physics A: Mathematical and General 27(12), L417–L421 (1994)
6. Field, A.J., Harder, U., Harrison, P.G.: Measurement and modelling of self-similar traffic in computer networks. IEE Proceedings Communications 151(4), 355–363 (2004)

7. Gavron, E.: NANOG traceroute (1995),
   ftp://ftp.login.com/pub/software/traceroute/
8. Georgatos, F., Gruber, F., Karrenberg, D., Santcroos, M., Susanj, A., Uijterwaal, H., Wilhelm, R.: Providing active measurements as a regular service for ISPs. In: Passive and Active Measurement Conference PAM, Springer, Heidelberg (2001)
9. Hyun, Y., Broido, A., Claffy, k.: On third-party addresses in traceroute paths. In: Passive and Active Measurement Workshop (April 2003)
10. Iannaccone, G., Chuah, C.-N., Mortier, R., Bhattacharyya, S., Diot, C.: Analysis of link failures in an IP backbone. In: Internet Measurement Conference (IMC), November 2002, pp. 237–242. ACM Press, New York (2002)
11. Kaulakys, B., Gontis, V., Alaburda, M.: Point process model of 1/f noise vs a sum of Lorentzians. Physical Review E 71(5), 51–105 (2005)
12. Kaulakys, B., Meškauskas, T.: Modeling 1/f noise. Physical Review E 58(6), 7013–7019 (1998)
13. Labovitz, C., Ahuja, A., Jahanian, F.: Experimental study of Internet stability and backbone failures. In: FCTS, June 1999, pp. 278–285. IEEE Computer Society, Los Alamitos (1999)
14. Markopoulou, A., Iannaccone, G., Bhattacharyya, S., Chuah, C.-N., Diot, C.: Characterization of failures in an IP backbone network. In: INFOCOM, March 2004, IEEE, Los Alamitos (2004)
15. Newman, M.E.J.: Power laws, Pareto distributions and Zipf's law. Contemporary Physics 46(5), 323–351 (2005)
16. Ohira, T., Sawatari, R.: Phase transition in computer network traffic model. Physical Review E 58(1), 193–195 (1998)
17. Oliveira, R.V., Zhang, B., Zhang, L.: Observing the evolution of Internet AS topology. In: SIGCOMM, August 2007, pp. 313–324. ACM Press, New York (2007)
18. Paxson, V.: End-to-end routing behavior in the Internet. IEEE/ACM Transactions on Networking 5(5), 601–615 (1997)
19. Paxson, V.: Measurements and Analysis of End-to-End Internet Dynamics. PhD dissertation, University of California, Lawrence Berkeley National Laboratory (April 1997)
20. Pucha, H., Zhang, Y., Mao, Z.M., Hu, Y.C.: Understanding network delay changes caused by routing events. In: SIGMETRICS, June 2007, ACM Press, New York (2007)
21. Solé, R.V., Valverde, S.: Information transfer and phase transitions in a model of Internet traffic. Physica A 289(3), 595–605 (2001)
22. Spring, N.T., Mahajan, R., Wetherall, D., Anderson, T.E.: Measuring ISP topologies with Rocketfuel. IEEE/ACM Transactions on Networking 12(1), 2–16 (2004)
23. Teixeira, R., Rexford, J.: Managing routing disruptions in Internet service provider networks. IEEE Communications Magazine 44(3), 160–165 (2006)
24. Toren, M.: Tcptraceroute (2001), http://michael.toren.net/code/tcptraceroute/
25. Wang, F., Feamster, N., Gao, L.: Measuring the contributions of routing dynamics to prolonged end-to-end Internet path failures. In: GLOBECOM, November 2007, IEEE Computer Society Press, Los Alamitos (2007)
26. Wang, F., Mao, Z.M., Wang, J., Gao, L., Bush, R.: A measurement study on the impact of routing events on end-to-end Internet path performance. In: SIGCOMM, September 2006, pp. 375–386. ACM Press, New York (2006)
27. Willinger, W., Taqqu, M.S., Sherman, R., Wilson, D.V.: Self-similarity through high-variability: Statistical analysis of Ethernet LAN traffic at the source level. IEEE/ACM Transactions on Networking 5(1), 71–86 (1997)
28. Yuan, J., Ren, Y., Shan, X.: Self-organized criticality in a computer network model. Physical Review E 61(2), 1067–1071 (2000)
29. Zhang, Y., Paxson, V., Shenker, S.: The stationarity of Internet path properties: Routing, loss, and throughput. ACIRI technical report, AT&T Centre for Internet Research at ICSI (2000)

# Non-Sticky Fingers: Policy-Driven Self-optimization for DHTs

Matti Siekkinen[1,2] and Vera Goebel[2]

[1] Helsinki University of Technology, Dept. of Computer Science and Engineering,
P.O. Box 5400, FI-02015 TKK, Finland
matti.siekkinen@tkk.fi

[2] University of Oslo, Dept. of Informatics, Postbox 1080 Blindern, 0316 Oslo, Norway
goebel@ifi.uio.no

**Abstract.** It is a common situation with distributed hash tables (DHT) that insertions and lookups frequently target only specific fractions of the entire value range. We present in this paper a self-optimization scheme for DHTs that optimizes the routing behavior in such situations. In our scheme, called Non-Sticky (NS) fingers, each node continuously measures the routing behavior and guides neighboring nodes to adjust their NS fingers (a subset of all the long distance links that the node establishes) accordingly in order to shortcut the most popular sections of routes. Our scheme enables self-optimization, which means that it adapts to the current system state and only operates when advantageous. It is also policy-driven, which means that the application can specify its policy on the tradeoff between performance and cost efficiency. We implemented the NS-fingers scheme for an existing order-preserving DHT and report the evaluation results. Our simulation results show that in a realistic application scenario, NS-fingers can halve the number of routing hops.

## 1 Introduction

Distributed hash tables (DHT) [1, 2, 3] provide efficient data exchange for completely distributed applications. These systems allow to lookup a node that stores a particular data value by specifying a key corresponding to that value.

In this paper, we present a generic self-optimization scheme for minimizing the length of routes in DHTs in cases where popularity of some value ranges in the key space is higher than others. The key observation is that in such a case, certain hops or sequences of hops (i.e. portions of entire route) become more utilized than others. Therefore, it makes sense to try to optimize these routes by making them pass through as few intermediate hops as possible. We call our scheme Non-Sticky fingers (NS-fingers) due to the way it functions: In DHTs, a node establishes a set of long distance links, a.k.a. fingers in Chord [1], in addition to "nearby" neighbors. The destination nodes of these fingers can be chosen in various ways. In Chord for instance, each node establishes fingers to nodes that are the powers of two distance in hops from the node (i.e. 2, 4, 8, etc. hops away from the node). These fingers are proven to enable efficient logarithmic routing performance. In NS-fingers, a subset of all fingers of a node are selected to be non-sticky. These fingers are continuously adjusted according to the estimation of

T. Spyropoulos and K.A. Hummel (Eds.): IWSOS 2009, LNCS 5918, pp. 89–100, 2009.
© IFIP International Federation for Information Processing 2009

demand for given links. In this way, we strive for shortening the most popular routes which may be shared portions of routes from many different source nodes to many different destination nodes.

In specific cases, reducing the number of routing hops with shortcuts may actually increase the end-to-end delay of the path. This is because the overlay topology does not necessarily reflect the underlying network-layer topology or geographical distance. Thus, neighbor nodes in the DHT may be located far away from each other geographically or many hops away in the network-layer topology. Therefore, we also present a simple extension of our scheme to consider also the delay when making decisions to shortcut routes.

Many optimizations for DHT routing exist today. What makes our scheme stand out from the crowd is that it proposes neither to grow the size of routing tables nor to add extra pointers or hints to optimize routing to the entire key space. We also do not change the forwarding procedure. We simply propose to dynamically adjust the routing table entries based on current demand. We show in later sections that such a straightforward modification in the routing table maintenance can deliver an impressive improvement in routing efficiency. The cost of this improvement is some additional load in terms of messages sent and memory used (see discussions in Section 3.5). Our scheme is an add-on that can be applied in principle to any kind of DHT. This kind of optimization approach is valuable especially for resource constrained devices which might not be able to afford to scale up the DHT routing table sizes.

Where do such non-uniform distributions occur that call for this kind of an optimization scheme? We give two example application scenarios that use *order-preserving* DHTs. Such DHTs (e.g. [4,5]) use order-preserving hash functions, or perform no hashing at all like in [4], in order to process range queries efficiently. First, consider locality aware applications using virtual network coordinates (e.g. [6]). In such applications, nodes store their coordinates and lookup nodes in their vicinity, that is, nodes having coordinates close to its own. These lookups can be efficiently expressed using range queries. As a second example, consider peer-to-peer video streaming where information about the blocks of video that a node currently has are inserted by the node and, consequently, looked up by other nodes. When a particular node is playing the stream, it queries a specific range of blocks at a time corresponding to its playout buffer. Note that using a standard DHT would in both cases require performing separately lookups corresponding to all the values within the ranges.

In both examples, insertions and lookups can have a high level of locality, that is, a given node often inserts and looks up similar value ranges: The lookups performed by a node discovering other nodes in its vicinity based on their coordinates exhibit a high degree of locality. Furthermore, if there are clusters of nodes, the value ranges corresponding to the coordinates of those regions become frequently addressed. As for the P2P video streaming example, the range of blocks that a particular node is interested in at a specific time instance exhibits temporally a high level of locality. Similarly, a node stores only information about pieces that it has just downloaded following the progress of viewing the video stream. In addition, a given video stream may experience a flash crowd phenomenon meaning that at a particular time (usually in the beginning) the stream becomes very popular. In such a case, the value ranges of the blocks following the progress of viewing of the nodes forming that flash crowd are very frequently addressed. DHTs without any optimizations treat ranges equally so that the expected number of hops to any range of values is the same. Our scheme would adapt in these examples to provide few-hop routes to the heavily used fraction of the range by increasing the expected number of hops to the other portions of the range that are little or not at all used.

Our scheme is *self-optimizing*: First of all, nodes continuously perform optimizations based on latest observed routing behavior in the system. In this way, the system adapts when the most popular range shifts, for instance. Second, the changing state of the system, e.g. the popular range size and the number of nodes, determines whether using the scheme is beneficial. NS-fingers can be made to adapt to this changing state. To enable this, we parameterize the scheme in order to provide "control knobs". By tuning these control knobs, its behavior can be adapted to the changing state of the system. For example, it can be turned off if the popularity distribution of the ranges is uniform.

The self-optimization is *policy-driven*: Tuning the control knobs allows to determine a tradeoff between how aggressively route optimizations are performed and how cost efficiently the scheme operates in terms of extra control messages routed. Thus, applications can specify their policy wrt. this tradeoff, as a result of which the scheme tunes its control knobs to comply with this policy, i.e. it *self-configures* its parameters based on estimations of the current state of the system and this policy.

The contributions of this paper are the following:

- We present a self-optimization scheme for routing in DHTs that is based on non-uniform distribution of popularity of value ranges. The self-optimization can be controlled through a set of parameters and, thus, allows customizing the routing behavior for a policy specified by a given application.
- We present an extension of the scheme that takes into account the delay, in addition to routing demand, of links when making decisions on shortcutting paths.
- We implement the scheme for an existing order-preserving DHT and show through simulations that in a realistic application scenario, NS-fingers can deliver up to 50% reduction in the average number of routing hops.

## 2   Related Work

Existing related work commonly focuses either on a growing the size of the actively managed routing table for increased performance or on relying on some additional lightweight, potentially obsolete information for routing. Considering proposals in the first category, Beehive [7] relies on Zipf-like query distributions and uses proactive replication to tradeoff resource consumption for improved lookup performance. Replication can be also problematic if data is volatile, i.e. has a relatively short lifetime. The work in [8] proposes to maintain complete routing tables using an aggressive hierarchical update protocol. Kelips [9] also achieves better lookup performance through increased routing table sizes and update traffic. EpiChord [10] maintains reactively a large routing state and copes with the possibly outdated routing state by using parallel lookups. As for the second category, "ShortCuts" approach [11] uses soft-state hints in local and global levels to improve the routing. Mercury incorporates simple route caching which can be used as a complement to our scheme. NS-fingers differs from these approaches in that we consider constant-size routing tables so that the applications using the DHT can set this size in order to control the resource requirements. Then, these allocated entries are dynamically adjusted according to the current lookup and insertion patterns. Furthermore, we provide the applications the control knobs to set the tradeoff between the aggressiveness and resource consumption of the optimization scheme itself.

"Interest-based shortcuts" presented in [12] is also based on creating shortcuts in P2P system based on interests. However, the scheme is targeted for Gnutella, an unstructured P2P system. Caching routes from source to destination as shortcuts in a DHT would be similar to the way their system works in Gnutella. Our approach is to continuously adjust routing table entries in order to shortcut one popular hop at a time.

In [5], the authors present a way to construct efficient routing tables in an order-preserving DHT that relies on estimates of hop counts between nodes. The scheme works also for skewed distributions of data values. The difference to our scheme is that we continuously adjust the routing based on recently measured behavior.

Yet another kind of approach is SkipNet which controls data placement by organizing data items according to their string names, which enables it to guarantee routing locality. SkipNet is useful when the data items stored by a node have certain persistent locality, e.g. content internal to an organization. However, this is not always the case. For example, while network coordinates exhibit temporarily locality (see examples in Section 1), nodes can move to another location in the coordinate system in which case the locality no longer holds. Also in the video streaming example, the locality of the data and lookups and the ranges that are most frequently addressed change continuously. Our scheme is able to optimize the routes even if the locality of the data and lookups change over time because the optimization is performed based on observed routing behavior at run time.

## 3   Non-Sticky Fingers

### 3.1   Overview

Our optimization scheme optimizes routing behavior when the popularity of the attribute value range is skewed. As we discussed in the introduction, such a case can occur when there is locality in the data or queries or when many nodes insert values that fall into the same ranges or query the same ranges resulting in a kind of flash crowd phenomenon.

We consider ring structured order-preserving DHTs where each node establishes links to predecessors and successors and, in addition, a number of long distance links, a.k.a. fingers in the literature. While we focus on a ring structured DHT in this paper, the scheme can be applied to other geometries as well (cf. Section 3.5). These fingers enable efficient routing within the overlay: simply passing data items and queries to successor or predecessor nodes would result in overall very inefficient routing (O(N)). With long distance links, it is possible to achieve logarithmic routing performance. The set of fingers can be static throughout the operation of a given node or can be periodically rebuilt. Nodes also maintain a set of reverse neighbors which are the nodes having a long distance link to the node.

We propose to have a set of dynamic long distance links, i.e. Non-Sticky fingers. The idea is that a node establishes $k$ fingers out of which it chooses a set of $l$ that will be Non-Sticky (NS). These NS fingers are adjusted continuously according to the most popular routes while the remaining $k - l$ fingers are static in the sense that they are only rebuilt periodically. The rationale is that when there is great demand for a particular multi-hop route within the ring, we make shortcuts to that route step by step in such a way that eventually the items are routed with as low number of hops as possible (with only one hop if possible) from the starting point to the end point of this popular route. The simplest case is when we shortcut a route so that data items are routed directly from the source node to the destination node. In such a case, using route caching at the source is a sufficient solution. However, note that a highly utilized route can also be an intermediate segment of many different routes from source to destination nodes. In such a case, caching a route directly from source to destination is not an optimal solution. We compare the performance of our scheme to route caching in Section 4.

Our scheme establishes the $k - l$ "normal" long distance links following the harmonic probability distribution function $(p_n(x) = 1/(x \ln n)$, when $x \in [1/n, 1])$ similarly to Symphony [13], which guarantees expected path length of $O(\frac{1}{k-1} \log^2 n)$ hops in a $n$ node network according to the Small World phenomenon [14]. This method gives us the flexibility to choose the number of NS-fingers to establish while still guaranteeing a certain level of routing performance in case of completely random data and query distributions. In this way, we sacrifice some of the routing performance bounds to little used ranges in order to optimize the routing towards the most popular ranges.

## 3.2   Adjusting Non-Sticky Fingers

The NS-fingers are adjusted according to the estimated service demand of links. Each node measures the service demands of each of its long distance and successor links to other nodes. Service demand $S_i(d)$ of the link of node $i$ to destination range $d$ is defined as the number of data items or queries forwarded per time unit using that link. Each node also keeps track of $S_i^j(d)$ which is the portion of service demand generated by reverse neighbor $j$ i.e. a node that has a successor or long distance link to node $i$. Those data items and queries that originate from the node itself are excluded from the computation of the service demands $(i \neq j)$. Hence, the service demands reflect the amount of services that the node provides as an intermediate hop for other nodes. Obviously, $\sum_j S_i^j(d) = S_i(d)$.

Each node checks periodically the service demands of its links and chooses the link to range $d_{max}$ which is the link with the highest demand. The node then sends adjustment requests to those reverse neighbors that contributed to the demand, i.e. node $i$ sends requests to nodes $j|S_i^j(d_{max}) > 0$. This request contains the ID of the node that is responsible for the target range $d_{max}$ and the service demand $S_i^j(d_{max})$. Figure 1 illustrates the scheme in a very simple scenario. In the figure, node 7 chooses the link to node 8 as the heaviest link because the computed service demand is 8 compared to 6 of the link to node 1. Note that the traffic originating from node 7 towards node 1 does not count. Node 7 then sends adjustment requests accordingly to nodes 5 and 3.

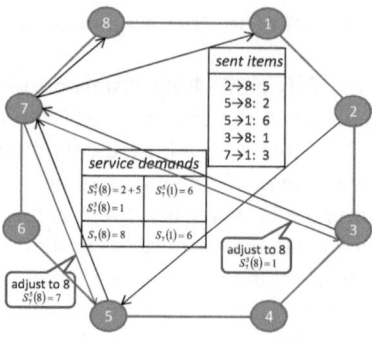

**Fig. 1.** Computing service demands

Upon receiving an adjustment request, a node stores it. Each node then adjusts its NS-fingers periodically: The node iterates through the received adjustment requests from largest service demand to the smallest. At each iteration, it compares the service demand included in the request to the smallest service demands of its currently established NS-fingers (i.e. the NS-finger with lightest load). If the demand in the request is larger, the node adjusts the NS-finger to the target range specified in the request. The iteration continues until each of the nodes NS-fingers has been adjusted or no more adjustments are necessary (i.e. the service demand of the current NS-fingers exceeds the demands in the remaining requests). It can happen that a node begins to adjust its NS-fingers right after it has checked its service demands and consequently reset the demands after sending necessary adjustment requests. In this case, the service demands of the NS-fingers are lower (close to zero) than what the true demand actually is. To avoid unnecessary adjustments in such cases, the node stores also the service demands of the NS-fingers from the previous period and uses those in addition to the current ones when making the choice whether to adjust a particular finger.

### 3.3 Delay-Aware Extension

As we pointed out in the introduction, in certain cases reducing the number of routing hops does not reduce the total end-to-end delay for the path. We present a simple extension to our scheme that takes the delay into account when making adjustments.

In the delay-aware extension of the scheme, nodes measure the delay to the nodes in their routing table while performing peer management (e.g. send keep-alive messages). Consider again Figure 1. Node 7 routinely measures the delay to nodes 8 and 1. It would then include the measured delay in the adjustment requests sent to nodes 5 and 3. When, for example, node 5 subsequently is about to adjust one of its fingers to point to node 8, it will first measure the delay to node 8 and then make the adjustment if the measured delay is less than the delay from node 7 to 8 (in the adjustment request) plus the delay to node 7 (measured during peer management). In this way, each adjustment is guaranteed to shorten the total delay in the forwarding path. The price to pay is the extra delay measurement to the new long distance neighbor candidate.

The scheme can be further extended to prioritize the shortcutting of longer delay links, as follows. In addition to adding the delay measurement to each corresponding adjustment request, all service demands of links are multiplied with the corresponding measured delay. Then this product, weighted service demand, is used in determining the "heaviest" link. In this way, when the "heaviest" link is chosen for shortcutting the delay is taken into account in addition to the amount of traffic passing through. The chosen link represents the one with largest estimated gain in total routing delay.

### 3.4   Parameters

Our scheme has three parameters: the two intervals that specify how often adjustment requests are sent ($I_r$) and how often the NS-fingers are adjusted ($I_a$), and the fraction of NS-fingers ($f = \frac{l}{k}$). The question is then how to choose these values.

In [15], we present detailed analysis of the impact of these parameter values through measurements from simulations. We first studied the tradeoff between fast route optimizations and cost efficiency of the scheme in terms of overhead messages routed. Then, we studied how a particular system setup (e.g. distribution of inserted data and number of nodes) affects the behavior of the system with a particular parameter configuration. Finally, we discuss how this information could allow us to design the scheme to be policy-driven in such a way that it self-configures its parameters depending on application specified policy (faster convergence vs. cost efficiency). Due to space constraints, we only present our main findings.

The experiments led us to conclude that the impact of the interval parameters $I_r$ and $I_a$ to the speed of convergence seems to be similar regardless of the other parameter values. The main observations are that 1) the values of the two interval parameters should be set to similar values and 2) the smaller the values, the faster the convergence. The results agree with common sense: a large adjustment interval intuitively slows down the convergence while a small adjustment interval does not help if requests are sent rarely, due to large request interval, because nodes do not know towards which range to adjust. The simulations also showed that many parameters affect the cost optimal configuration of the scheme, which suggests that analytically determining $I_a$ and $I_r$ to achieve cost optimal routing behavior for any given setup is a complex problem. Addressing this challenge is currently left for future work. Nevertheless, the results gave us a good idea on the trend of the behavior with different interval lengths which we report in [15]. It is intuitive that the scheme reduces the overall number of routing hops only when the popular range is small enough compared to the number of NS-fingers established by each node. Thus, we compared the converged mean routing hops when $f = 1$ (all fingers are non-sticky) to the average number of routing hops when $f = 0$ (i.e. no NS-fingers) with different values of $R$. We observed that, regardless of the absolute number of fingers, when $R$ is roughly larger than 0.6 the scheme no longer provides benefits. Of course, in order to check whether this result can really be generalized, we would need an analytical model of the system which we do not have at the moment.

### 3.5   Discussions

We briefly discuss in this section the extra cost imposed by the scheme, impact of churn to the scheme, applicability of the scheme to different kinds of DHTs, and the impact to load balancing.

The price to pay for the optimization using the NS-fingers scheme is increased memory demands and traffic. Each node needs to keep track of the service demands. Given that each node maintains $k$ long distance links and a few successor links, the number of reverse neighbors for a node is also around $k$. This means that maintaining the service demands requires in the worst case storing and updating $k \times k$ of state information, $k$ being typically $log_2 n$. The extra traffic introduced includes the adjustment requests and adjustments themselves, i.e. neighbor requests sent to the new target of adjusted finger. Since adjustment requests are sent periodically, they can be piggybacked into heartbeat messages in which case they come almost for free.

Churn is a common issue with DHTs. The impact of churn is similar to a DHT equipped with NS-fingers than to one without. Thus, the strategies described in [16] can be used. In fact, NS-fingers can help choose suitable timeout values (one of the important performance factors under churn according to [16]) through the delay-aware extension.

We focus in this paper on ring structured DHTs. However, the concept of NS-fingers is generic and the scheme can be applied to other geometries as well. For example, in Pastry [2], which is a kind of hybrid combining ring and tree geometries, NS-fingers could be used to shortcut routes on specific rows of the routing table (i.e. level of the tree) that shares the same ID prefix. Similarly, in CAN [3], which relies on a hypercube geometry, NS-fingers can be used to adjust the routing tables with the help of neighboring nodes considering the routing demand in order to maximize the number of "corrected" bits on each hop.

The scheme has also an impact on the load balancing of the system. Indeed, the routing load is implicitly driven via the adjustments towards the nodes that are responsible for a popular range. The advantage is that there are fewer nodes that become loaded but the drawback is that this load can be higher. To deal with this load imbalance, we can leverage existing explicit load balancing mechanisms such as the one of Mercury (see [4] for details).

### 3.6 Implementation

We implemented our self-optimization scheme on top of Mercury. We made this choice because the separation of concerns via the concept of attribute hubs allows applying our scheme flexibly to the desired hubs and customizing it to each hub separately if necessary. In addition, we can leverage some of Mercury's mechanisms for the self-configuration of the scheme (see [15] for details). Routing of items is similar in Mercury to Chord except that, thanks to the order preservation, range queries can be expressed and routed as one lookup instead of making multiple point queries.

The Mercury program code is open source. The code can be compiled to run on a simple discrete event-based simulator. This simulator does not model any queuing delays or packet losses, which enables the simulation of thousands of nodes. Such a simulation environment is sufficient for us to evaluate and analyze the behavior of our optimization scheme and we used it for all the results presented in Sections 3.4 and 4. Note that with this simulator we cannot simulate the delay-aware extension presented in Section 3.3. However, since the extension is an optimization of the scheme and the basic mechanism does not change with it, this evaluation approach is still valid.

# 4 Evaluation

In this section, we evaluate the performance of NS-fingers in different situations. Our goal is to understand the level of performance improvement the scheme can deliver in different scenarios. Unless explicitly mentioned, the simulations in this section were run with $n = 5000$ and $r = n$. In reality, it is commonly the case that some lookups and insertions also fall outside of the popular range. Thus, in all of the simulations performed for the results presented in this section, a given data item was inserted to the popular range (modelled with $R$) with 90% probability and, consequently, to the remaining non-popular range with 10% probability.

## 4.1 Stable Popular Range

We first evaluate the gain with NS-fingers scheme in terms of reduction in routing hops compared to the case without NS-fingers. In order to further put the numbers in perspective, we include the case of using simple route caching. Route caching being a complementary mechanism that can be used together with NS-fingers, we merely want to show that it alone does not provide the same benefits as our scheme.

Figure 2 shows how the average number of routing hops evolves when the size of the network grows. For the NS-fingers cases, we computed the converged mean which represents a lower bound below which the average number of hops does not go even if further adjustments are made (please, refer to [15] for a more detailed explanation and example) and for the other cases we computed the average over all routed items. In all cases $R = 0.05$ and the total number of fingers is $log_2 n$. The cache size was set to the same as the number of fingers, i.e. $log_2 n$. Note that each cache entry implies same extra maintenance cost as an additional normal finger would do. We observe that the additional cache delivers only a marginal improvement in the performance compared to NS-fingers. Furthermore, the figure shows that the number of hops do not scale similarly for the scenarios with route caching than with NS-fingers or no optimization. This observation suggests that the cache size should be increased more than logarithmically with the number of nodes in the system in order to have similarly scaling performance in terms of number of routing hops when the number of nodes in the system increases.

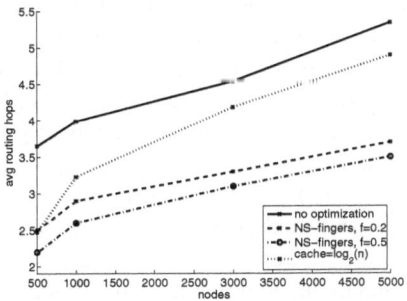
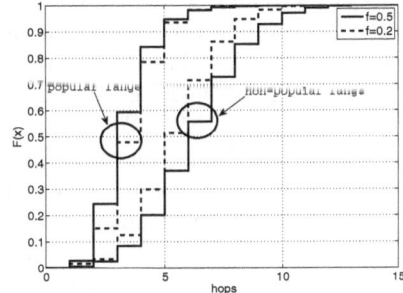
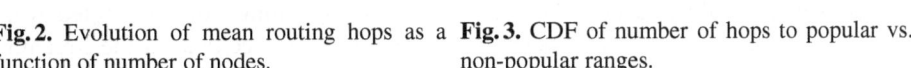

**Fig. 2.** Evolution of mean routing hops as a function of number of nodes.

**Fig. 3.** CDF of number of hops to popular vs. non-popular ranges.

Figure 3 plots a CDF of the number of hops separately for the insertions to the popular range and outside of the popular range. The figure illustrates the main tradeoff, i.e. how much the scheme penalizes the non-popular range. We can see that this tradeoff can be effectively controlled by adjusting the $f$ parameter: with a smaller number of NS-fingers (smaller $f$) the difference in number of hops is smaller between the popular and non-popular ranges.

We also looked at what happens in a likely case where we have many smaller distinct popular ranges instead of just one bigger popular range. We simulated cases having from one to five popular ranges and the sum of the range sizes being the same in each case. We observed that the fewer partitions of the popular range there are, the better the scheme works. The difference in average number of hops between having one or five ranges is approximately one. There is intuitively an advantage of having a contiguous popular range because each of the NS-fingers direct traffic towards the popular range, and within that range, routing takes very few hops.

## 4.2   Unstable Popular Range

It is a likely scenario that the popular range is not stable but instead changes with time. Remember, for instance, the video streaming example from Section 1 where the popular range of video blocks shifts continuously following the progress of the stream. In the following, we investigate what is the routing behavior in such a case. We choose the parameters according to the example: We consider a two hours long movie stored into a 1.3GByte file which is divided into roughly 5000 pieces of 256KBytes each (like in BitTorrent). Each piece has a sequence number and all the sequence numbers together form the entire range. Nodes lookup peers having a copy of specific pieces using a range of sequence numbers as a key and insert sequence numbers of the pieces they have downloaded (note that the pieces themselves are not inserted into the DHT). Since we have 5000 pieces, each of the 5000 simulated nodes is responsible of one piece. We set the popular range so that it corresponds to 1 minute's worth of pieces, i.e. $R = \frac{1}{120} = 0.0083$. The rate at which the popular range shifts is equal to $\frac{\text{whole range}}{7200}$ per second. Finally, nodes request a range of 10 pieces at a time and make an insertion for each downloaded piece yielding a following total rate of routed items: $r = \frac{\text{insertions+lookups}}{\text{duration}} = \frac{5000n+500n}{7200} \approx 0.76n$ items/s where $n$ is number of nodes.

Now consider the same scenario but with progressive download, i.e. nodes download the video file at full speed to a buffer while playing it locally. A similar situation would occur with large software updates which naturally lead to a flash crowd phenomenon. We assume a generous 15Mbit/s average download rate, which yields a download duration of 650s. We set the popular range to a window of 10s which gives us $R = \frac{10}{650} = 0.0154$, the rate at which the popular range shifts equal to $\frac{\text{whole range}}{650}$, and $r = \frac{5000n+500n}{650} \approx 8.46n$ items/s. We set $I_a = I_r = 500$ in both scenarios.

Table 1 compares the average number of hops and total cost in hops (lookups and insertions) resulting from the simulations of these two scenarios. For the cases where $f > 0$, we computed again the converged mean and included finger adjustment requests and adjustments to the total cost. Thus, the cost is the total number of hops routed by the DHT until each node has downloaded the entire movie. We see that for the streaming case, with $f = 0.5$ the average number of hops is reduced to half and the total cost is

**Table 1.** Avg hop count/total cost for the video streaming example

| Scenario | $f = 0$ | $f = 0.2$ | $f = 0.5$ |
|---|---|---|---|
| streaming | 5.2/143M | 2.9/92M | 2.6/84M |
| progressive dl | 5.3/146M | 3.3/100M | 2.8/89M |

reduced to 59% compared to the case without NS-fingers ($f = 0$). With $f = 0.2$ the improvement is slightly smaller. When downloading at full steam, there improvement is almost similar for both values of $f$.

## 5  Conclusions and Future Work

In this paper, we presented NS-fingers, a self-optimization scheme for order-preserving DHTs, which performs route optimizations in the case of non-uniform popularity distribution of data or lookup value ranges. Our future work includes further studies regarding the relationship between the current system state and cost optimal parameter configuration in order to facilitate the configuration of the parameters of the scheme, esp. for specification of the self-configuration rules. In addition, we want to be able to express the expected number of routing hops with NS-fingers for a given configuration. We would also like to evaluate the delay-aware extension by, for example, deploying a set of nodes in PlanetLab. Our simulations revealed that the routing behavior somewhat oscillates when there are not enough NS-fingers per node to cover the whole popular region. We intend to investigate whether simple schemes such as using a weighted average of the service demands can alleviate this issue. Our evaluations focused on a specific order-preserving DHT, but the NS-fingers can be applied to any DHT. While it is really the application workload that in the end determines how big performance improvement our scheme can provide, it would still be interesting to try the scheme with another kind of DHT. We would also like to study whether in certain situations some particular nodes relying on the most popular paths experience overload, and if so, how to prevent it from happening.

## Acknowledgments

The authors would like to thank Thomas Plagemann, Sasu Tarkoma, and Ovidiu Drugan for their helpful comments. This work has been funded by the Autonomic Network Architecture (ANA) project No. FP6-IST-27489 of the EU 6th Framework Programme, Situated and Autonomic Communications (SAC).

## References

1. Stoica, I., Morris, R., Karger, D., Kaashoek, M.F., Balakrishnan, H.: Chord: A scalable peer-to-peer lookup service for internet applications. In: Proceedings of SIGCOMM 2001, pp. 149–160 (2001)
2. Rowstron, A.I.T., Druschel, P.: Pastry: Scalable, decentralized object location, and routing for large-scale peer-to-peer systems. In: Guerraoui, R. (ed.) Middleware 2001. LNCS, vol. 2218, pp. 329–350. Springer, Heidelberg (2001)

3. Ratnasamy, S., Francis, P., Handley, M., Karp, R., Schenker, S.: A scalable content-addressable network. In: Proceedings of SIGCOMM 2001, pp. 161–172 (2001)
4. Bharambe, A.R., Agrawal, M., Seshan, S.: Mercury: supporting scalable multi-attribute range queries. In: Proceedings of SIGCOMM 2004, pp. 353–366 (2004)
5. Klemm, F., Girdzijauskas, S., Boudec, J.Y.L., Aberer, K.: On routing in distributed hash tables. In: P2P 2007: Proceedings of the Seventh IEEE International Conference on Peer-to-Peer Computing, pp. 113–122. IEEE Computer Society, Los Alamitos (2007)
6. Dabek, F., Cox, R., Kaashoek, F., Morris, R.: Vivaldi: a decentralized network coordinate system. In: Proceedings of SIGCOMM 2004, pp. 15–26. ACM, New York (2004)
7. Ramasubramanian, V., Sirer, E.G.: Beehive: O(1)lookup performance for power-law query distributions in peer-to-peer overlays. In: NSDI 2004: Proceedings of the 1st conference on Symposium on Networked Systems Design and Implementation, p. 8 (2004)
8. Gupta, A., Liskov, B., Rodrigues, R.: One hop lookups for peer-to-peer overlays. In: Ninth Workshop on Hot Topics in Operating Systems (HotOS-IX), Lihue, Hawaii, pp. 7–12 (2003)
9. Gupta, I., Birman, K., Linga, P., Demers, A., van Renesse, R.: Kelips: Building an efficient and stable P2P DHT through increased memory and background overhead. In: Kaashoek, M.F., Stoica, I. (eds.) IPTPS 2003. LNCS, vol. 2735, Springer, Heidelberg (2003)
10. Leong, B., Liskov, B., Demaine, E.: EpiChord: parallelizing the chord lookup algorithm with reactive routing state management. Proceedings of ICON 1, 270–276 (2004)
11. Tati, K., Voelker, G.M.: ShortCuts: Using Soft State to Improve DHT Routing. In: Chi, C.-H., van Steen, M., Wills, C. (eds.) WCW 2004. LNCS, vol. 3293, pp. 44–62. Springer, Heidelberg (2004)
12. Sripanidkulchai, K., Maggs, B., Zhang, H.: Efficient content location using interest-based locality in peer-to-peer systems. In: Proceedings of INFOCOM 2003, vol. 3, pp. 2166–2176 (2003)
13. Manku, G., Bawa, M., Raghavan, P.: Symphony: Distributed hashing in a small world. In: Proceedings of the USITS 2003 (2003)
14. Kleinberg, J.: The small-world phenomenon: an algorithm perspective. In: STOC 2000: Proceedings of the 32nd annual ACM symposium on Theory of computing, pp. 163–170. ACM, New York (2000)
15. Siekkinen, M., Goebel, V.: Non-sticky fingers: Policy-driven self-optimization for order-preserving dhts. Technical report, University of Oslo / Helsinki University of Technology (2009), http://www.tkk.fi/~siekkine/pub/siekkinen09nsfingers.pdf
16. Rhea, S., Geels, D., Roscoe, T., Kubiatowicz, J.: Handling churn in a DHT. In: ATEC 2004: Proceedings of the annual conference on USENIX Annual Technical Conference, Berkeley, CA, USA, USENIX Association, p. 10 (2004)

# Passive/Active Load Balancing with Informed Node Placement in DHTs

Mikael Högqvist and Nico Kruber

Zuse Institute Berlin
Takustr. 7, 14195, Berlin, Germany
hoegqvist@zib.de, kruber@zib.de

**Abstract.** Distributed key/value stores are a basic building block for large-scale Internet services. Support for range queries introduces new challenges to load balancing since both the key and workload distribution can be non-uniform.

We build on previous work based on the power of choice to present algorithms suitable for active and passive load balancing that adapt to both the key and workload distribution. The algorithms are evaluated in a simulated environment, focusing on the impact of load balancing on scalability under normal conditions and in an overloaded system.

## 1 Introduction

Distributed key/value stores [1,2,3] are used in applications which require high throughput, low latency and have a simple data model. Examples of such applications are caching layers and indirection services. Federated key/value-stores, where the nodes are user contributed, require minimal management overhead for the participants. Furthermore, the system must be able to deal with large numbers of nodes which are often unreliable and have varying network bandwidth and storage capacities. We also aim to support both exact-match and range queries to increase flexibility for applications and match the functionality of local key/value-stores such as Berkeley DB and Tokyo Cabinet.

Ring-based Structured Overlay Networks (SONs) provide algorithms for node membership (join/leave/fail) and to find the node responsible for a key within $O(\log N)$ steps, where $N$ is the number of nodes. One of the main advantages of SONs for large-scale services is that each node only has to maintain state of a small number of other nodes, typically $O(\log N)$. Most SONs also define a static partitioning strategy over the data items where each node is responsible for the range of keys from itself to its predecessor.

At first glance SONs may therefore seem to be a good fit for distributed key/value stores. However, the static assignment of data items to nodes in combination with the dynamic nature of user-donated resources make the design of the data storage layer especially challenging in terms of reliability [4] and load balancing.

The goal of load balancing is to improve the fairness regarding storage as well as network and CPU-time usage between the nodes. Imbalance mainly occurs

T. Spyropoulos and K.A. Hummel (Eds.): IWSOS 2009, LNCS 5918, pp. 101–112, 2009.

due to: 1) non-uniform key distribution, 2) skewed access frequency of keys and 3) node heterogeneity. First, by supporting range-queries, an order-preserving hash function is used to map keys to the overlay's identifier space. With a non-uniform key distribution a node can become responsible for an unfair amount of items. Second, keys are typically accessed with different popularity which creates uneven workload on the nodes. The third issue, node capacity differences, also impacts the imbalance. For example, a low capacity node gets overloaded faster than a high capacity node. We assume that nodes are homogeneous or have unit size, where a single physical node can run several overlay nodes.

Our main contribution is a self-adaptive balancing algorithm which is aware of both the key distribution and the item load, i.e. used storage and access-frequency. The algorithm has two modes: *active*, which triggers a node already part of the overlay to balance with other nodes and *passive*, which places a joining node at a position that reduces the overall system imbalance. In both the passive and active mode, a set of nodes are sampled and the algorithm balance using the node with the highest load.

Our target application is a federated URL redirection service. This service allow users to translate a long URL, from for example Google Maps, to a short URL. The redirection service supports look-ups of single URLs as well as statistics gathering and retrieval over time which motivates the need for range queries to execute aggregates. Popular URL redirection providers such as `tinyurl.com` have over 60 million requests per day and close to 300 million indirections.

Section 2 contains the model, assumptions and definitions that are used for the load balancing algorithm presented in Section 3. In Section 4, we evaluate the system using a simulated environment. Results from the simulation show that the algorithm improves the load imbalance within a factor 2-3 in a system with 1000 nodes. In addition, we also show that load balancing reduces the storage capacity overhead necessary in an overloaded system from a factor 10 to 8.

## 2   System Model

A ring-based DHT consists of $N$ nodes and an identifier space in the range $[0, 1)$. This range wraps around at 1.0 and can be seen as a ring. A node, $n_i$, at position $i$ has an identifier $n_i^{ID}$ in the ID space. Each node $n_i$ has a *successor*-pointer to the next node in clockwise direction, $n_{i+1}$, and a *predecessor*-pointer to the first counter-clockwise node, $n_{i-1}$. The last node, $n_{N-1}$, has the first node, $n_0$ as successor. Thus, the nodes and their pointers create a double linked list where the first and last node are linked. We define the distance between two identifiers as $d(x, y) = |y - x| \bmod 1.0$.

Nodes can fail and join the system at any time. When a node joins, it takes over the range from its own ID to the predecessor of its successor. Similarly, when a node $n_i$ fails, its predecessor becomes predecessor of $n_i$'s successor. We model churn by giving each node a mean time to failure (MTTF). To maintain the system size, a failed node is replaced after a recovery time-out.

*Storage:* When a key/value-pair or item is inserted in the system it is assigned an ID using an order-preserving hash-function in the same range as the node IDs, i.e. $[0, 1)$. Each node in the system stores the subset of items that falls within its responsibility range. That is, a node $n_i$ is *responsible* for a key iff it falls within the node's key range $(n_{i-1}^{ID}, n_i^{ID}]$.

Each item is replicated with a replication factor $f$. The replicas are assigned replica keys according to symmetric replication where the identifier of an item replica is derived from the key and the replica factor using the formula $r(k, i) = k + (i - 1) * \frac{1}{f} \mod N$, $k$ is the item ID and $i$ is the $i$th replica [5]. An advantage of symmetric replication is that the replica keys are based on the item key. This makes it possible to look-up any replica by knowing the original key. In other approaches such as successor-list replication [6] the node responsible for the key must first be located in order to find the replicas.

A replica maintenance protocol ensures that a node stores the items and the respective replicas it is responsible for. The protocol consist of two phases; the synchronization phase and the data transfer phase. In the synchronization phase, a node determines which items should be stored at the node using the symmetric replication scheme. And if they are not stored or not up-to-date, which replicas need to be retrieved. The retrieval is performed during the data transfer phase by issuing a read for each item.

*Load and Capacity:* Each node has a workload and a storage capacity. The workload can be defined arbitrarily, but for a key/value-store this is typically the request rate. Each stored item has a workload and a storage cost. A node cannot store more items than its storage capacity allows. The workload, on the other hand, is limited by for example bandwidth, and a node can decide if a request should be ignored or not. We model the probability of a request failure as $P(fail) = 1 - \frac{1}{\mu}$, where $\mu$ is the current node utilization, i.e. the measured workload divided by the workload capacity.

*Imbalance:* We define the system imbalance of a load attribute (storage or workload) as the ratio between the highest loaded node and the system average. For example, for the storage, the imbalance is calculated as $\frac{L_{max}}{L_{avg}}$. $L_{max}$ is the maximum number of items stored by a node and $L_{avg}$ is the average number of items per node.

## 3    Load Balancing Algorithm

The only way to change the imbalance in our model is to change the responsibility of the nodes. A node's responsibility changes either when another node joins between itself and its predecessor, or when the predecessor fails. Thus, we can balance the system either *actively* by triggering a node to fail and re-join or *passively* by placing a new node at an overloaded node when joining. Passive balancing uses the system churn, while active induces churn and extra data transfers. We first present the passive/active balancing algorithm followed by the placement function.

```
1   def placement ():
2       balanced_ID = ⊥
3       current_distance = ∞
4       for item in (n^{ID}_{i-1}, n^{ID}_i]:
5           distance = f(item^{ID}) # the placement function
6           if distance < current_distance:
7               balanced_ID = item^{ID} + d(item^{ID}, next(item^{ID}))/2
8               current_distance = distance
9
10      return balanced_ID
11
12  def sample ():
13      samples = [(n.load(), n)
14                    for n in random_nodes(k)]
15      return max(samples)
16
17  def passive ():
18      (n_load, n) = sample()
19      join(n)
20
21  def active ():
22      (n_load, n) = sample()
23      if n_load > local_load * ε:
24          leave()
25          join(n.placement())
```

**Fig. 1.** Passive and Active load balancing

The passive/active balancing algorithm presented in Figure 1 uses only local knowledge and can be divided into three parts. 1) sample a set of $k$ random nodes to balance with using e.g. [7], 2) decide the placement of a potential new predecessor and 3) select one of the $k$-nodes that reduce the imbalance the most. We assume that there is a join function which is used to join the overlay given an ID. passive is called before a node is joining and active is called periodically. active is inspired by Karger's [8] balancing algorithm, but we only consider the case where the node has a factor $\epsilon$ less load than the remote node. The $\epsilon$ is used to avoid oscillations by creating a relative load range where nodes do not trigger a re-join. sample calls a function random_nodes that uses a random walk or generates random IDs to find a set of $k$ nodes. The node with the highest load is returned.

## Placement Function

The goal of the placement function is to find the ID in a node's responsibility range that splits the range in two equal halves considering both workload and key distribution. When defining the cost for a single load attribute, it is optimal to always divide the attribute in half [9]. We use this principle for each attribute by calculating the ratio between the range to the left of the identifier $x$ and the remaining range up to the node's ID. The optimal position is where this ratio approaches 1. A ratio therefore increases slowly from 0 towards 1 until the optimal value of $x$ is reached, and after 1 the value approaches the total cost for the attribute.

First, let $l_r(a, b) = \sum_{i=0}^{items \in (a,b]} l(item_i)$ be a function returning the load of the items in the range $(a, b]$. $l(item_i)$ is the load of a single item and is defined arbitrarily depending on the load attribute. Second, let $n_i$ be the node at which we want to find the best ID, then the ratio function is defined as follows

$$r(x) = \frac{l_r(n_{i-1}^{ID}, x)}{l_r(x, n_i^{ID})}$$

The workload ratio, $r_w(x)$, could for example be defined using $l(item_i) = weight(item_i) + (rate_{access}(item_i) \times weight(item_i))$. The weight is the total bytes of the item and the access rate is estimated with an exponentially weighted moving mean. For the key distribution ratio, $r_{ks}(x)$, the load function is $l(item_i) = 1$. This means that $r_{ks}(x) = 1$ for the median element in $n_i$'s responsibility range. An interesting aspect of the ratio definitions is that they can be weighted in order to ignore load attributes that changes fast or taking on extreme values.

In order to construct a placement function acknowledging different load attributes, we calculate the product of their respective ratio function. The point $x$ where this product is closest to 1 is where all attributes are being balanced equally. Note that when it equals 1, it means that the load attributes have their optimal point at the same ID.

The placement function we use here considers both the key-space and workload distribution and is more formally described as

$$f(x) = |1 - r_w(x) \times r_{ks}(x)|$$

where $x$ is the ID and $n_j$ is the joining node. The ratio product value is subtracted from 1 and the absolute value of this is used since we are interested in the ratio product value "closest" to 1. Finally, when the smallest value of $f(x)$ is found, a node is placed at the ID between the item, $item_i$ preceding $x$ and the subsequent item, $item_{i+1}$. That is, the resulting ID is $item_i^{ID} + d(item_i^{ID}, item_{i+1}^{ID})/2$.

## 4   Evaluation

This section present simulation results of the passive and active algorithms. The goal of this section is to 1) show the effects of different access-load and key distributions, 2) show the scalability of the balancing strategies when increasing the system size and 3) determine the impact of imbalance in a system close to its capacity limits. Table 1 summarizes the parameters used for the different experiments.

**Effect of Workloads:** In this experiment, we quantify the effect that different access-loads and key distributions have on the system imbalance. The results from this experiment motivate the use of a multi-attribute placement function. Specifically, we measure the imbalance of the nodespace (ns), keyspace (ks) and the access workload (w).

**Table 1.** Parameters of the different experiments

|  | Nodes | Items | Replicas | $k$ | MTTF | Storage | Item Size |
|---|---|---|---|---|---|---|---|
| Effect of Workloads | 256 | 32768 | 7 | 7 | $\infty$ | $\infty$ | 1 |
| Network costs | 256 | 8192 | 7 | 7 | 1h | $\infty$ | 1-1MB |
| Size of $k$ | 256 | 8192 | 7 | 0-20 | 1h | $\infty$ | 1 |
| System size | 64-1024 | $2^{15}$-$2^{18}$ | 3 | 7 | 1h | $\infty$ | 1 |
| Churn | 256 | 8192 | 7 | 7 | 1h-1d | $\infty$ | 1 |
| Overload | 256 | 8192 | 7 | 7 | 1h | $128*7$-$1024*7$ | 1 |

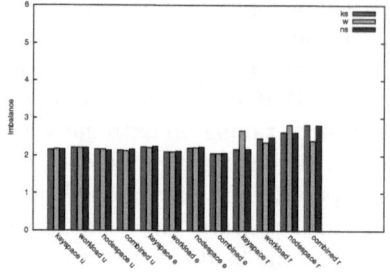
(a) Uniform key distribution

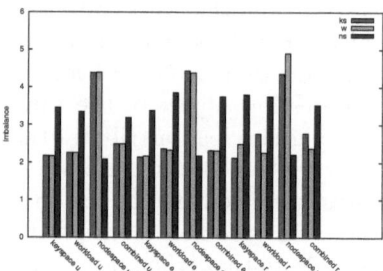
(b) Dictionary key distribution

**Fig. 2.** The effect of different access workloads and key distributions

Four different placement functions are used (x-axis in Fig. 2)

**nodespace** places a new node in the middle between the node and its predecessor, i.e. $n_i + \frac{d(n_{i-1}, n_i)}{2}$.

**keyspace** places the node according to the median item, $f(x) = |1 - r_{ks}(x)|$.

**workload** halves the load of the node, i.e $f(x) = |1 - r_w(x)|$

**combined** uses the placement function defined in section 3.

The simulation is running an active balancing algorithm with $\epsilon = 0.15$.

Workload is generated using three scenarios; uniform (u), exponential (e) and range (r). In the uniform and exponential cases, the items receive a load from either a uniform or exponential distribution at simulation start-up. The range workload is generated by assigning successive ranges of items with random loads taken from an exponential distribution. We expect this type of workload from the URL redirection service when, for example, summarizing data of a URL for the last week.

From the results shown in Figure 2, we can see that the imbalance when using the different placement strategies are dependent on the load type. Figure 2(a) clearly shows that a uniform hash-function is efficient to balance all three metrics under both uniform and exponential workload. In the latter case, this is because the items are assigned the load independently. However, for the range workload, the imbalances are showing much higher variation depending on the placement function. We conclude that in a system supporting range queries, the placement function should consider several balancing attributes for fair resource usage.

**Size of $k$:** In this experiment, we try to find a reasonable value of the number of nodes to sample, $k$. A larger $k$ implies more messages used for sampling, but also reduces the imbalance more. The results in figure 3 imply that the value of $k$ is important for smaller values of between 2-10. However, the balance improvement becomes smaller and smaller for each increase of $k$, similar to the law of diminishing returns. In the remaining experiments we use $k = 7$.

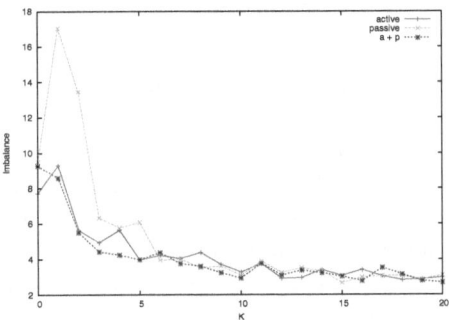

**Fig. 3.** Imbalance when increasing the number of sampled nodes

**Network costs:** We define cost as the total amount of data transferred in the system up to a given iteration. This cost is increased by the item size each time an item is transferred. Since there is no application traffic in the simulation environment, the cost is only coming from replica maintenance. That is, item transfers are used to ensure that replicas are stored according to the current node responsibilities. Active load balancing creates traffic when a node decides to leave and re-join the system.

We measure the keyspace imbalance and the transfer cost at the end of the simulation, which is run for 86400s (1 day). Each simulation has 8192 items with 7 replicas and the size of the items is increased from $2^{10}$ to $2^{20}$. The item size has minor impact on the imbalance (Fig. 4(a)). Interestingly, the overhead when using the hash-based balancing strategy as a reference, of active and passive (a+p in the figure) and active only is 5-15% (Fig. 4(b)). The passive strategy does not show a significant difference. Noteworthy is also that in a system storing around 56 GB of total data (including replicas), over 1 TB aggregated data is transferred. This can be explained with the rather short node lifetime of 3600s.

**Churn:** A node joining and leaving (churn) changes the range of responsibility for a node in the system. Increasing the rate of churn influences the cost of replica maintenance since item repairs are triggered more frequently. In this experiment, we quantify the impact of churn on transferred item cost and the storage imbalance.

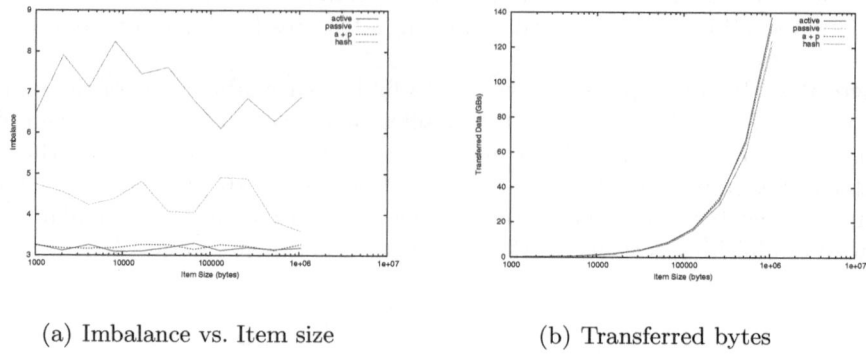

(a) Imbalance vs. Item size          (b) Transferred bytes

**Fig. 4.** Imbalance and cost of balancing for increasing item size

In figure 5(a) the node MTTF is varied from 1 to 24 hours. As expected the amount of data transferred is decreasing when the MTTF is increasing. Also as noted in the network costs experiment, the different schemes for load balancing have a minor impact on the total amount of transferred data. Figure 5(b) shows that churn has in principle no impact on the imbalance for the different strategies. This is also the case for the passive approach which only relies on churn to balance the system.

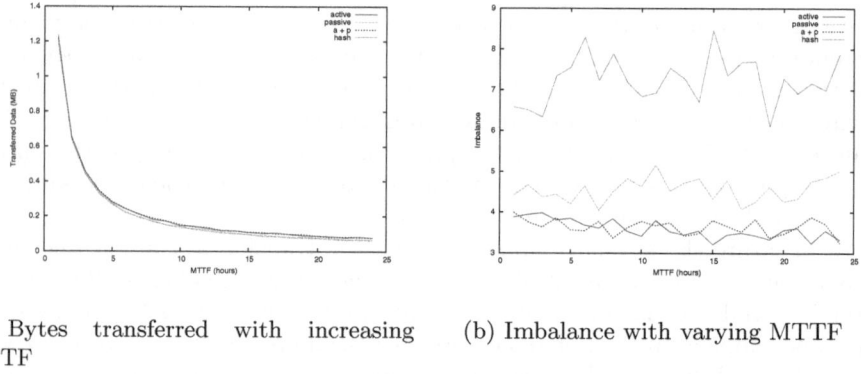

(a) Bytes transferred with increasing    (b) Imbalance with varying MTTF
MTTF

**Fig. 5.** Imbalance and network cost for varying levels of churn (MTTF)

**System size:** The imbalance in a system with hash-based balancing was shown theoretically to be bounded by $O(\log N)$, where $N$ is the number of nodes in the system [10]. However, this assumes that both the nodes and the keys are assigned IDs from a uniform hash-function. In this experiment, we try to determine the efficiency of the placement function with an increasing number of nodes and items.

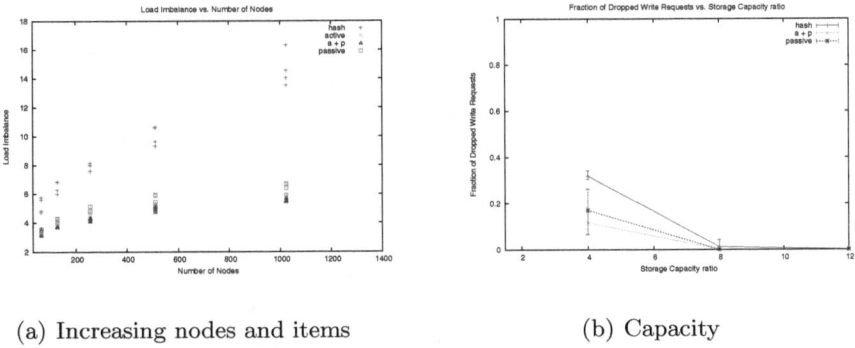

(a) Increasing nodes and items  (b) Capacity

**Fig. 6.** Imbalance of the system using different balancing strategies while increasing the system size. The right figure shows the influence of load balancing in an overloaded system.

We measure the keyspace imbalance for an increasing number of nodes between $2^5$ and $2^{10}$. In addition, for each system size we vary the number of items from $2^{15}$ to $2^{18}$. Keys are generated from a dictionary and nodes are balanced using the combined placement function. Four different balancing strategies are compared; 1) IDs generated by a uniform hash-function 2) active without any passive placement, 3) passive without any active and 4) active and passive together (a+p). For the last three, 7 nodes are sampled when selecting which node to join at or whether to balance at all.

Figure 6(a) shows that the hash-based approach performs significantly worse with an imbalance up to 2-3 times higher compared to the other balancing strategies. Interestingly, the difference in load imbalance when varying the number of items is also growing slightly with larger system sizes. All three variants of the passive/active algorithm show similar performance. The imbalance grows slowly with increasing system size and the difference for different number of items is small. Thus, we draw the conclusion that these strategies are only minimally influenced by system size and number of items. However, note that we need to perform further experiments varying other parameters such as $k$ to validate these results.

**Overload:** In a perfectly balanced system where at most one consecutive node can fail, nodes can use at most up to 50% of their capacity to avoid becoming overloaded when a predecessor fails. This type of overload leads to dropped write requests when there is insufficient storage capacity and dropped read request with insufficient bandwidth and processing capacity. Since a replica cannot be recreated when a write is dropped, this influences the data reliability. The goal of this experiment is to better understand the storage capacity overhead to avoid dropped writes.

We start the experiment such that the sum of the item weights equals the aggregated storage capacity of all nodes. Then by increasing the node's storage capacity we decrease their fill-ratio and thereby the probability of a dropped write. The system is under churn and lost replicas are re-created using a replica maintenance algorithm executed periodically at each node. The y-axis in Figure 6(b) shows the fraction of dropped write requests and the x-axis shows the storage capacity ratio. We do not add any data to the system which means that a write request is dropped when a replica cannot be created at the responsible node because of insufficient storage capacity. We measured the difference with hash-based balancing vs. the active and active + passive with 7 sampled nodes and the combined placement function.

Figure 6(b) shows that a system must have at least 10x the storage capacity over the total storage load to avoid dropped write requests when using hash-based balancing. Active and active-passive delays the effect of overload and a system with at least 8x storage capacity exhibits a low fraction of dropped requests.

## 5  Related Work

Karger et al. [8] and Ganesan et al. [11] both present active algorithms aiming at reducing the imbalance of item load. Karger uses a randomized sampling-based algorithm which balances when the relative load value between two nodes differs by more than a factor $\epsilon$. Ganesan's algorithm triggers a balancing operation when a node's utilization exceeds (falls below) a certain threshold. In that case, balancing is either done with one of its neighbors or the least (most) loaded node found. Aspnes at al. [12] describe an active algorithm that categorizes nodes as closed or open depending on a threshold and groups them in a way so that each closed node has at least one open neighbor. They balance load when an item is to be inserted into a closed node that cannot shed some of its load to an open neighbor without making it closed as well. A rather different approach has been proposed by Charpentier et al. [13] who use mobile agents to gather an estimate of the system's average load and to balance load among the nodes. Those algorithms however do not explicitly define a placement function or use a simple "split loads in half" approach which does not take several load attributes into account.

Byers et. al. [14] proposed to store an item at the $k$ least loaded nodes out of $d$ possible. Similarly, Pitoura et al. [15] replicate an item to $k$ of $d$ possible identifiers when a node storing an item becomes overloaded (in terms of requests). This technique, called the "power of two choices" was picked up by Ledlie et. al [16] who apply it to node IDs and use it to address workload skew, churn and heterogeneous nodes. With their algorithm, k-Choices, they introduce the concept of passive and active balancing. However, their focus is on virtual server-based systems without range-queries. Giakkoupis and Hadzilacos [17] employ this technique to create a passive load balancing algorithm including a weighted version for heterogeneous nodes. There, joining nodes contact a logarithmic (in system size) number of nodes and choose the best position to join at. Their focus on the other hand is on balancing the address-space partition rather than arbitrary loads. Manku [18] proposes a similar algorithm issuing one random probe and contacting a logarithmic number of its neighbors. An analysis of such algorithms using $r$ random probes each followed by a local probe of size $v$ is given by Kenthapadi and Manku [19]. However, only the nodespace partitioning is examined.

In Mercury [20] each node maintains an approximation of a function describing the load distribution through sampling. This works well for simple distributions, but as was shown in [21] it does not work for more complex cases such as filenames. Instead, [21] introduces OSCAR where the long-range pointers are placed by recursively halving the traversed peer population in each step. Both OSCAR and Mercury balance the in/out-degree of nodes. While this implies that the routing load in the overlay is balanced, it does not account for the placement of nodes according to item characteristics.

## 6    Conclusions

With the goal of investigating load balancing algorithms for distributed key/value-stores, we presented an active and a passive algorithm. The active algorithm is triggered periodically, while the passive algorithm uses joining nodes to improve system imbalance. We complement these algorithms with a placement function that splits a node's responsibility range according to the current key and workload distribution. Initial simulation results are promising showing that the system works well under churn and scales with increasing system sizes. Ongoing work include quantifying the cost of the algorithms within a prototype implementation of a key/value-store.

**Acknowledgments.** This work is partially funded by the European Commission through the SELFMAN project with contract number 034084.

## References

1. DeCandia, G., Hastorun, D., Jampani, M., Kakulapati, G., Lakshman, A., Pilchin, A., Sivasubramanian, S., Vosshall, P., Vogels, W.: Dynamo: amazon's highly available key-value store. In: SOSP, pp. 205–220. ACM, New York (2007)
2. Rhea, S.C., Godfrey, B., Karp, B., Kubiatowicz, J., Ratnasamy, S., Shenker, S., Stoica, I., Yu, H.: Opendht: a public dht service and its uses. In: SIGCOMM, pp. 73–84. ACM, New York (2005)
3. Reinefeld, A., Schintke, F., Schütt, T., Haridi, S.: Transactional data store for future internet services. Towards the Future Internet - A European Research Perspective (2009)
4. Blake, C., Rodrigues, R.: High availability, scalable storage, dynamic peer networks: Pick two. In: HotOS, USENIX, pp. 1–6 (2003)
5. Ghodsi, A., Alima, L.O., Haridi, S.: Symmetric replication for structured peer-to-peer systems. In: DBISP2P, pp. 74–85 (2005)
6. Stoica, I., Morris, R., Karger, D.R., Kaashoek, M.F., Balakrishnan, H.: Chord: A scalable peer-to-peer lookup service for internet applications. In: SIGCOMM, pp. 149–160 (2001)

7. Vishnumurthy, V., Francis, P.: A comparison of structured and unstructured p2p approaches to heterogeneous random peer selection. In: USENIX, pp. 309–322 (2007)
8. Karger, D.R., Ruhl, M.: Simple efficient load balancing algorithms for peer-to-peer systems. In: Voelker, G.M., Shenker, S. (eds.) IPTPS 2004. LNCS, vol. 3279, pp. 131–140. Springer, Heidelberg (2005)
9. Wang, X., Loguinov, D.: Load-balancing performance of consistent hashing: asymptotic analysis of random node join. IEEE/ACM Trans. Netw. 15(4), 892–905 (2007)
10. Karger, D., Lehman, E., Leighton, T., Levine, M., Lewin, D., Panigrahy, R.: Consistent hashing and random trees: Distributed caching protocols for relieving hot spots on the world wide web. In: ACM Symposium on Theory of Computing, May 1997, pp. 654–663 (1997)
11. Ganesan, P., Bawa, M., Garcia-Molina, H.: Online balancing of range-partitioned data with applications to peer-to-peer systems. In: VLDB, pp. 444–455. Morgan Kaufmann, San Francisco (2004)
12. Aspnes, J., Kirsch, J., Krishnamurthy, A.: Load balancing and locality in range-queriable data structures. In: PODC, pp. 115–124 (2004)
13. Charpentier, M., Padiou, G., Quéinnec, P.: Cooperative mobile agents to gather global information. In: NCA, pp. 271–274. IEEE Computer Society, Los Alamitos (2005)
14. Byers, J.W., Considine, J., Mitzenmacher, M.: Simple load balancing for distributed hash tables. In: Kaashoek, M.F., Stoica, I. (eds.) IPTPS 2003. LNCS, vol. 2735, pp. 80–87. Springer, Heidelberg (2003)
15. Pitoura, T., Ntarmos, N., Triantafillou, P.: Replication, load balancing and efficient range query processing in dhts. In: Ioannidis, Y., Scholl, M.H., Schmidt, J.W., Matthes, F., Hatzopoulos, M., Böhm, K., Kemper, A., Grust, T., Böhm, C. (eds.) EDBT 2006. LNCS, vol. 3896, pp. 131–148. Springer, Heidelberg (2006)
16. Ledlie, J., Seltzer, M.I.: Distributed, secure load balancing with skew, heterogeneity and churn. In: INFOCOM, pp. 1419–1430. IEEE, Los Alamitos (2005)
17. Giakkoupis, G., Hadzilacos, V.: A scheme for load balancing in heterogenous distributed hash tables. In: PODC, pp. 302–311. ACM, New York (2005)
18. Manku, G.S.: Balanced binary trees for id management and load balance in distributed hash tables. In: PODC, pp. 197–205 (2004)
19. Kenthapadi, K., Manku, G.S.: Decentralized algorithms using both local and random probes for p2p load balancing. In: SPAA, pp. 135–144. ACM, New York (2005)
20. Bharambe, A.R., Agrawal, M., Seshan, S.: Mercury: supporting scalable multi-attribute range queries. In: SIGCOMM, pp. 353–366. ACM, New York (2004)
21. Girdzijauskas, S., Datta, A., Aberer, K.: Oscar: Small-world overlay for realistic key distributions. In: Moro, G., Bergamaschi, S., Joseph, S., Morin, J.-H., Ouksel, A.M. (eds.) DBISP2P 2005 and DBISP2P 2006. LNCS, vol. 4125, pp. 247–258. Springer, Heidelberg (2006)

# Optimal TCP-Friendly Rate Control for P2P Streaming: An Economic Approach

Jinyao Yan[1,2], Martin May[3], and Bernhard Plattner[1]

[1] Computer Engineering and Networks Laboratory, Swiss Federal Institute of Technology, ETH Zurich, CH-8092, Switzerland
{jinyao,plattner}@tik.ee.ethz.ch
[2] Computer and Network Center, Communication University of China, 100024, Beijing, China
jyan@cuc.edu.cn
[3] Thomson Paris Research Lab, Thomson, France
martin.may@thomson.net

**Abstract.** TCP and TCP-friendly rate control protocols, designed for unicast, do not take neighbor connections into account in P2P networks. In this paper, we study the topic of distributed and optimal rate control for scalable video streams in P2P streaming applications. First, we propose a fully distributed and TCP-friendly network analytical model for rate control and formulate an optimization problem to maximize the aggregate utility for the P2P streams. In the model, we further extend the definition of TCP-friendliness for P2P network. Second, we propose a shadow price-based distributed algorithm for P2P Streaming that solves the optimization problem. Finally, we evaluate the performance of the proposed algorithm in terms of streaming quality and messaging overhead. Extensive simulations show that the proposed algorithms generate very small overhead and that they are optimal in terms of overall quality for scalable streams.

## 1 Introduction

Multimedia streaming over Internet has been a hot topic both in academia and in industry for two decades. Since the emergence of peer-to-peer architectures, there has been significant interest in streaming applications over peer-to-peer overlay networks [4] [5] [6]. P2P streaming does not require support from Internet routers compared to IP layer multicast, therefore, it is easy to deploy and also scaleable to very large group sizes.

Rate control is one of key technologies in multimedia communications to deal with the diverse and constantly changing conditions of the Internet. TCP, the dominant congestion protocol designed for client-server unicast communication in the Internet, is also used as rate/congestion control protocol in most of P2P streaming systems. However, using TCP for P2P streaming also has some disadvantages. Streaming applications are usually sensitive to delay. TCP adopts an Additive-Increase Multiplicative-Decrease (AIMD) strategy to react to packet losses and retransmits packets lost in congestion, therefore it introduces long delay and jitters and hence is not well suited for real-time streaming applications. By contrast, UDP is an unreliable and connection-less protocol without integrated rate/congestion control. Without congestion control however, non-TCP traffic can cause starvation or even congestion collapse to TCP traffic [12]. To

T. Spyropoulos and K.A. Hummel (Eds.): IWSOS 2009, LNCS 5918, pp. 113–124, 2009.

overcome the disadvantages of TCP and to handle competing dominant TCP flows in a friendly manner, TCP-Friendly Rate Control (TFRC) was introduced for streaming applications in [1] .

On the other hand, existing P2P streaming systems using rate control send data flows without considering the structure of the overlay tree (e.g., TCP in [5] and TFRC in [4]). TCP and TFRC/UDP, both being client/server (unicast) protocols, prevent applications from either overloading or under-utilizing the available bandwidth of their *local* connections. Moreover, they do not take neighbor connections and the quality of media stream into account.

With the goal to optimize the aggregate utility (video-quality) for P2P streaming application, we develop a fully distributed and optimal TCP-friendly rate control model in Section 2 and propose a shadow price-based distributed algorithm to solve the optimization problem in Section 3. The proposed algorithm is distributed with very small messaging overhead to allow P2P streaming systems to scale up to very large sizes while being TCP-friendly to coexisting traffic outside of the P2P session. We further extend the definition of TCP-friendliness to P2P network. With the help of extensive simulations, we evaluate the performance of the proposed TCP-friendly algorithm in terms of streaming quality and messaging complexity in Section 4. In Section 5, we discuss the implementation issues and conclude the paper.

# 2   Network Model and Rate Control Problem Formulation

## 2.1   Network Model

A large number of approaches have emerged in recent years for P2P streaming systems ([6] and its references). The vast majority of systems to date are tree-based P2P streaming, where peers are organized in trees to deliver data. Consider a P2P overlay tree of $n+1$ end hosts, denoted as $H = \{h_0, h_1, \ldots, h_n\}$. End host $h_0$ is the source of the P2P multicast channel. The structure of the overlay tree is given by the used P2P streaming approach. Non-leaf nodes are forwarding streaming data to its children and are able to scale-down the streams, fulfilling the constraint of the flow data. For our model, we assume that streams are fine-grained scalable [7]. The P2P streaming channel consists of $n$ end-to-end unicast flows, denoted as $F = \{f_1, \ldots, f_n\}$. Flow $f_i$ is the flow that terminates at $h_i$. Flow $f_i \in F$ has a rate $x_i$. We collect all the $x_i$ into a rate vector $x = (x_i, i = 1, 2..., n)$. We denote $U(x_i)$ as the utility of flow $f_i$, when $f_i$ transmits at rate $x_i$. We assume that $U(x)$ is strictly increasing and concave, and twice continuously differentiable. We measure the utility $U(x)$ for streams in section 4. $F_h'$ is the set of flows sent from $h$. If a host $h_i$ is the destination of a flow $f_i$ and the source of another flow $f_i' \in F_{h_i}'$, then $f_i'$ is the child flow of $f_i$, denoted as $f_i \to f_i'$. We denote $h'$ as the child of $h$ and $h^p$ is the parent node of $h$, i.e., $h^p \to h \to h'$. Let us define [16],

**Definition 1.** *A rate control algorithm is **TCP-friendly for P2P multicast**, if and only if the coexisting TCP traffic outside of the P2P channel achieves not less throughput than what it would achieve if all flows of the overlay channel were using TCP as rate control algorithm.*

Based on the fact that the backbone links of today's Internet are usually overly provisioned [13], we assume that the bottleneck of a unicast flow $f_i$ only appears at access links, namely upload link $l_u(f_i)$ and download link $l_d(f_i)$.

**Assumption 1.** *Access links (download and upload links) of end hosts are the only bottleneck links of a unicast path.*

Moreover, three possible bottleneck links between host $h$ and its children of a subtree $(h_1, h_2, h_3 \in H'_h)$ are presented in Fig. 1.

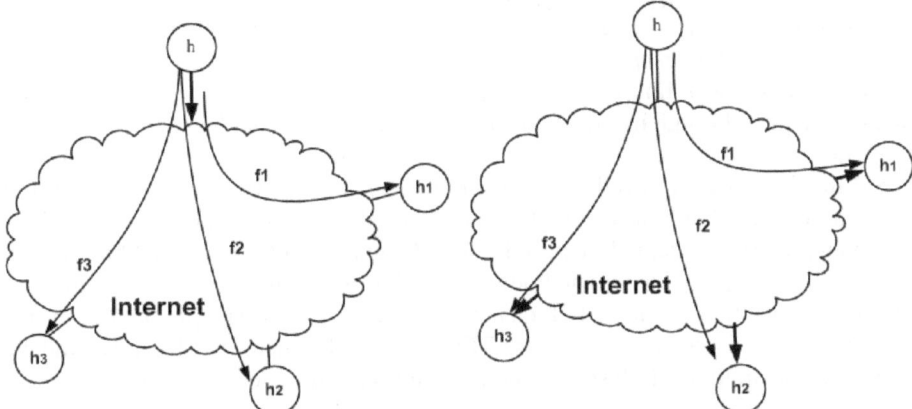

(a) Bottleneck link at the upload link and host $h$ (Case $I1$)

(b) Bottleneck links at download links and host $h$ (Case $I2$)

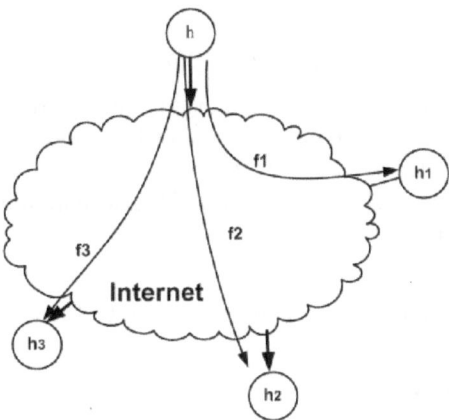

(c) Bottleneck links at the upload link and download links, and host $h$ (Case $I3$)

**Fig. 1.** Locations of Bottleneck links (Note: Bold lines are bottleneck links in the arrow direction. All links are directed)

**Proposition 1.** *Only sibling flows in the tree may share bottlenecks.*

Here, sibling flows are flows sent from the same end host $h$, i.e., all $f \in F'_h$ are sibling flows. Proposition 1 is straightforwardly provable with the locations model of bottleneck links shown in Fig.1. Therefore, non-sibling flows have independent bottleneck links and the overlay tree can be *fully decomposed* into subtrees with independent bottleneck links.

Let $t_i$ be the TCP-friendly available bandwidth for the unicast flow $f_i$ at the bottleneck links determined by end-to-end TFRC algorithm. We measure all $t_i$ for flows $f_i \in F$. Hence, we get the TCP-friendly available bandwidth for P2P multicast channel at bottleneck links, $c_{l_d(f_i)}$ and $c_{l_u(f_i)}$. For cases $I2 \cup I3$ where bottleneck links locate at download links $l_d(f_i)$: $c_{l_d(f_i)} = t_i$. For cases $I1 \cup I3$ where bottleneck links locate at upload links $l_u(f_i)$: $c_{l_u(f_i)} = \sum_{f_i \in F(l_u)} t_i$.

For each bottleneck link $l$, $F(l) = \{f \in F \mid l(f) = l\}$ is the set of flows in the channel that pass through it and $l(f)$ is the bottleneck link through which $f$ goes.

We define the constraints for rate control as follows: Flow rate of $f_i$ should not exceed the TCP-friendly available bandwidth $c_{l_d(f_i)} = t_i$ when the bottleneck link locates at download link of $h_i$. On the other hand, the sum of all flow rates in one direction and the same channel that go through the upload link of $h_i$ should not exceed $c_{l_u(f_i)} = \sum_{f_i \in F(c_{l_u})} t_i$, when the bottleneck link is at the upload-link. Therefore, co-existing TCP traffic outside of P2P channel obtain no less throughput than what they would achieve if all streams would use TFRC. Formally, such TCP-friendly available bandwidth constraint for P2P streaming rate control is expressed as follows:

$$\sum_{h_i^p \to h_i} x_i \leq c_{l_u(f_i)} = \sum_{h_i^p \to h_i} t_i, \quad \forall h_i^p \in I1 \cup I3. \tag{1}$$

$$x_i \leq c_{l_d(f_i)} = t_i, \quad \forall h_i^p \in I2 \cup I3 \tag{2}$$

Moreover, the downstream rate is constrained by the upstream rate, namely, if $f_i \to f_j$ then $x_j \leq x_i$. We define the data constraint or flow preservation $F \times F$ matrix B. $B_{f1,f2} = -1$, if $f2 \to f1$,i.e., $f1 = f2'$; $B_{f1,f2} = 1$, if $f1 = f2$, and $f1$ has a parent flow; Otherwise $B_{f_1 f_2} = 0$. Hence, given the P2P distribution tree, the data constraint can be formalized as follows:

$$B \cdot x \leq 0 \tag{3}$$

A summary of the notations used in the model can be found in Table 1.

## 2.2   Problem Formulation

Our objective is to devise a distributed rate control algorithm that maximizes the aggregate utility, *i.e.*, the overall video quality of all streams in the P2P streaming channel:

$$\max \sum_{i=1,2\dots n} U(x_i) \tag{4}$$

**Table 1.** Summary of Notations in the Model

| Notation | Definition |
|---|---|
| $h \in H = \{h0, h1, \ldots, hn\}$ | End Host |
| $h^p \to h \to h' \in H'_h$ | $h^p$ is the parent node of $h$, $h'$ is a child of $h$ |
| $H'_h$ | Set of child of h |
| $f \in F = \{f1, f2, \ldots, fn\}$ | Unicast flow in P2P streaming channel |
| $f_i \to h_i$ | Flow $f_i$ terminated at $h_i$ |
| $f_h$ | Flow terminated at $h$ |
| $x = (x_i, i = 1, 2, \ldots, n)$ | Flow rate set of $fi \in F$ |
| $l \in \Gamma = 1, 2, \ldots, L$ | Bottleneck Link $l$ (download link or upload link) |
| $c_l \in C, l \in \Gamma$ | TCP-friendly available bandwidth |
|  | for the channel at bottleneck link $l$ |
| $fi \to fi' \in F'_{hi}$ | $fi'$ is a child flow of $fi$ |
| $F'_{hi}$ | Set of flow sent from $hi$ in the channel |
| $l_u(fi) \in \Gamma$ | The upload link that $fi$ goes through |
| $l_d(fi) \in \Gamma$ | The download link that $fi$ goes through |
| $F(l)$ | Set of siblings flows that go through bottleneck link $l$ |
| $B = (B_{fi,fj})_{F \times F}$ | Data constraint matrix |
| $t_i$ | TCP-friendly available bandwidth for unicast for $fi$ at bottleneck |
| $U(x_i)$ | Utility Function of streams at rate $x_i$ |

fulfilling the following constraints:

$$\begin{cases} \sum_{h^p_i \to h_i} x_i \leq c_{l_u(f_i)} = \sum_{h^p_i \to h_i} t_i, & \forall h^p_i \in I1 \cup I3 \\ x_i \leq c_{l_d(f_i)} = t_i, & \forall h^p_i \in I2 \cup I3 \\ B \cdot x \leq 0 \end{cases}$$

## 3   Algorithm

In this section, we propose a distributed rate control algorithm for P2P streaming based on a shadow price concept that solves the convex optimization problem (4). Compared with the dual approaches proposed in [3][16][2], our primal algorithm is a feasible direction algorithm [9] which is applied to the original problem (primal problem) directly by searching the feasible region in the direction of improving the aggregate utility for an optimal solution. Please note that the proposed primal algorithm is different from the primal algorithm introduced by Kelly' in [8] or other penalty algorithms. Thanks to our fully distributed model, solving the optimization program (4) directly requires the coordination among those sibling flows only sharing bottleneck links. In order to find the direction for improving the aggregate utility, we define,

**Definition 2.** *The **Data shadow price** of a flow is the change in the aggregate utility of the flow itself and its subtree by relaxing the data constraint by one unit (a small move).*

By moving the bandwidth of the bottleneck link from children flows with lower shadow prices to children flows with higher shadow prices, the aggregate utility is improved.

We call a flow a data constrained flow when it is actively constrained by its parent flow, *i.e.*, $x_i = x_j$ (where $f_j = f_i'$); otherwise it is a data unconstrained flow, *i.e.*, $x_i > x_j$ (where $f_j = f_i'$) and actively constrained by the bottleneck link.

For a data constrained leaf flow $f_i$, its data shadow price is:

$$p_{f_i} = \Delta U(x_i)/\Delta x_i = U'(x_i) \tag{5}$$

For a data constrained intermediate flow $f_i$:

$$p_{f_i} = U'(x_i) + \sum_{f_j \in F'_{h_i}} p_{f_j} \tag{6}$$

When a flow is not constrained by its parent flow, its data shadow price is zero ($p_{f_i} = 0$). For example, all dashed flows in Fig. 2 have data shadow price zero. We call a node data constrained node (see Fig. 2(b)) when its incoming flow is a data constrained flow, otherwise it is a data unconstrained node (as in Fig. 2(a)). Each end host $h_i$ is assumed to

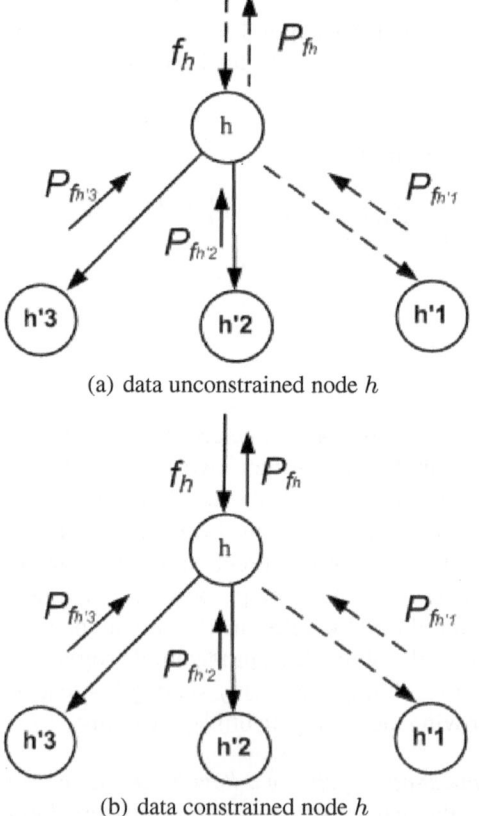

(a) data unconstrained node $h$

(b) data constrained node $h$

**Fig. 2.** Nodes and flows in the algorithm (dashed line means data unconstrained flow, constrained flow otherwise. $h_1', h_2', h_3' \in H_h'$)

**Table 2.** Algorithm of End Host $h_i$

---

**Initialization**
  Sending data with the TFRC unicast rate for each flow.
**Update the data shadow price from children**
  Get shadow price for children flows $f_i': p_{f_j}(t), f_j \in F_{h_i}'$
  Compute the median shadow price of children flows: $pj_{med}(t)$
**Update information from the parent node**
  Get the flow rate $x_i(t)$ and the data constraint information
  **Re-allocate the rate among the children flows for** $f_j \in F_{h_i}'$
  for $h_i \in I1 \cup I3$
    if $p_{f_j}(t) > pj_{med}(t)$
      $x_j(t+1) = x_j(t) + \gamma$
    else if $p_{f_j} < pj_{med}$
      $x_j(t+1) = x_j(t) - \gamma$
    end if
  end if
**Update Data Shadow Price to parent node**
  For data constrained node $h_i : p_{f_i}(t+1) = U'(x_i)$
  for $f_j \in F_{h_i}' := 1$ to n do
    if $x_j(t) \geq x_i(t)$
      $x_i(t+1) = x_j(t); p_{f_i}(t+1) = p_{f_i}(t+1) + p_{f_j}(t)$
    else $x_{f_j}(t+1) = x_{f_j}(t)$
    end if
  For data unconstrained node $h_i: p_{f_i} = 0$
  Send $p_{f_i}(t+1)$ up to parent node $h_i^p$
**Update stream rates and inform the children**
  for $f_j \in F_{h_i}'$
    Stream media to child $j$ with updated rate $x_j(t+1)$
    Update the data constraint information and $x_j(t+1)$ to $h_j$

---

be capable of communicating with neighbors, to determine the locations of bottleneck links, $t_j$ (where $f_j \in F_{h_i}'$) and to compute and adapt the sending rate for each flow $f_h'$ (*i.e.*, sender-based flow).

We present the algorithm of an intermediate peer in Table 2. We choose the TCP-friendly available rate of unicast flows as the initial rate, *i.e.*,$x_j(0) = t_j$.The algorithm purely depends on the coordination of end nodes. Each node receives the data shadow prices from its children nodes for children flows at each step. The algorithm *(i)* real-locates the bandwidth of the bottleneck link with stepsize $\gamma$ ($\gamma > 0$) from children flows with lower shadow prices to children flows with higher shadow prices such that the TCP-friendly available bandwidth constraints are not violated and flows with higher data shadow price get more bandwidth; and *(ii)* the algorithm obtains a better rate allo-cation after each step with an improved aggregate streaming quality. Thus, we have the following theorem,

**Theorem 1.** *For any P2P multicast streaming session, the rate allocation by the algorithm in Table 2 with sufficient small stepsize $\gamma(\gamma > 0)$ converges to the optimal allocation.*

*Proof.* For the subtree rooted at end host $h_i$, each allocation generated in the algorithm process is feasible and flows with higher data shadow price get more bandwidth. Therefore, the value of the aggregate utility of the subtree $\sum U(x_j(t)) < \sum U(x_j(t+1))$ improves constantly. Given the receiving rate $f_i$, as there is a limit for the aggregate utility of the subtree, the algorithm will finally converge to a maximum. For a convex optimization problem, the convergent rate allocation is the global maximum (the optimality) of the subtree(Chapter 11.1 in [9]).By each subtree converging to the optimal allocation for a given receiving rate iteratively, the optimal allocation of the entire multicast tree with its root at $h_0$ will be eventually reached.                                    □

Unlike the fluctuating convergence procedure in dual approaches [3][16], the feasible direction algorithm steadily converges to the optimal allocation .

## 4   Performance Evaluation

### 4.1   The Utility Function of Scalable Streams

The utility function used in [3] was $U(x_i) = \ln(x_i)$, which did not reflect the application quality of video streams. To tailor the utility function to the application quality, we use the rate-distortion function as the utility of our algorithm for each flow. The classic rate-distortion function for Gaussian distribution video source with mean $\mu = 0$ and variance $\sigma^2$ [15] is,

$$D(x_i) = \sigma^2 \cdot 2^{-\alpha x_i} \qquad (7)$$

We decided to use MPEG-4 fine-grained Scalable video (FGS) steams [7] in our performance evaluation, due to its ability to be sent at any given rate determined by a rate control algorithm at server side or any intermediate peer in the tree.

To measure the quality function of FGS coded P2P streams, we first use the Microsoft MPEG-4 software encoder with FGS functionality to encode the stored raw video streams. Then, we cut the corresponding FGS enhancement layer at the increasing and equally spaced bit rates (step size = 100kbps). For each compressed and cut bit-stream, we specify the distortion $D$ after decoding. Subsequently, we generate the rate-distortion curve of the FGS video stream using these sample points and finally we estimate the parameters in the classic video rate-distortion function that fit the rate-distortion traces [14]. We compute these parameters values for video sequences and keep them constant throughout the entire streaming process. Parameters we measured for some typical streams are presented in Table 3. All streams measured are CIF format,

Table 3. Measurement of Rate-distortion function for streams

| Video streams | $a$ | $\sigma^2$ | Fitting Goodness($SSE/sum(D)$) |
|---|---|---|---|
| Forman | -0.8625 | 100.915 | 0.0355 |
| Akiyo | -1.728 | 38.827 | 0.0970 |
| Mobile | -0.4917 | 256.711 | 0.0656 |
| Highway | -1.514 | 41.041 | 0.0611 |
| Tempete | -0.6177 | 167.963 | 0.0728 |
| Container | -1.098 | 82.196 | 0.0967 |

30 fps and 300 frames in length. A value of $SSE/sum(D)$ closer to 0 indicates a better fit, where $SSE$ is the sum of squares due to error and distortion $D$ is measured by the average MSE of a truncated video sequence.

We use the utility (video quality) function for *Forman*(CIF, 30fps, 300frames) in the experiments:

$$U(x_i) = -D(x_i) = -100.915 * 2^{-0.8625x_i} \qquad (8)$$

where $D(x_i)$ stands for the distortion of the stream and *mbit/s* is used as unit for streaming rate $x_i$. The utility function (8) is strictly increasing and concave, and twice continuously differentiable. It follows that solving problem (4) is equivalent to maximizing the overall video quality or minimizing the overall video distortion.

The primary concept of incorporating the rate-distortion function of a video encoding scheme into congestion control is directly applicable to other video-encoding schemes beyond FGS. As a matter of fact, we can use the same model with a different utility function (namely the utility function of TCP [10] or TFRC) for any other TCP-like P2P application.

## 4.2 Simulation Setting

While we have carried out simulations on various network topologies, we present here only the representative results of simulations on a topology generated with Brite [11] in the router level topology model with 1000 routers. The average time interval of shadow price updates and constraint information updates is 10ms. The bandwidths of all links are randomly distributed between 100Mbps and 1000Mbps with 0.6ms average delay. To investigate the message overhead of the algorithm in difference size of network, we set up other two smaller topologies with 20 and 100 routers of the same average link delay and bandwidth properties (part of the 1000 routers topology). We build the P2P streaming sessions consisting of various number of peers, each with a random access link bandwidth from 1Mbps to 100Mbps.

*Tree construction mechanism:* Since streaming applications are very time sensitive, we design a delay-based tree construction mechanism for P2P streaming systems. A new peer selects the closest peer in the tree as its parent node in terms of end to end delay. We further constrain that each peer has at most four children.

Assumption 1 holds in our experiments. By examining all bottleneck link constraint matrixes in our experiments, it was confirmed that "only sibling flows in the tree may share bottlenecks", i.e, namely on-sibling flows are with independent bottlenecks. Therefore the TCP-friendly bandwidth constraints at bottleneck links are fully decomposed for each subtree (1)(2), i.e., the network model and algorithm are fully distributed.

## 4.3 Rate Allocation

First, we compare the rate allocation results of our proposed algorithms with a standard unicast algorithm. We generate various P2P streaming systems sizes from 5 to 200 peers. In our simulations, the stepsize $\gamma$ is set to 0.0001. Fig. 3 shows that the

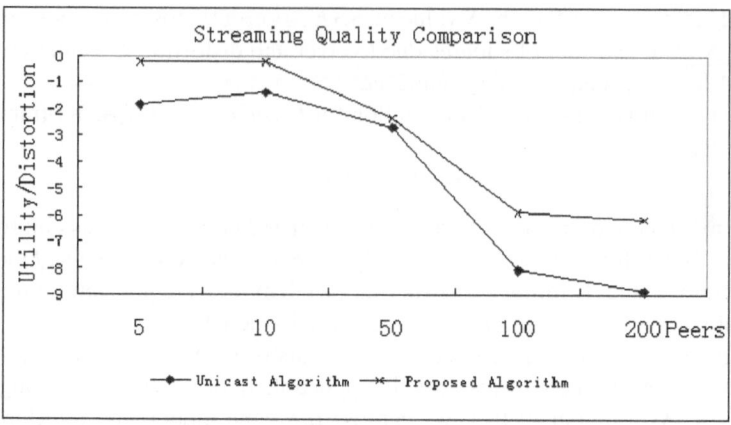

**Fig. 3.** Comparison of Average Streaming Quality

proposed algorithm is optimal in terms of average utility for various number of peers. If we first allocate the rates independently as unicast flows using the TCP/TFRC algorithm and then apply the data constraint at the same time, we get a set of rates with average/aggregate utility lower than the average/aggregate utility allocated by our algorithm.

### 4.4   Messaging Overhead

Next, we investigate the messaging overhead of the algorithm in various size of network topology. Fig. 4 shows the average number of messages sent by all peers per time interval. The results show that the larger P2P session the more messages are produced. Moreover, the number of messages increases with the number of peers in the session

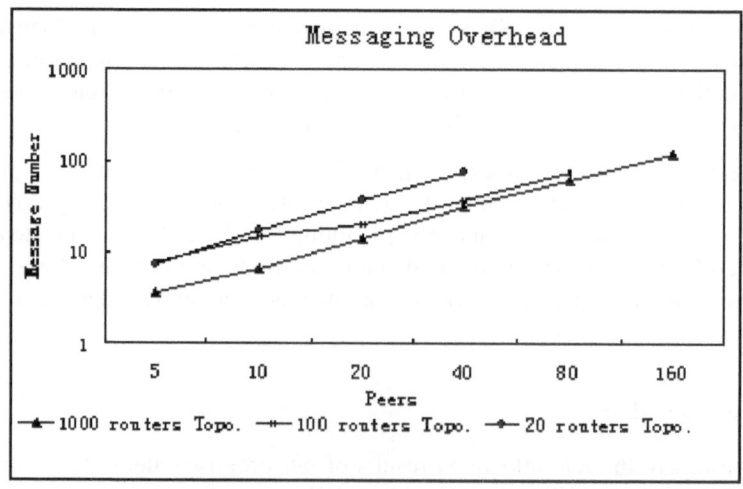

**Fig. 4.** Comparison of Messaging Overhead

linearly, i.e., each peer produces the same and small amount of message no matter the number of simultaneous P2P sessions. Therefore, our proposed algorithm can scale up to very large sizes and produces a small messaging overhead. Hence, we conclude that our algorithm is a fully distributed algorithm with small messaging overhead while maximizing the aggregate utility of P2P multicast tree.

## 5    Concluding Remarks

In this paper, we have proposed a fully distributed and TCP-friendly rate control model which maximizes the social utility for the P2P streams. The proposed algorithm works very well when bottleneck links are not access links. It is TCP-friendly to cross traffic outside the P2P session, while the rate allocation is proportionally fair in the P2P distribution tree [3]. In particular, the average time interval of the data shadow price updates and rate updates in the algorithm are much smaller than that of the TCP-friendly available bandwidth measurement so that the algorithm converges fast while the TCP-friendly available bandwidth measurement overhead is very small. Concerning future work, we are about to implement the algorithm in a real, large-scale P2P streaming system and will present more measurement results in upcoming publications.

## Acknowledgments

The first author is supported in part by Swiss National Science Foundation under grant No.200020-121753 and National Science Foundation of China under grant No.60970127.

## References

1. Floyd, S., Handley, M., Padhye, J.: Equation-Based Congestion Control for Unicast Application. In: ACM SIGCOMM 2000 (September 2000)
2. Low, S., Lapsley, D.E.: Optimization Flow Control, I: Basic Algorithm and Convergence. IEEE/ACM Trans. on Networking 7(6) (December 1999)
3. Cui, Y., Xue, Y., Nahrstedt, K.: Optimal Resource Allocation in Overlay Multicast. IEEE Transactions on Parallel and Distributed Systems 17(8) (2006)
4. Chu, Y., Rao, S.G., Zhang, H.: A case for End System Multicast. In: Proc. ACM Sigmetrics (June 2000)
5. PPlive, http://www.pplive.com/
6. Liu, J., Rao, S.G., Li, B., Zhang, H.: Opportunities and Challenges of Peer-to-Peer Internet Video Broadcast. IEEE JSAC 96(1) (January 2008)
7. Li, W.: Overview of Fine Granularity Scalability in MPEG-4 Video Standard. IEEE Trans.on CSVT 11(3) (March 2001)
8. Kelly, F.P., Maulloo, A.K., Tan, D.K.H.: Rate Control for Communication Networks:Shadow Prices, Proportional Fairness and Stability. Journal of the Operational Research Society 49 (1998)
9. Luenberger, D.: Linear and Nonlinear Programming. Addison-Wesley, Reading (1984)
10. Low, S.H., Paganini, F., Doyle, J.C.: Internet congestion control. IEEE Control Systems Magazine 22(1) (Febuary 2002)

11. http://www.cs.bu.edu/BRITE/
12. Floyd, S., Fall, K.: Promoting the Use of End-to-End Congestion Control in the Internet. IEEE/ACM Transactions on Networking (1999)
13. Barakat, C., Thiran, P., Iannaccone, G., Diot, C., Owezarski, P.: Modeling Internet backbone traffic at the flow level. IEEE Trans. on Signal Processing 51(8) (August 2003)
14. Yan, J., May, M., Plattner, B.: Media and TCP-Friendly Congestion Control for Scalable Video Streams. IEEE Trans. on Multimedia 8(2) (2006)
15. Cover, T.M., Thomas, J.A.: Elements of Information Theory. Wiley, New York (1991)
16. Yan, J., May, M., Plattner, B.: Distributed and Optimal Congestion Control for Application-layer Multicast: A Synchronous Dual Algorithm. In: IEEE CCNC, Las Vegas, US (2008)

# The Degree of Global-State Awareness in Self-Organizing Systems

Christopher Auer, Patrick Wüchner, and Hermann de Meer

Faculty of Informatics and Mathematics,
University of Passau
94032 Passau, Germany
{auerc,wuechner,demeer}@fim.uni-passau.de

**Abstract.** Since the entities composing self-organizing systems have direct access only to information provided by their vicinity, it is a non-trivial task for them to determine properties of the global system state. However, this ability appears to be mandatory for certain self-organizing systems in order to achieve an intended functionality.

Based on Shannon's information entropy, we introduce a formal measure that allows to determine the entities' degree of global-state awareness. Using this measure, self-organizing systems and suitable system settings can be identified that provide the necessary information to the entities for achieving the intended system functionality.

Hence, the proposed degree supports the evaluation of functional properties during the design and management of self-organizing systems. We show this by applying the measure exemplarily to a self-organizing sensor network designed for intrusion detection. This allows us to find preferable system parameter settings.

**Keywords:** Self-organizing systems, Mathematical modeling, Quantitative evaluation, Information theory, System design, Sensor networks.

## 1 Introduction

Self-organization is foreseen to enable efficient, scalable, and robust large-scale distributed systems, like the future Internet. However, the design of self-organizing systems (SOSs) that fulfill a certain intended functionality is a difficult task. Three different design approaches are sketched in [1]: the trial-and-error approach, the bio-inspired design, and the design by learning from an omniscient entity (see [2]).

In SOSs, the entities can only observe events that happen in their immediate vicinity. This makes it a non-trivial task to design the entities such that they foster the desired functionality of the SOS. Entity design can be simplified if the entities have access to information on the SOS' global state. This, however, presumes that the system entities are provided with the necessary global state

T. Spyropoulos and K.A. Hummel (Eds.): IWSOS 2009, LNCS 5918, pp. 125–136, 2009.

information to foster the preferred system behavior. A quantitative characterization of the entities' ability to derive such global state information could help to identify such SOSs that can then be further investigated (e.g., by the method in [2]) towards how the entities can foster the intended functionality. In this paper, we are proposing such a quantitative characterization.

Several formal measures have been proposed for describing certain properties of SOSs, like autonomy [3,4], emergence [3,4,5], adaptivity, homogeneity, and resilience [6]. However, these measures do not evaluate to which extent the entities can derive the necessary global-state information from the information provided by their vicinity.

In this paper, we propose a novel measure of the entities' degree of global-state awareness. By evaluating this degree, systems can be identified in which, over time, the necessary information is communicated to the entities, i.e., the entities become *aware* of important properties of a former global system state. These system candidates can then be further investigated, e.g., by the method proposed in [2], towards how the entities can use the provided information in a purposeful, target-oriented manner. The resulting system can finally be evaluated by using the measure of target orientation proposed in [6].

Hence, the main contribution of this paper is proposing an answer to the question *if*, and to which degree, the SOS under investigation is able, in principle, to provide the necessary information to the system's entities. For instance, in an SOS where global consensus has to be found in a completely decentralized manner, a low degree of global-state awareness indicates that the system entities are not provided with sufficient information. Hence, the system is not suitable to fulfill the task and should be redesigned.

We derive the measure of the degree of the entities' global-state awareness utilizing Shannon's information entropy (see [7]) and are hence in line with the previously proposed measures of autonomy [3,4], emergence [3,4,5], and homogeneity [6].

As an illustrative example, we apply the proposed measure to find suitable system parameters for a sensor network (adopted from [8]) which is designed for intrusion detection. It can be shown that the simulated sensor network indeed is able to fulfill the desired functionality in a self-organizing manner if the system parameters are chosen such that the degree of global-state awareness is maximized. The sensors are then able to reach a global consensus on whether an alarm triggered by a subset of the sensors was a false positive.

The remainder of the paper is organized as follows: In Sec. 2, we provide a reminder of Shannon's information entropy, which is essential to the understanding of following sections. We also introduce the model representing the SOS. As the main contribution of this paper, we introduce the measure of the entities' degree of global-state awareness in Sec. 3. In Sec. 4, we apply the proposed measure to discuss a sensor network for intrusion detection as an illustrative example. How the degree of global-state awareness can assist during the design of SOSs is sketched in Sec. 5. A conclusion and directions for future work are given in Sec. 6.

## 2  Background

This section briefly recapitulates the basic concepts of information theory and introduces the system model.

### 2.1  Entropy of Information

In this paper, a random variable $X$ is denoted by an upper-case letter and the realization of $X$ is denoted by the lower-case letter $x$. For the range of $X$ we use the bold-face character $\mathbf{X}$. In our case, the range of any random variable is finite, i.e., $\#\mathbf{X} < \infty$.

As a measure of uncertainty, a useful tool is Shannon's information entropy as defined in [7]: given some discrete random variable $X$ with finite range $\mathbf{X}$, the entropy of $X$ is defined as:

$$\mathrm{H}[X] = - \sum_{x \in \mathbf{X}} P[X = x] \cdot \log_2 P[X = x] . \tag{1}$$

The entropy is 0 iff $X$ almost surely takes a value $x \in \mathbf{X}$, i.e., $P[X = x] = 1$ for a single $x \in \mathbf{X}$. $\mathrm{H}[X]$ takes its maximum iff $X$ is uniformly distributed. Generally, the lower the value of the entropy, the more certain the outcome of a random variable can be predicted.

If $X, Y$ are two discrete random variables with finite ranges $\mathbf{X}, \mathbf{Y}$, then knowing the outcome of $Y$ might reduce the uncertainty of the outcome of $X$. When the outcome of $Y$ is known, the remaining entropy of $X$ is measured by the conditional entropy $\mathrm{H}[X|Y] = \mathrm{H}[X, Y] - \mathrm{H}[Y]$ (cf. [7,9]). It can be shown (cf. [9]) that $\mathrm{H}[X|Y] = \mathrm{H}[X]$ iff $X$ and $Y$ are independent and $\mathrm{H}[X|Y] = 0$ iff $X = f(Y)$, where $f$ is a non-stochastic function.

### 2.2  System Model

In this paper, we focus on the class of technical SOSs that can be modeled as discrete-event systems consisting of a finite set $\mathbf{N}$ of entities. An example of such a system is depicted in Figure 1.

At each time step $t \in \mathbb{N}_0$, each entity $n \in \mathbf{N}$ receives input $i_{t,n}$ from its vicinity that contains neighboring entities and possibly comprises parts of the environment. Entity $n$ then produces the output $o_{t,n}$ which is received in the next time step $t + 1$ by entity $n$'s neighbors and possibly also influences the environment. For example, in Figure 1, $o_{t,n_{2,2}} = i_{t+1,n_{2,1}}$.

We assume that the SOS' entities can be modeled as deterministic finite-state Mealy automatons: The finite state space of entity $n$ is denoted by $\mathcal{S}_n$ and its transition function by $\zeta_n$ that maps the input $i_{t,n}$ and current state $s_{t,n} \in \mathcal{S}_n$ of entity $n$ to its output $o_{t,n}$ and its successor state $s_{t+1,n} \in \mathcal{S}_n$.

The tuple $\gamma_{t,n} = (i_{t,n}, s_{t,n})$ is called *local configuration* of entity $n$ at time step $t$: At each time step, each entity can only observe its local configuration, i.e., its own state and the input provided by its vicinity. Up to some time step $t$, the sequence of local configurations $(\gamma_{t',n})_{t'=0...t}$ constitutes the *local history* $\overleftarrow{\gamma}_{t,n}$

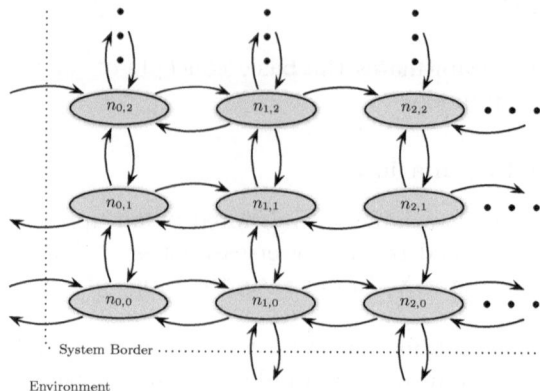

**Fig. 1.** Model of a self-organizing system: The system consists of several entities (nodes) and can be distinguished from its environment (system border). The entities may exchange information (arrows) with their vicinity that might contain parts of the environment.

of entity $n$. The state of the whole SOS at time step $t$ can be fully described by the local configurations of all entities: We call $\gamma_t = (\gamma_{t,n})_{n \in \mathbf{N}}$ the *configuration* at time step $t$. The set of all possible configurations is called *state space* of the SOS and is denoted by $\mathbf{\Gamma}$.

The configuration $\gamma_0$ at time step 0 is called *initial configuration (IC)*. The IC is modeled by the random variable $\Gamma_0$ with range $\mathbf{\Gamma}_0 \subseteq \mathbf{\Gamma}$, where $\mathbf{\Gamma}_0$ is the set of all possible ICs which all have a non-zero probability. Since the IC is chosen randomly, all subsequent configurations can also be described by random variables even if there is no random input from the environment. We denote the random variable of the SOS' configuration $\gamma_t$ by the capital letter $\Gamma_t$. Accordingly, we denote with $\Gamma_{t,n}$ and $\overleftarrow{\Gamma}_{t,n}$ the random variables of entity $n$'s local configuration $\gamma_{t,n}$ and of the local history $\overleftarrow{\gamma}_{t,n}$, respectively.

In principle, our system model assumptions can be relaxed by generalizing the model using the modeling approaches presented in [3] (asynchronous communication and stochastic transition functions) and [4] (continuous time and continuous state space). However, due to space limitations and to preserve clarity, we refrain from applying these generalizations here.

## 3   The Degree of Global-State Awareness

In this section, we present the main contribution of this paper. Our main goal is to evaluate an entity's ability to obtain global-state information when only the information provided by the entity's vicinity is directly available.

### 3.1   Classification Problem

To define the measure of the entities' degree of global-state awareness as general as possible, we introduce the notion of classification problems. In general, it is unsuitable, unnecessary, and often even impossible to distribute the full information on the system's global state to each single entity.

Hence, we aggregate global states that share a common property of interest to the same state class. For an entity it is then sufficient to figure out the state class, i.e., to solve the classification problem by only taking the information into account that is provided by its vicinity.

We assume that the entities need to determine the state class of the SOS at some time step $t$. Without loss of generality, we refer to this time step $t$ as the initial time step $t = 0$. Therefore, we aggregate initial system states of the set of possible ICs $\Gamma_0$ to state classes to define the *classification problem* as follows:

**Definition 1 (Classification Problem).** *Let* $\mathbf{L}$ *be a partition of* $\Gamma_0$, *i.e.:*

$$\bigcup_{l \in \mathbf{L}} l = \Gamma_0 \wedge \forall l, l' \in \mathbf{L}, l \neq l' : l \cap l' = \emptyset \wedge \emptyset \notin \mathbf{L}.$$

$\vartheta : \Gamma_0 \to \mathbf{L}$ *is a function which maps any initial configuration to its corresponding state class in* $\mathbf{L}$: $\forall \gamma_0 \in \Gamma_0 : \forall l \in \mathbf{L} : \vartheta(\gamma_0) = l \iff \gamma_0 \in l$.

*Applying function* $\vartheta$ *to random variable* $\Gamma_0$ *produces the random variable* $L = \vartheta(\Gamma_0)$, *which is the random variable of the state class of* $\Gamma_0$. *The realization of* $L$ *is denoted by* $l$. $L$ *naturally inherits the probability distribution from* $\Gamma_0$.

*The problem of determining the state class of the initial configuration* $l \in \mathbf{L}$ *(i.e., the outcome of* $L$*) given only the local history of an entity is called* classification problem $\mathbf{L}$.

In the following, we assume that a classification problem is non-trivial in the sense that $\mathbf{L}$ contains at least two state classes.

The mapping of the IC $\gamma_0$ to the property of interest can be represented in form of a function $\psi : \Gamma_0 \to \mathbf{P}$, i.e., $\gamma_0$ has the property $\psi(\gamma_0)$, where $\mathbf{P}$ is the set of global-state properties of interest that depend on the specific application scenario. This implies the partition $\mathbf{L}_\psi$ in which all ICs are aggregated that map to the same value of $\psi$. Hence, the system entities can obtain the value of $\psi(\gamma_0)$ after determining the corresponding state class of $\mathbf{L}_\psi$, i.e., after solving the classification problem $\mathbf{L}_\psi$.

## 3.2   Defining the Degree of Global-State Awareness

As a reminder, $L = \vartheta(\Gamma_0)$ denotes the random variable of the state class of the random IC $\Gamma_0$. To solve a classification problem $\mathbf{L}$, the entities of an SOS have to decrease their uncertainty about the random variable $L$, given their local history. In order to measure the uncertainty about $L$ when the local history is given, we use Shannon's information entropy (see Sec. 2.1):

**Definition 2 (Degree of Global-State Awareness).** *The* degree of global-state awareness $\omega_{t,n}(\mathbf{L})$ *observable by entity* $n \in \mathbf{N}$ *at time step* $t$ *is defined by:*

$$\omega_{t,n}(\mathbf{L}) = 1 - \frac{\mathrm{H}[L | \overleftarrow{\Gamma}_{t,n}]}{\mathrm{H}[L]}. \tag{2}$$

*The system's* overall degree of global-state awareness *is then defined as the limiting value for $t \to \infty$, averaged over all entities:*

$$\omega(\mathbf{L}) = \lim_{t\to\infty} \frac{1}{\#\mathbf{N}} \sum_{n\in\mathbf{N}} \omega_{t,n}(\mathbf{L}) . \tag{3}$$

Note that $\mathrm{H}[L]$ measures the uncertainty of predicting $L$ when no additional information is given. Since $\mathrm{H}[L|\overleftarrow{\Gamma}_{t,n}] \leq \mathrm{H}[L]$, we have $\omega_{t,n}(\mathbf{L}) \in [0,1]$ and $\omega(\mathbf{L}) \in [0,1]$.

By the definition of the entropy (Eq. (1)), the denominator in Eq. (2) is strictly greater than 0 since $\mathbf{L}$ consists of at least two state classes that have a non-zero probability.

Intuitively, $\omega_{t,n}(\mathbf{L})$ measures how certain entity $n$ can determine the outcome of $L$ at time step $t$ by taking its local history $\overleftarrow{\Gamma}_{t,n}$ into account. If $\omega_{t,n}(\mathbf{L}) \approx 1$, then $\mathrm{H}[L|\overleftarrow{\Gamma}_{t,n}]$ is small compared to $\mathrm{H}[L]$ and, hence, $n$ can use its local history to determine the outcome of $L$ with a high certainty. As a matter of fact, $\omega_{t,n}(\mathbf{L}) = 1$ iff there exists a non-stochastic function $f$ with $L = f(\overleftarrow{\Gamma}_{t,n})$ (see Sec. 2.1). At the other extreme, if $\omega_{t,n}(\mathbf{L}) \approx 0$, then $\mathrm{H}[L|\overleftarrow{\Gamma}_{t,n}] \approx \mathrm{H}[L]$, which implies that the local history contributes almost no information about the outcome of $L$. Again, $\omega_{t,n}(\mathbf{L}) = 0$ iff $L$ and $\overleftarrow{\Gamma}_{t,n}$ are independent random variables (see Sec. 2.1).

The system's overall degree of global-state awareness $\omega(\mathbf{L})$ (Eq. (3)) measures to which extent the information about $L$ is distributed within the SOS in the long-term. Note that a high overall degree of global-state awareness indicates the distribution of the information about the IC's state class $L$ among most entities since all entities equally contribute to $\omega(\mathbf{L})$.

## 4    Intrusion Detection in Sensor Networks

To show the applicability of the proposed measure, we now utilize it to find suitable parameter settings for a sensor network designed for intrusion detection. It can be shown that the sensor network indeed is able to fulfill the desired functionality in a self-organizing manner as long as suitable system parameters are chosen. The degree of global-state awareness helps to find these system parameters.

### 4.1    Scenario and Problem Description

Consider the distributed sensor network sketched in Figure 2(a). Suppose that the group $\mathbf{G}$ of sensors detects an intrusion. In absence of an omniscient central entity and with a non-zero probability of single sensors giving a false alarm, a consensus needs to be achieved in a completely decentralized and self-organized manner whether the alarm was false positive. Due to hard resource constraints, this should also be accomplished by exchanging a minimum amount of information. In the following, we use the degree of global-state awareness to find

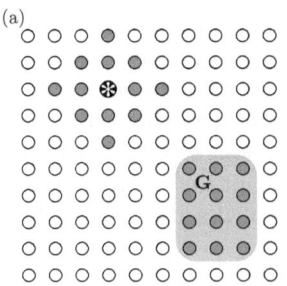 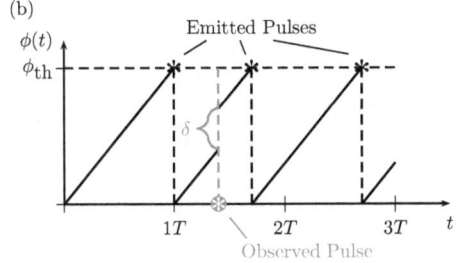

**Fig. 2.** Sensor network (a) and phase function $\phi(t)$ of pulse-coupled oscillator (b)

system candidates for which the sensors are able to find a global consensus on the detected intrusion.

In [8], a mechanism is described how a sensor network can reach global consensus on an intrusion detected by a sensor group **G**. Each sensor periodically emits pulses (illustrated by black asterisk in Fig. 2(a)) with a fixed identical period $T$ but varying phases. These pulses can be observed by all sensors in its neighborhood (gray dots) only.

In the following, we neglect transmission delays which are discussed in, e.g., [10]. With the help of [10], the following discussion can similarly be applied to more realistic scenarios where delays cannot be neglected, e.g., in the internet.

Before intrusions can be detected, all sensors have to synchronize their phases. According to [11], which is based on the theoretical model of pulse-coupled oscillators presented in [12], synchronization can be achieved using the following mechanism: Each entity calculates a local phase $\phi$ following the phase function illustrated in Figure 2(b). The phase is increased linearly over time. If the entity receives no pulse, it periodically emits a pulse with a period of $T$. If an entity observes a pulse from another entity, it additionally increases its current phase by $\delta = (\alpha - 1)\phi + \beta$, where $\alpha > 1$ and $\beta > 0$ are constant system parameters (see also [12, 11] for details). If the phase reaches a threshold value $\phi_{\text{th}}$, the entity emits a pulse and resets its phase to $\phi = 0$. It is proved in [12] that using this mechanism, synchronization can be reached almost surely if an entity's pulse can be directly observed by all other entities. Moreover, it is shown in [13] that pulse-coupled oscillators can also synchronize if each entity communicates only with its nearest neighbors.

After synchronization, the sensor network is ready for detecting intrusions using the method described in [8]: On detection of an intruder, the sensors of group **G** shift their phase by a predefined amount of $\Delta\phi$. This results in a partitioning of the network into two groups where intra-group synchronization is still given. Inter-group synchronization, however, is no longer given. The groups automatically start to resynchronize until, at some time instant, all sensors are again synchronized and all phases have been shifted by some $\Delta\Theta$ compared to the phase before the intrusion was detected. According to [8], if $\Delta\phi$ is chosen appropriately and if the sensor network is fully connected, i.e., every sensor receives pulses from all other sensors, then each sensor is able to infer from $\Delta\Theta$

whether group $\mathbf{G}$ was large enough to exclude false alarm. However, in [8], no concrete advice is given how to find appropriate system parameters (i.e., $\alpha$, $\beta$, $\Delta\phi$, and $\phi_{th}$) for this mechanism, despite focusing on fully connected networks. Additionally, [8] neglects that each sensor can observe more than just $\Delta\Theta$: Each sensor might use the whole history of its local observations to infer whether a false alarm has occurred. In Section 4.2, we now derive suitable system parameters by evaluating the degree of global-state awareness of a sensor network that is not even fully connected.

## 4.2   Application of Proposed Measure to Sensor Network

Without loss of generality, we assume that each entity $n \in \mathbf{N}$ ($\mathbf{N} = 1, \ldots, 100$) of the sensor network should conclude that an intrusion has in fact occurred, if more than one fifth of all sensors detected the intrusion, i.e., $\#\mathbf{G} > \#\mathbf{N}/5$.

We model the sensor network at discrete time steps that are defined at all time instants where at least one sensor emits a pulse. At these time steps, each sensor can be modelled as a finite state automaton: As input from its neighboring sensors, each sensor either perceives a pulse or not. Additionally, at time step 0, the detection of an intrusion by a sensor is modelled as an input from the sensor's environment. To keep the model simple, we assume that after time step 0 no further intrusions are detected. The state $s_{t,n}$ of sensor $n$ at time step $t$ is given by the sensors' phase value $\phi_n(t)$. As output, the sensor may or may not emit a pulse to its neighbors. In order for the state space of the sensor network to be discrete, we assume, without loss of generality, that at time step 0 all sensors are synchronized and have a phase value of 0 except for the sensor group $\mathbf{G}$ that have an identically shifted phase value of $\Delta\phi$. The resulting configuration of the sensor network is referred to as IC $\gamma_0$. This implies a discrete set of possible ICs $\Gamma_0$ and, since the sensor network is investigated at discrete time steps, also the set of possible system configurations $\Gamma$ and sensor states $\mathcal{S}_n$ are discrete.

To keep the model concise, we assume that intrusions are always detected by compact rectangular groups of sensor nodes, i.e., $\mathbf{G}$ forms a rectangular subset with width $w$ and height $h$ within the sensor grid of size $10 \times 10$ nodes (see Fig. 2(a)). The width and height of $\mathbf{G}$ are described by independent random variables $W$ and $H$, respectively. Both random variables have the range $\{0, \ldots, 10\}$, where $W = 0$ or $H = 0$ implies that no sensor has triggered an alarm. Given the width $w$ and height $h$, the position of $\mathbf{G}$ is then chosen according to a uniform distribution on all possible positions within the sensor grid.

According to the notation introduced in Section 3.1, there are two equivalence classes on the IC $\gamma_0 \in \Gamma_0$ of interest. Class $l_{>1/5}$ contains all ICs where $\#\mathbf{G} > \#\mathbf{N}/5$ and intrusion should be recognized by all sensors. Class $l_{\leq 1/5}$ contains all ICs where $\#\mathbf{G} \leq \#\mathbf{N}/5$ and should be treated as false alarm. It hence has to be checked for which system parameters each sensor is able to solve the classification problem $\mathbf{L} = \{l_{>1/5}, l_{\leq 1/5}\}$.

By simulating the sensor network for all possible $\mathbf{G}$ and weighing the outcomes with the according probabilities, exact values for the degree of global-state awareness $\omega(\mathbf{L})$ can be obtained by applying Eq. (2). In particular, we investigate the

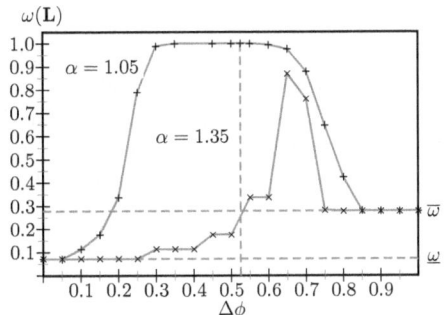

**Fig. 3.** The degree of global-state awareness $\omega(\mathbf{L})$ (y-axis) depending on the phase shift $\Delta\phi \in [0, 0.05, \ldots, 1.0]$ (x-axis) for $\alpha = 1.05$ and $\alpha = 1.35$.

influence of the system parameters $\alpha$ and $\Delta\phi$ on the system's ability to solve the classification problem $\mathbf{L}$. For $\beta$ we choose a value of 0.01 and $\phi_{th} = 1$, which are reasonable values according to [11].

Figure 3 shows the influence of $\Delta\phi$ on the degree of global-state awareness $\omega(\mathbf{L})$ (y-axis) for $\alpha = 1.05$, $\alpha = 1.35$, and $\Delta\phi \in [0, 0.05, 0.1, \ldots, 1]$ (x-axis). It can be seen that with increasing $\Delta\phi$, both curves initially increase, reach a maximum, and then drop. For both extremes, $\Delta\phi$ close to 0 and 1, $\omega(\mathbf{L})$ is small which implies that the sensors cannot decide certainly whether $L = l_{\leq 1/5}$ or $L = l_{>1/5}$. This is because the difference between the phases of $\mathbf{G}$ and $\mathbf{N} \setminus \mathbf{G}$ is small. Hence, the two sensor groups almost immediately resynchronize and the information about $L$ has no time to spread within the network.

For $\Delta\phi \to 0$, $\omega(\mathbf{L})$ tends to the limiting value $\underline{\omega}$. In this case, the sensors in $\mathbf{G}$ detect an intrusion but do not shift their phase. Each sensor in $n \in \mathbf{G}$ knows that at least one sensor, namely $n$ itself, has detected an intrusion. This reduces the uncertainty about $L$ for $n$ and leads to the limiting value $\underline{\omega}$. At the other extreme, for $\Delta\phi \to 1$, the sensors in $\mathbf{G}$ detect an intrusion, do not shift their phase, but emit a pulse which is also perceived by the sensors neighboring $\mathbf{G}$. All these sensors can then reduce their uncertainty about $L$, resulting in the limiting value $\overline{\omega}$ which is greater than $\underline{\omega}$.

For $\alpha = 1.05$, the degree of global-state awareness is close to 1 for $\Delta\phi \in [0.3, 0.6]$. Indeed, for $\Delta\phi = 0.525$ (vertical dashed line in Fig. 3), $\omega(\mathbf{L}) = 1$, which has an important implication. It follows that $H[L|\overleftarrow{\Gamma}_{t,n}] = 0$ for a sufficiently large $t$ and all $n \in \mathbf{N}$. From information theory (see Sec. 2.1), it is then known that there exists a non-stochastic function $f_n : \overleftarrow{\Gamma}_{t,n} \to \mathbf{L}$ for every sensor $n \in \mathbf{N}$ such that $f_n(\overleftarrow{\Gamma}_{t,n}) = L$, almost surely. Each sensor $n \in \mathbf{N}$ can then apply $f_n$ to obtain $L$. Such a mapping can, for instance, be obtained by the simulation process we used to calculate the degree of global-state awareness. Another approach to implement $f_n$ is discussed in Section 5.

It can also be seen in Fig. 3 that the curve for $\alpha = 1.35$ remains well below the curve for $\alpha = 1.05$. In [12, 13] it is shown that the time until the sensors reach synchronization is inversely proportional to $\alpha$. Intuitively, a reduced convergence time prevents the information about state class $L$ from spreading within the whole network since the sensors within the vicinity of $\mathbf{G}$ resynchronize too quickly. Hence, suitable choices for $\alpha$ and $\Delta\phi$ maximize the degree of global-state awareness while minimizing the convergence time of the network to assure that the information about an intrusion is spread quickly within the network.

## 5  Effects on Entity Design

Remember that, by using the degree of global-state awareness, it is possible to characterize *if* it is, in principle, possible for the entities to obtain the desired global-state information. However, the degree does not indicate *how* the system entities can use the information provided by their vicinity to derive the desired global-state information and *how* this information can be used to foster the desired functionality of the SOS. In this section, we sketch how suitable local interaction strategies can be found.

A degree of global-state awareness of 1 implies that there exists a non-stochastic function that maps any local history to the corresponding global state class. In principle, such a mapping could be obtained by a simulation process similar to the one we used to produce the results shown in Sec. 4. However, the resulting table that maps a local history to the corresponding state class would be very large and, hence, cumbersome, if not impossible, to implement in devices with limited memory and computation power, e.g., wireless sensors. Furthermore, in a real-world scenario, events may happen that are not encountered by the simulation process. This could lead to an undefined input value for the obtained mapping.

In [2], we presented a method to derive local interactions strategies for entities of an SOS by learning from an omniscient entity, called the *Laplace's Daemon* (LD). We now discuss the application of LDs to obtain a mapping from the local history to the corresponding state class of the IC. In a simulation environment, the LD of each entity is equipped with information about the IC's state-class. At each time step, as input, each LD receives the local configuration of the corresponding entity and outputs the current state-class of the IC. The sequence of local inputs to and outputs from each LD generated during the simulation is investigated by a time series analysing algorithm (CSSR algorithm; cf. [14, 15]) to obtain a minimal Markov chain description of each LD. The obtained Markov chains can be effectively implemented in the entities and are minimal in the sense that they only use the relevant information from the local history to predict the global-state class. Furthermore, the derived mappings even produce reasonably reliable results when they encounter situations that were not faced during the simulation. This information about the global-state class can then be used by the entities to foster the desired functionality of the SOS. A high degree of global-state awareness indicates that it is worthwhile to apply the approach presented in [2] to optimize the given SOS.

# 6  Conclusion

In order for the entities of a self-organizing system to optimize the overall system performance, it is useful for them to know certain aspects of the system's global state. To model such aspects of the system's global state, we introduced the classification problem where system global states that share common properties of interest are aggregated to state classes. As the main contribution of this paper, the degree of global-state awareness was introduced that evaluates to which extent the entities of a self-organizing systems are able to find out the respective state class while using only the local information provided by the entities' vicinity. By using our measure, it is possible to find preferable system parameters that enable the entities to adjust their behavior for an optimized overall system performance.

As an illustrative example, a sensor network for intrusion detection was used in this paper. By applying the measure of the entities' degree of global-state awareness, we were able to find system parameter settings that allow all sensors to determine whether an intrusion was detected by a significant number of sensors in a completely decentralized manner. The sensors can use this global-state information to exclude false alarms.

By using the found system parameter settings, it is then possible to derive local interaction strategies that can rely on this global information. We also described how to obtain such interaction strategies in general, i.e., by learning from an omniscient entity, an approach we presented in [2].

In near future, we intend to apply our measure of global-state awareness to other application scenarios. We also plan to investigate further properties of self-organizing systems, e.g., decentralization, while also providing suitable and generally applicable formal measures of these properties.

## Acknowledgements

The authors thank Dr. Richard Holzer for fruitful discussions. This research is partially supported by the AutoI project (STREP, FP7 Call 1, ICT-2007-1-216404), by the ResumeNet project (STREP, FP7 Call 2, ICT-2007-2-224619), and by the Network of Excellence EuroNF (FP7, IST 216366).

## References

1. Elmenreich, W., De Meer, H.: Self-organizing networked systems for technical applications: A discussion on open issues. In: Hummel, K.A., Sterbenz, J.P.G. (eds.) IWSOS 2008. LNCS, vol. 5343, pp. 1–9. Springer, Heidelberg (2008)
2. Auer, C., Wüchner, P., De Meer, H.: A method to derive local interaction strategies for improving cooperation in self-organizing systems. In: Hummel, K.A., Sterbenz, J.P.G. (eds.) IWSOS 2008. LNCS, vol. 5343, pp. 170–181. Springer, Heidelberg (2008)

3. Holzer, R., de Meer, H., Bettstetter, C.: On autonomy and emergence in self-organizing systems. In: Hummel, K.A., Sterbenz, J.P.G. (eds.) IWSOS 2008. LNCS, vol. 5343, pp. 157–169. Springer, Heidelberg (2008)
4. Holzer, R., de Meer, H.: On modeling of self-organizing systems. In: Proc. of Autonomics 2008, Turin, Italy (September 2008)
5. Mnif, M., Müller-Schloer, C.: Quantitative emergence. In: Proc. of the 2006 IEEE Mountain Workshop on Adaptive and Learning Systems (SMCals 2006), July 2006, pp. 78–84. IEEE, Piscataway (2006)
6. Holzer, R., de Meer, H.: Quantitative modeling of self-organizing properties. In: Plattner, B., Spyropoulos, T., Hummel, K.A. (eds.) Proc. of the 4th International Workshop of Self-Organizing Systems (IWSOS 2009), Zurich, Switzerland, December 2009. LNCS, Springer, Heidelberg (2009)
7. Shannon, C.E.: A mathematical theory of communication. Bell System Technical Journal 27, 379–423 (1948)
8. Hong, Y., Scaglione, A.: Distributed change detection in large scale sensor networks through the synchronization of pulse-coupled oscillators. In: Proc. of the IEEE International Conference on Acoustics, Speech, and Signal Processing (ICASSP 2004), Montreal, Canada, May 2004, vol. 3, pp. 869–872 (2004)
9. Cover, T.M., Thomas, J.A.: Elements of information theory. Wiley-Interscience, Hoboken (1991)
10. Ernst, U., Pawelzik, K., Geisel, T.: Synchronization induced by temporal delays in pulse-coupled oscillators. Phys. Rev. Lett. 74(9), 1570–1573 (1995)
11. Tyrrell, A., Auer, G., Bettstetter, C.: In: Biologically Inspired Synchronization for Wireless Networks. Vol. 69/2007 of Studies in Computational Intelligence, pp. 47–62. Springer, Heidelberg (2007)
12. Mirollo, R.E., Strogatz, S.H.: Synchronization of pulse-coupled biological oscillators. SIAM Journal on Applied Mathematics 50(6), 1645–1662 (1990)
13. Lucarelli, D., Wang, I.: Decentralized synchronization protocols with nearest neighbor communication. In: Proc. of the 2nd International Conference on Embedded Networked Sensor Systems, pp. 62–68. ACM, New York (2004)
14. Shalizi, C.R.: Causal Architecture, Complexity and Self-Organization in Time Series and Cellular Automata. PhD thesis, University of Wisconsin, Supervisor: Martin Olsson (2001)
15. Shalizi, C.R., Shalizi, K.L.: Blind construction of optimal nonlinear recursive predictors for discrete sequences. In: AUAI 2004: Proceedings of the 20th Conference on Uncertainty in Artificial Intelligence, pp. 504–511. AUAI Press, Arlington (2004)

# Revisiting the Auto-Regressive Functions of the Cross-Entropy Ant System

Laurent Paquereau and Bjarne E. Helvik

Centre for Quantifiable Quality of Service in Communication Systems*,
Norwegian University of Science and Technology, Trondheim, Norway
{laurent.paquereau,bjarne}@q2s.ntnu.no

**Abstract.** The Cross-Entropy Ant System (CEAS) is an Ant Colony Optimization (ACO) system for distributed and online path management in telecommunication networks. Previous works on CEAS have enhanced the system by introducing new features. This paper takes a step back and revisits the auto-regressive functions at the core of the system. These functions are approximations of complicated transcendental functions stemming from the Cross-Entropy (CE) method for stochastic optimization, computationally intensive and therefore not suited for online and distributed operation. Using linear instead of hyperbolic approximations, new expressions are derived and shown to improve the adaptivity and robustness of the system, in particular on the occurrence of radical changes in the cost of the paths sampled by ants.

**Keywords:** Ant-based optimization, Cross-Entropy method, CEAS, linear approximations.

## 1 Introduction

Ant Colony Optimization (ACO) [1] systems are self-organizing systems inspired by the foraging behaviour of ants and designed to solve discrete combinatorial optimization problems. More generally, ACO systems belong to the class of Swarm Intelligence (SI) systems [2]. SI systems are formed by a population of agents, which behaviour is governed by a small set of simple rules and which, by their collective behaviour, are able to find good solutions to complex problems. ACO systems are characterized by the indirect communication between agents - (artificial) ants - referred to as stigmergy and mediated by (artificial) pheromones. In nature, pheromones are a volatile chemical substance laid by ants while walking that modifies the environment perceived by other ants. ACO systems have been applied to a wide range of problems [1]. The Cross-Entropy Ant System (CEAS) is such a system for path management in dynamic telecommunication networks.

The complexity of the problem arises from the non-stationary stochastic dynamics of telecommunication networks. A path management system should adapt

---

* "Centre for Quantifiable Quality of Service in Communication Systems, Centre of Excellence" appointed by The Research Council of Norway, funded by the Research Council, NTNU, UNINETT and Telenor. http://www.q2s.ntnu.no

T. Spyropoulos and K.A. Hummel (Eds.): IWSOS 2009, LNCS 5918, pp. 137–148, 2009.

to changes including topological changes, e.g. link/node failures and restorations, quality changes, e.g. link capacity changes, and traffic pattern changes. The type, degree and time-granularity of changes depend on the type of network. For instance, the level of variability in link quality is expected to be higher in a wireless access network than in a wired core network.

Generally, the performance of an ACO system is related to the number of iterations required to achieve a given result. Specific to the path management problem in telecommunication networks are the additional requirements put on the system in terms of time and overhead. On changes, the system should adapt, i.e. converge to a new configuration of paths, in short time and with a small overhead. In addition, finding a good enough solution in short time is at least as important as finding the optimal solution, and there is a trade-off between quality of the solution, time and overhead.

Previous works on CEAS have improved the performance of the system by introducing new features. See for instance [3,4,5]. This paper follows the same objective, but takes a different approach. Rather than adding yet another mechanism, it takes a step back and revisits the auto-regressive functions at the core of the system.

The rest of this paper is organized as follows. Section 2 presents CEAS. Section 3 addresses the auto-regressive functions at the core of the system and motivates the introduction of a new set of functions. The performance of the system applying these new functions is then evaluated in Section 4. Finally, Section 5 concludes the paper.

## 2    Cross Entropy Ant System (CEAS)

CEAS is an online, distributed and asynchronous ACO system designed for adaptive and robust path management in telecommunication networks. Ants cooperate to collectively find and maintain minimal cost paths, or sets of paths, between source and destination pairs. Each ant performs a random search directed by the pheromone trails to find a path to a destination. Each ant also deposits pheromones so that the pheromone trails reflect the knowledge acquired by the colony thus enforcing the stigmergic behaviour characterizing ACO systems. Similarly to what happens in nature, good solutions emerge as the result of the iterative indirect interactions between ants.

Contrary to other ACO systems, the random proportional rule used by ants to decide about their next-hop and the pheromone update rule used in CEAS are formally founded and stem from the Cross-Entropy (CE) method for stochastic optimization [6]. A brief overview of these formal foundations is given below before CEAS is described in more details.

### 2.1    Cross-Entropy (CE) Method

The CE method is a Model-Based Search (MBS) [7] procedure applying importance sampling techniques to gradually bias a probability distribution over the solution space (probabilistic model) towards high-quality solutions and almost surely converge to the optimal solution. An outline of the method applied to the shortest path problem is given below. For further details and proofs, the reader is referred to [6].

Let $\mathbf{G} = (\mathbf{V}, \mathbf{E})$ denote a bidirectional weighted graph where $\mathbf{V}$ is the set of vertices (nodes) and $\mathbf{E}$ the set of edges (links), and let $\mathbf{p}_t = [p_{t,vi}]_{\|\mathbf{V}\| \times \|\mathbf{V}\|}$ denote the probability distribution after $t$ updates. Finding the shortest path between nodes $s$ and $d$ consists in solving the minimization problem $(\Omega, L)$ where $\Omega$ is the set of feasible solutions (paths) and $L$ is the objective function, which assigns a cost $L(\boldsymbol{\omega})$ to each path $\boldsymbol{\omega} = \langle (s, v_1), (v_1, v_2), \ldots, (v_{h-1}, d) \rangle \in \Omega$. $(v, i) \in \mathbf{E}$ denotes the link connecting node $v$ to node $i$ and $L$ is an additive function, $L(\boldsymbol{\omega}) = \sum_{\forall (v,i) \in \boldsymbol{\omega}} L((v, i))$. The CE method works as follows. At each iteration $t$, a sample of $m$ paths $\{\boldsymbol{\omega}_1, \ldots, \boldsymbol{\omega}_m\}$ is drawn from $\mathbf{p}_{t-1}$, and $\mathbf{p}_t$ is obtained by minimizing the cross entropy between $\mathbf{p}_{t-1}$ multiplied by a quality function $Q$ of the cost values and $\mathbf{p}_t$, which is equivalent to solving

$$\mathbf{p}_t = \arg\max_{\mathbf{p}} \frac{1}{m} \sum_{k=1}^{m} Q(L(\boldsymbol{\omega}_k)) \sum_{(v,i) \in \boldsymbol{\omega}_k} \ln p_{vi} \ . \tag{1}$$

By choosing $Q(L(\boldsymbol{\omega}_k)) = H(L(\boldsymbol{\omega}_k), \gamma_t)$, the solution to (1) is

$$p_{t,vi} = \frac{\tau_{t,vi}}{\sum_{j \in \mathbf{N}_v} \tau_{t,vj}} \ , \tag{2}$$

where $\mathbf{N}_v = \{i \in \mathbf{V} \mid (v, i) \in \mathbf{E}\}$ and

$$\tau_{t,vi} = \sum_{k=1}^{m} I((v, i) \in \boldsymbol{\omega}_k) H(L(\boldsymbol{\omega}_k), \gamma_t) \tag{3}$$

and $I(x) = 1$ if $x$ is true, 0 otherwise.

$$H(L(\boldsymbol{\omega}_k), \gamma_t) = e^{-\frac{L(\boldsymbol{\omega}_k)}{\gamma_t}} \tag{4}$$

is the Boltzmann function, see Fig. 1. $\gamma_t > 0$ is an internal parameter (*temperature*) determined by minimizing it subject to $h_t(\gamma_t) \geqslant \rho$, where

$$h_t(\gamma_t) = \frac{1}{m} \sum_{l_0=1}^{m} H(L(\boldsymbol{\omega}_k), \gamma_t) \tag{5}$$

is the overall performance function and $\rho$ is a configuration parameter (*search focus*) close to 0, typically 0.01. Both $\gamma_t$ and $\rho$ control the relative weights given to solutions and thereby the convergence of the system. $\gamma_t$ is self-adjusting and depends on the sampled solutions. $\rho$ decides the absolute value of the temperature.

## 2.2    Online Distributed Asynchronous Operation

CEAS is an implementation of the CE method as an ACO system where updates are performed online by ants at each node along their way. The CE method,

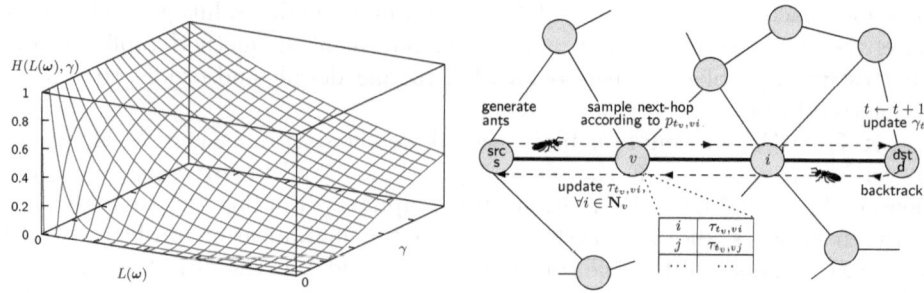

**Fig. 1.** Boltzmann function          **Fig. 2.** Behaviour of ants in CEAS

as described in Section 2.1, is a centralized and batch-oriented procedure and updates are performed offline and synchronously at the end of each iteration. As such, it is therefore not suited for online, distributed and asynchronous operation. To achieve this, CEAS substitutes (3) and (5) with auto-regressive functions. Reusing the notations defined in Section 2.1, the behaviour of ants in CEAS is now detailed. See Fig. 2 for an illustration.

Starting from node $s$, an ant incrementally builds a path to node $d$ by moving through a sequence of neighbour nodes. At each node, the ant samples its next-hop according to $\mathbf{p}_t$ (*biased exploration*) where $t$ now represents the number of paths sampled. At node $v$, the probability that an ant decides to move to node $i$ is given by the *random proportional rule*

$$p_{t_v,vi} = \frac{\tau_{t_v,vi}}{\sum_{j \in \mathcal{N}_v} \tau_{t_v,vj}}, \quad \forall i \in \mathcal{N}_v \tag{6}$$

where $\tau_{t_v,vi}$ is the pheromone trail value at node $v$ for the link $(v,i)$ after $t_v$ updates[1] at node $v$, see below, and $\mathcal{N}_v \subseteq \mathbf{N}_v$ is the set of neighbours of node $v$ not yet visited by the ant. $\mathcal{N}_v = \mathbf{N}_v$ if the ant has visited none or all of the nodes in $\mathbf{N}_v$. To bootstrap the system, ants do not apply (6) but a *uniformly distributed proportional rule* (*uniform exploration*) $p_{t_v,vi} = \frac{1}{|\mathcal{N}_v|}$, $\forall i \in \mathcal{N}_v$. During normal operation, a given percentage of such *explorer ants* is maintained to allow the system to adapt to changes, e.g. to discover new solutions.

Immediately after the ant has arrived at the destination, $t$ is incremented and the temperature $\gamma_t$ is computed. The auto-regressive counterpart of (5) is

$$h_t(\gamma_t) = \beta h_{t-1}(\gamma_t) + (1 - \beta) H(L(\omega_t), \gamma_t) \ , \tag{7}$$

which is approximated by

$$h_t(\gamma_t) \approx \frac{1 - \beta}{1 - \beta^t} \sum_{k=1}^{t} \beta^{t-k} H(L(\omega_k), \gamma_t) \tag{8}$$

---

[1] $t_v = \sum_{x=1}^{t} I((v, \cdot) \in \omega_x)$.

where $\beta \in [0,1)$ is the memory factor and $\boldsymbol{\omega}_k$ is the path sampled by the $k^{\text{th}}$ ant arrived at the destination. $\gamma_t$ is obtained by minimizing it subject to $h_t(\gamma_t) \geqslant \rho$, that is[2]

$$\gamma_t = \left\{ \gamma \mid \frac{1-\beta}{1-\beta^t} \sum_{k=1}^{t} \beta^{t-k} H(L(\boldsymbol{\omega}_k), \gamma) = \rho \right\} . \tag{9}$$

The ant then backtracks and updates pheromones trail values (online delayed pheromone update) along $\boldsymbol{\omega}_t$. The pheromone values are calculated according to the auto-regressive counterpart of (3)

$$\tau_{t_v, vi} = \sum_{k=1}^{t_v} I((v,i) \in \boldsymbol{\omega}_k) \beta^{t_v - k} H(L(\boldsymbol{\omega}_k), \gamma_{t_v}), \quad \forall i \in \mathbf{N}_v, \ \forall v \in \boldsymbol{\pi}_t \tag{10}$$

where $\boldsymbol{\pi}_t = \langle s, v_1, v_2, \ldots, v_{h-1}, d \rangle$ denotes the sequence of nodes traversed by the ant on its way forward.

# 3    Auto-Regressive Functions

Both (8) and (10) are complicated transcendental functions. The exact evaluation of these functions is both processing and storage intensive. The entire path cost history $\mathbf{L}_t = \{L(\boldsymbol{\omega}_k) \mid k = 1, \ldots, t\}$ must be stored and, for each update, the weights for all the costs must be recalculated. Such requirements are impractical for the online operation of a network node. Instead, assuming that the temperature does not radically change, each term in (8) and (10) is approximated by a Taylor expansion, and these functions are replaced by a set of auto-regressive functions with limited computational and memory requirements.

Since CEAS has been first introduced in [8], $H(L(\boldsymbol{\omega}_k), \gamma)$ has been approximated using a Taylor expansion of $H(L(\boldsymbol{\omega}_k), \frac{1}{\gamma})$ around $\frac{1}{\gamma_k}$ (hyperbolic approximation). However, for a similar computational complexity, approximating $H(L(\boldsymbol{\omega}_k), \gamma)$ using a Taylor expansion around $\gamma_k$ (linear approximation) provides a more accurate and more robust approximation. Details on the resulting auto-regressive schemes and their stepwise derivation are given in Appendix A.

## 3.1    Linear vs. Hyperbolic Approximations

Since [8], first order Taylor expansions of $H(L(\boldsymbol{\omega}_k), \frac{1}{\gamma})$, around $\frac{1}{\gamma_k}, \forall k < t$, and around $\gamma_{t-1}$ for $k = t$, have been used to compute the temperature $\gamma_t$. This amounts to approximating $H(L(\boldsymbol{\omega}_k), \gamma)$ by a hyperbola. Now, choosing $\rho < e^{-2}$ yields $\gamma_k < \frac{L(\boldsymbol{\omega}_k)}{2}$, unless the occurrence of a radical change in the costs of the sampled paths, see Section 3.2. Hence, $H(L(\boldsymbol{\omega}_k), \gamma)$ is convex[3] around $\gamma_k$ and a linear approximation is always more accurate than a hyperbolic approximation, see Fig. 3(a). Fig. 4 shows the temperature values obtained by the exact

---

[2] For $\gamma > 0$, $h_t(\gamma)$ is strictly increasing and $h_t(\gamma) \in (0,1)$. Therefore,
$\forall \rho \in (0,1)$, $\exists! \ \gamma_t$, $h_t(\gamma_t) = \rho$ and $\min \gamma$ s.t. $h_t(\gamma) \geqslant \rho \implies \gamma = \gamma_t$.

[3] $H(L(\boldsymbol{\omega}_k), \gamma)$ is convex for $\gamma \leqslant \frac{L(\boldsymbol{\omega}_k)}{2}$ and concave for $\gamma \geqslant \frac{L(\boldsymbol{\omega}_k)}{2}$.

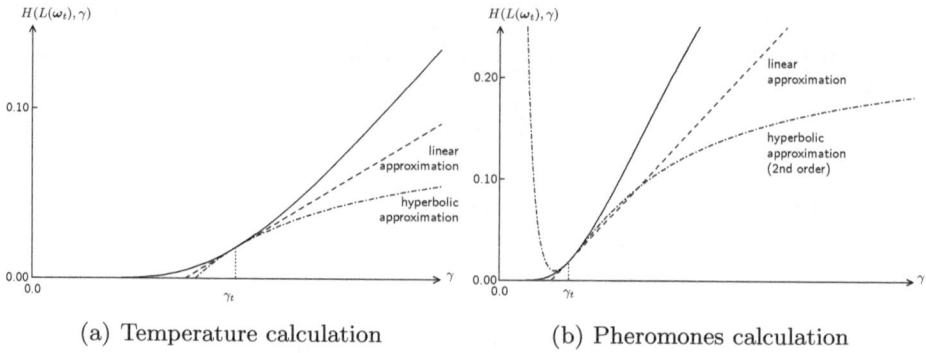

(a) Temperature calculation                    (b) Pheromones calculation

**Fig. 3.** Linear vs. hyperbolic approximations

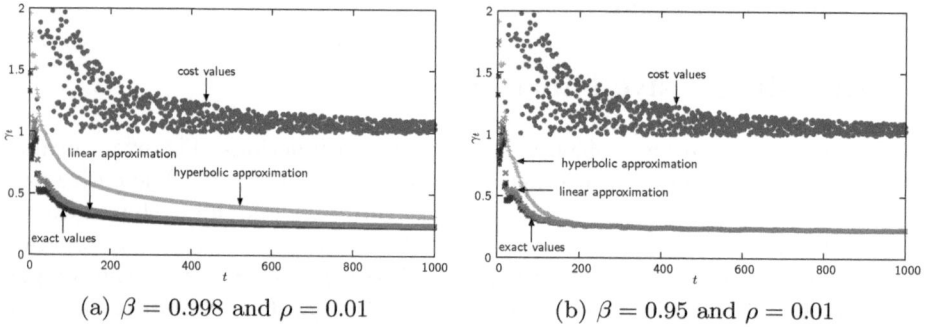

(a) $\beta = 0.998$ and $\rho = 0.01$            (b) $\beta = 0.95$ and $\rho = 0.01$

**Fig. 4.** Temperature obtained by numerical solution of (9) and by using hyperbolic and linear approximations

evaluation of (9) and by applying hyperbolic and linear approximations given the same sequence of cost values. The series of costs imitates the convergence of the system. The $t^{\text{th}}$ cost value is $L_t = L_{\min} + U_t$ where $L_{\min} = 1.0$ and $U$ is a random variable uniformly distributed between 0 and $\frac{100}{t+1}$. The temperature values obtained using linear approximations are much closer to the exact values than the values obtained using hyperbolic approximations.

Since [8], pheromones values are computed using second order Taylor expansions around $\frac{1}{\gamma_k}$ to avoid negative values in case of a rapid decay of the temperature. For $\gamma$ close to $\gamma_k$, the second order hyperbolic expansion of $H(L(\omega_k), \gamma_t)$ provides a better approximation than a linear approximation, see Fig. 3.(b). However, when $\gamma_t < \gamma_k$ $H(L(\omega_k), \gamma_t)$ is over-estimated[4], and when $\gamma_t > \gamma_k$ $H(L(\omega_k), \gamma_t)$ is under-estimated. Hence, using second order expansions tends to smoothen the difference in weights given to costs as the temperature varies. As a result, it may slow down convergence as poor quality solutions (high costs)

---

[4] When $\gamma_t < \gamma_k$ and $\left. \frac{d\tau_{vi,t}(\gamma)}{d\gamma} \right| < 0$, the resulting over-estimation of $\tau_{vi,t}$ is only limited by $\min_{\forall \gamma} \tau_{vi,t}(\gamma)$.

receive relatively higher weights. A first order approximation, and a fortiori a linear approximation, is therefore preferable as it provides a better approximation over a larger interval and better discriminates between good and poor solutions.

## 3.2   Radical Changes

This section addresses the adaptivity and robustness of the approximations to radical changes, i.e. when the assumption of small changes in the temperature does not hold anymore. Radical changes in temperature are caused by radical changes in the costs of the sampled paths, either because paths become unavailable or degraded, or because new paths are discovered. At intermediate nodes, radical temperature differences may also be observed when a node is seldom visited. For pheromone calculations, the observations made above apply, and the larger the change, the better the linear approximations compared to the hyperbolic approximations. In the following, the challenges posed by large variations in the costs of the sampled paths on the calculation of the temperature are considered. We distinguish between radically lower and radically higher costs.

Radically lower costs are observed when a radically better path is discovered. A radically better path is a path such that $L(\omega_t) < 2\gamma_{t-1}$. In this case, $H(L(\omega_t), \gamma)$ is concave around $\gamma_{t-1}$. A linear approximation may then not provide a good approximation of $H(L(\omega_t), \gamma)$ and (9) may not have a positive solution. However, the shape of $H(L(\omega_t), \gamma)$ can easily be exploited to provide a better approximation and ensure that a positive solution is found. See Fig. 5(a) and Appendix A for details. Using a hyperbolic approximation, (9) always have a solution, although it may be quite far from the exact value, see Fig. 5(a). It could be significantly improved by exploiting the shape of $H(L(\omega_t), \gamma)$, but would remain less accurate than a linear approximation.

Radically higher costs are obtained for instance on the degradation of the path the system has converged to. In this case, $H(L(\omega_t), \gamma)$ is convex around $\gamma_{t-1}$. Therefore, a linear approximation is more accurate than a hyperbolic approximation. In addition, a linear approximation is more robust than a hyperbolic approximation. The main problem with using a hyperbolic approximation is that, contrary to a linear approximation, it may be significantly smaller than $\rho, \forall \gamma > 0$. See Fig. 5(b). As a result, (9) may output a very high value or may not have a positive solution. In either case, there is no good alternative; not updating the temperature results in very low pheromone updates and therefore slow convergence, resetting the temperature in possibly premature convergence as pheromone values at intermediate nodes may be strongly directed towards a poor quality path.

The effect of radical changes in the observed costs on the approximations of the temperature is shown in Fig. 6. The settings are identical to those used for the example presented in Fig. 4, except that radical changes are introduced, namely $L_t = 0.01, \forall t \in (450, 700)$. $\beta$ and $\rho$ are chosen to illustrate the potential pitfalls described above when using hyperbolic approximations and radically higher costs are sampled, i.e. very high values (Fig. 6(a)) and no positive solution in which case $\gamma_t$ is not updated (Fig. 6(b)). On the other hand, in both cases,

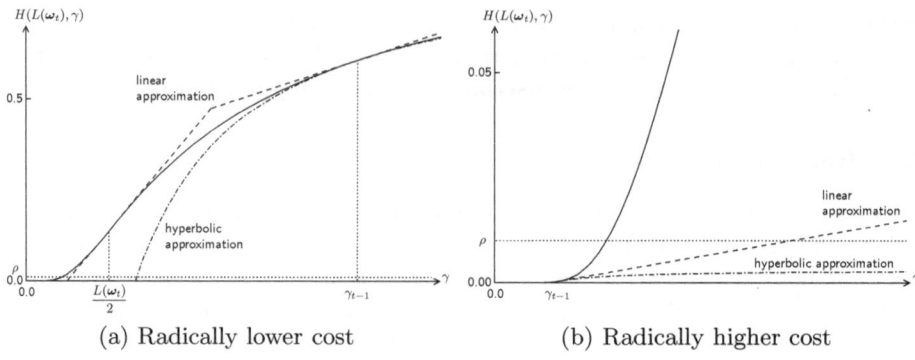

(a) Radically lower cost          (b) Radically higher cost

**Fig. 5.** Approximation of $H(L(\omega_t), \gamma)$ on radical changes

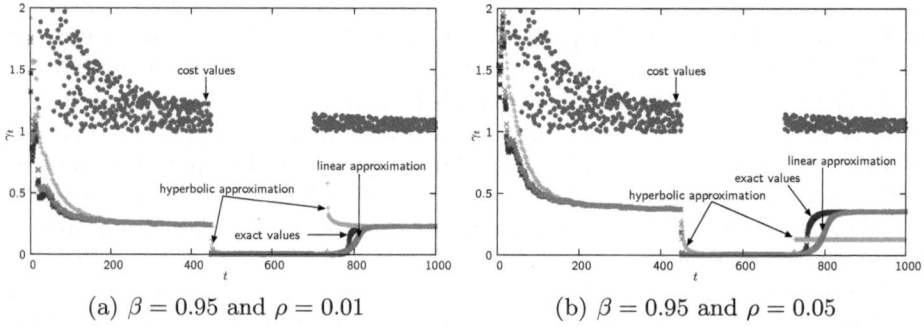

(a) $\beta = 0.95$ and $\rho = 0.01$          (b) $\beta = 0.95$ and $\rho = 0.05$

**Fig. 6.** Effect of radical changes on the temperature obtained by numerical solution of (9) and by using hyperbolic and linear approximations

the temperature obtained applying linear approximations correctly converges to the exact value.

## 4    Case Study

In this section, the effect of replacing the original auto-regressive update schemes obtained using hyperbolic approximations by the new schemes derived from applying linear approximations on the performance of the system is evaluated by simulation. The effect is more pronounced as good solutions are hard to find. Hence, it is demonstrated by applying CEAS to solve the `fri26` symmetric static Traveling Salesman Problem (TSP) taken from TSPLIB[5] and also used in [3]. Note that CEAS has not been specifically designed to solve the TSP. Such a hard (NP-complete) problem is chosen to stress the performance of the system.

Fig. 7 shows the mean value of the cost $L(\omega)$ of the sampled paths with respect to the number of tours completed by ants, averaged over 13 runs. Error

---

[5] http://www.iwr.uni-heidelberg.de/groups/comopt/software/TSPLIB95

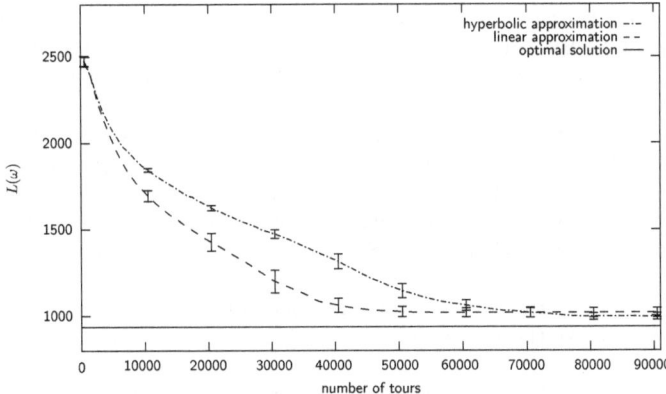

**Fig. 7.** 26 node TSP example

bars indicate 95% confidence intervals. Parameter settings are similar to those used in [3] and elite selection is applied, see [3]. It is observed that applying linear approximations improves the convergence speed of the system in terms of number of iterations ("tours" here). This improvement is due to the combined effect of the faster convergence of the temperature, hence the higher relative weights given to good solutions, and the better differentiation between paths when updating pheromone levels.

## 5    Conclusion

CEAS is an online, distributed and asynchronous ACO system designed for adaptive and robust path management in telecommunication networks. The performance of such a system is related to the number of iterations, the time and the management overhead required to converge after a change. Previous works on CEAS have improved the performance of the system by adding new mechanisms. This paper takes a step back and revisits the auto-regressive functions at the core of the system. These functions are approximations of functions stemming from the CE method for stochastic optimization, processing and storage intensive and therefore not suited for online and distributed operation. The functions used so far were based on hyperbolic approximations. In this paper, new functions are derived applying linear approximations instead. For a similar computational complexity, linear approximations are shown to be both more accurate and more robust to radical changes. Results show that the performance of the system is also improved.

## References

1. Dorigo, M., Di Caro, G., Gambardella, L.M.: Ant Algorithms for Discrete Optimization. Artificial Life 5(2), 137–172 (1999)
2. Bonabeau, E., Dorigo, M., Theraulaz, G.: Swarm Intelligence: From Natural to Artificial Systems. Oxford University Press, Oxford (1999)

3. Heegaard, P.E., Wittner, O.J., Nicola, V.F., Helvik, B.E.: Distributed Asynchronous Algorithm for Cross-Entropy-Based Combinatorial Optimization. In: Rare Event Simulation and Combinatorial Optimization (RESIM/COP 2004), Budapest, Hungary (2004)
4. Heegaard, P.E., Wittner, O.J.: Overhead Reduction in a Distributed Path Management System. In: Computer Networks, Elsevier, Amsterdam (in press, 2009)
5. Paquereau, L., Helvik, B.E.: Ensuring Fast Adaptation in an Ant-Based Path Management System. In: 4th International Conference on Bio-Inspired Models of Network, Information, and Computing Systems (BIONETICS 2009), Avignon, France (2009)
6. Rubinstein, R.Y.: The Cross-Entropy Method for Combinatorial and Continuous Optimization. In: Methodology and Computing in Applied Probability, pp. 127–190 (1999)
7. Zlochin, M., Birattari, M., Meuleau, N., Dorigo, M.: Model-Based Search for Combinatorial Optimization: A Critical Survey. Annals of Operations Research 131, 373–395 (2004)
8. Helvik, B.E., Wittner, O.J.: Using the Cross Entropy Method to Guide/Govern Mobile Agent's Path Finding in Networks. In: Goos, G., Hartmanis, J., van Leeuwen, J. (eds.) MATA 2001. LNCS, vol. 2164, pp. 255–268. Springer, Heidelberg (2001)

## Appendix A: Approximations Details

### Temperature Calculation

Assuming that the temperature does not change radically, $H(L(\boldsymbol{\omega}_k), \gamma_t)$ is approximated by its first order Taylor expansion around $\gamma_k$ for $k < t$

$$H(L(\boldsymbol{\omega}_k), \gamma_t) \approx H(L(\boldsymbol{\omega}_k), \gamma_k) + H'(L(\boldsymbol{\omega}_k), \gamma_k)(\gamma_t - \gamma_k)$$

$$= e^{-\frac{L(\boldsymbol{\omega}_k)}{\gamma_k}} + \frac{L(\boldsymbol{\omega}_k)}{\gamma_k^2} e^{-\frac{L(\boldsymbol{\omega}_k)}{\gamma_k}} (\gamma_t - \gamma_k)$$

$$= e^{-\frac{L(\boldsymbol{\omega}_k)}{\gamma_k}} \left( 1 - \frac{L(\boldsymbol{\omega}_k)}{\gamma_k} + \gamma_t \frac{L(\boldsymbol{\omega}_k)}{\gamma_k^2} \right)$$

and $H(L(\boldsymbol{\omega}_t), \gamma_t)$ around $\gamma_{t-1}$

$$H(L(\boldsymbol{\omega}_t), \gamma_t) = e^{-\frac{L(\boldsymbol{\omega}_t)}{\gamma_t}} \approx e^{-\frac{L(\boldsymbol{\omega}_t)}{\gamma_{t-1}}} \left( 1 - \frac{L(\boldsymbol{\omega}_t)}{\gamma_{t-1}} + \gamma_t \frac{L(\boldsymbol{\omega}_t)}{\gamma_{t-1}^2} \right)$$

Eq.(9) can then be rewritten

$$\rho \frac{1 - \beta^t}{1 - \beta} = \overbrace{\sum_{k=1}^{t-1} \beta^{t-k} e^{-\frac{L(\boldsymbol{\omega}_k)}{\gamma_k}} \left( 1 - \frac{L(\boldsymbol{\omega}_k)}{\gamma_k} \right)}^{a_{t-1}} + \gamma_t \overbrace{\sum_{k=1}^{t-1} \beta^{t-k} e^{-\frac{L(\boldsymbol{\omega}_k)}{\gamma_k}} \frac{L(\boldsymbol{\omega}_k)}{\gamma_k^2}}^{b_{t-1}}$$

$$+ e^{-\frac{L(\boldsymbol{\omega}_t)}{\gamma_{t-1}}} \left( 1 - \frac{L(\boldsymbol{\omega}_t)}{\gamma_{t-1}} + \gamma_t \frac{L(\boldsymbol{\omega}_t)}{\gamma_{t-1}^2} \right)$$

$$= a_{t-1} + e^{-\frac{L(\boldsymbol{\omega}_t)}{\gamma_{t-1}}} \left( 1 - \frac{L(\boldsymbol{\omega}_t)}{\gamma_{t-1}} \right) + \gamma_t \left( b_{t-1} + \frac{L(\boldsymbol{\omega}_t)}{\gamma_{t-1}^2} e^{-\frac{L(\boldsymbol{\omega}_t)}{\gamma_{t-1}}} \right)$$

Hence, $\gamma_t$ is obtained as

$$\gamma_t = -\frac{a_{t-1} + e^{-\frac{L(\boldsymbol{\omega}_t)}{\gamma_{t-1}}}\left(1 - \frac{L(\boldsymbol{\omega}_t)}{\gamma_{t-1}}\right) - \rho\frac{1-\beta^t}{1-\beta}}{b_{t-1} + \frac{L(\boldsymbol{\omega}_t)}{\gamma_{t-1}^2}e^{-\frac{L(\boldsymbol{\omega}_t)}{\gamma_{t-1}}}}$$

where $a_t$ and $b_t$ can be expressed as auto-regressive functions:

$$a_t \leftarrow \beta\left(a_{t-1} + e^{-\frac{L(\boldsymbol{\omega}_t)}{\gamma_t}}\left(1 - \frac{L(\boldsymbol{\omega}_t)}{\gamma_t}\right)\right)$$

$$b_t \leftarrow \beta\left(b_{t-1} + e^{-\frac{L(\boldsymbol{\omega}_t)}{\gamma_t}}\frac{L(\boldsymbol{\omega}_t)}{\gamma_t^2}\right)$$

and the initial values are $a_0 = b_0 = 0$ and $\gamma_0 = -\frac{L(\boldsymbol{\omega}_1)}{\ln \rho}$.

Now when $\gamma_{t-1} > \frac{L(\boldsymbol{\omega}_t)}{2}$ [6], the first order Taylor expansion around $\gamma_{t-1}$ may not be a good approximation of $H(L(\boldsymbol{\omega}_t), \gamma_t)$ and may even result in a negative value for $\gamma_t$. A better approximation may be obtained by approximating $H(L(\boldsymbol{\omega}_t), \gamma_t)$ by its tangent at the inflection point $(\frac{L(\boldsymbol{\omega}_t)}{2}, e^{-2})$. The value obtained using this approximation is used as a lower bound for $\gamma_t$ when $\rho < e^{-2}$ and $\gamma_{t-1} > \frac{L(\boldsymbol{\omega}_t)}{2}$. See Fig. 8 for an illustration. Hence, if $\gamma_{t-1} > \frac{L(\boldsymbol{\omega}_t)}{2}$, $\gamma_t$ is calculated as

$$\gamma_t = \max\left(-\frac{a_{t-1} - e^{-2} - \rho\frac{1-\beta^t}{1-\beta}}{b_{t-1} + \frac{4}{L(\boldsymbol{\omega}_t)}e^{-2}}, -\frac{a_{t-1} + e^{-\frac{L(\boldsymbol{\omega}_t)}{\gamma_{t-1}}}\left(1 - \frac{L(\boldsymbol{\omega}_t)}{\gamma_{t-1}}\right) - \rho\frac{1-\beta^t}{1-\beta}}{b_{t-1} + \frac{L(\boldsymbol{\omega}_t)}{\gamma_{t-1}^2}e^{-\frac{L(\boldsymbol{\omega}_t)}{\gamma_{t-1}}}}\right)$$

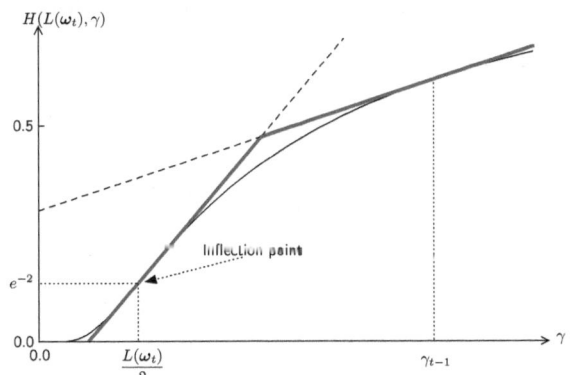

**Fig. 8.** Approximation of $H(L(\boldsymbol{\omega}_t), \gamma_t)$ for $\gamma_{t-1} > \frac{L(\boldsymbol{\omega}_t)}{2}$

---

[6] This happens when a radically better path is discovered.

**Pheromones Calculation**

Eq.(10) can be rewritten

$$\tau_{t_v,vi} = I((v,i) \in \omega_{t_v})H(L(\omega_{t_v}),\gamma_{t_v}) + \underbrace{\sum_{k=1}^{t_v-1} I((v,i) \in \omega_k)\beta^{t_v-k}H(L(\omega_k),\gamma_{t_v})}_{\tau_{t_v,vi}|_{(v,i)\notin\omega_{t_v}}}$$

Approximating $\tau_{t_v,vi}|_{(v,i)\notin\omega_t}$ by a weighted sum of first order Taylor expansions gives

$$\tau_{t_v,vi}|_{(v,i)\notin\omega_{t_v}} \approx \overbrace{\sum_{k=1}^{t_v-1} I((v,i) \in \omega_k)\beta^{t_v-k}e^{-\frac{L(\omega_k)}{\gamma_k}}\left(1-\frac{L(\omega_k)}{\gamma_k}\right)}^{A_{t_v-1,vi}}$$

$$+ \gamma_{t_v}\overbrace{\sum_{k=1}^{t_v-1} I((v,i) \in \omega_k)\beta^{t_v-k}e^{-\frac{L(\omega_k)}{\gamma_k}}\frac{L(\omega_k)}{\gamma_k^2}}^{B_{t_v-1,vi}}$$

where $A_{t_v,vi}$ and $B_{t_v,vi}$ can be expressed as auto-regressive functions:

$$A_{t_v,vi} \leftarrow \beta\left(A_{t_v-1,vi} + I((v,i) \in \omega_{t_v})e^{-\frac{L(\omega_{t_v})}{\gamma_{t_v}}}\left(1-\frac{L(\omega_{t_v})}{\gamma_{t_v}}\right)\right)$$

$$B_{t_v,vi} \leftarrow \beta\left(B_{t_v-1,vi} + I((v,i) \in \omega_{t_v})e^{-\frac{L(\omega_{t_v})}{\gamma_{t_v}}}\frac{L(\omega_{t_v})}{\gamma_{t_v}^2}\right)$$

and where the initial values are $A_{0,vi} = B_{0,vi} = 0$. In addition, $\tau_{t_v,vi}|_{(v,i)\notin\omega_{t_v}}$ is bounded by 0 and $\sum_{k=1}^{t-1}\beta^{t-k} = \beta\frac{1-\beta^{t-1}}{1-\beta}$.

# Quantitative Modeling of Self-organizing Properties*

Richard Holzer and Hermann de Meer

Faculty of Informatics and Mathematics, University of Passau, Innstrasse 43, 94032
Passau, Germany
{holzer,demeer}@fim.uni-passau.de

**Abstract.** For analyzing properties of complex systems, a mathematical model for these systems is useful. In this paper we give quantitative definitions of adaptivity, target orientation, homogeneity and resilience with respect to faulty nodes or attacks by intruders. The modeling of the system is done by using a multigraph to describe the connections between objects and stochastic automatons for the behavior of the objects. The quantitative definitions of the properties can help for the analysis of existing systems and for the design of new systems. To show the practical usability of the concepts, the definitions are applied to a slot synchronization algorithm in wireless sensor networks.

**Keywords:** Self-Organziation, Mathematical modeling, Systems, Adaptivity, Target orientation, Homogeneity, Resilience.

## 1 Introduction

A goal for networking systems is to reduce administrative requirements for users and operators. A technical system should be able to manage itself as much as possible without requiring human effort. Failures and malfunctioned modules should be detected automatically and auto-corrected if possible. This leads to the concept of self-organization, which is a topic that have become more and more important in the last few years. Much research has been done for the design and analysis of autonomous and self-organizing systems.

To assist these technological advances, a mathematical model is useful for a better understanding of complex systems and to improve the design of new systems. The main goal of this paper is to contribute to this issue. We give formal definitions of adaptivity, target orientation, homogeneity and resilience. We also apply these concepts to a practical example, namely the self-organized slot-synchronization in wireless networks.

In this paper, Section 2 gives an overview of the related work and Section 3 gives a mathematical model for complex systems. Sections 4-7 define quantitatively the concepts adaptivity, target orientation, homogeneity and resilience.

---

* This research is partially supported by the SOCIONICAL project (IP, FP7 Call 3, ICT-2007-3-231288), by the ResumeNet project (STREP, FP7 Call 2, ICT-2007-2-224619) and by the Network of Excellence EuroNF (IST, FP7, ICT-2007-1-216366).

T. Spyropoulos and K.A. Hummel (Eds.): IWSOS 2009, LNCS 5918, pp. 149–161, 2009.

In Section 8, we apply our definitions to model slot synchronization in wireless networks. Section 9 contains some discussions, applications and advanced conclusions of the formal model described in this paper.

## 2   Related Work

In the last few years, much research has been done in the topic of self-organizing systems. Unfortunately, there is no generally accepted meaning of self-organization. According to [1] and [2] some typical features of self-organization are autonomy, emergence, self-maintenance, adaptivity, decentralization, and optimization.

A non-technical overview of self-organization can be found in [2]. Other definitions and properties of self-organizing systems can be found in thermodynamics [3], information theory [4], cybernetics [5], [6], [7] and synergetics [8]. An extensive description about the design of self-organizing systems is in [9]. A good overview about modeling complex systems can be found in [10]. [11] gives a survey about practical applications of self-organization. For modeling continuous self-organizing systems and a comparison between discrete and continuous modeling see [12]. Quantitative definitions of autonomy, emergence and global-state awareness, which use the information theoretical concept of entropy, can be found in [13], [14] and [15]. The methods of [13] for modeling a system with stochastic automatons are also used in Section 3 of this paper. While [13] gives quantitative definitions only for the concepts autonomy and emergence, this paper contains the definitions of adaptivity, target orientation, homogeneity and resilience.

## 3   Discrete Systems

For modeling discrete systems, we use the methods of [13], which are based on the ideas of [2]: A multigraph describes the connections between objects and stochastic automatons describe the behavior of the objects. These concepts allow the modeling of a wide variety of complex systems of the real world, e.g. systems that appear in biology, physics, computer science or any other field.

In the real world, not all properties are known in all detail (e.g. it would be very difficult to describe a deterministic behavior of an animal), but there are many things, that can better be described by probabilities. Therefore, a stochastic behavior is more adequate than a deterministic one. In this section, we use directed multigraphs to describe the communication channels for the interaction between objects: Each node in the multigraph corresponds to an object and each edge of the multigraph is used to model the interaction (e.g. transfer of data) between the objects. For modeling the external influence of the environment we use special vertices (external nodes) in the multigraph, where the edges from these vertices represent the channels for the input into the system, and the edges to these vertices represent the output of the system. We distinguish between user data (data from the environment that is processed by the system) and control data (data from the environment to change the behavior of the system). The behavior of the objects is modeled by finite stochastic automatons.

**Definition 1.** *A discrete system[1]* $\mathcal{S} = (G, E, C, A, a)$ *consists of*

- *a finite directed multigraph* $G = (V, K, \tau)$, *where* $V$ *is the set of vertices,* $K$ *is the set of directed edges (loops are also allowed), and* $\tau : K \to V^2$ *assigns to each edge* $k \in K$ *the corresponding vertices* $\tau(k)$, *where the starting vertex is also denoted by* $k-$ *and the ending vertex is denoted by* $k+$. *Therefore we have* $\tau(k) = (k-, k+)$ *for each* $k \in K$. *For a vertex* $v \in V$ *the set of edges ending in* $v$ *is denoted by* $v- := \{k \in K \mid k+ = v\}$ *and the set of edges starting at* $v$ *is denoted by* $v+ := \{k \in K \mid k- = v\}$. *Analogously for* $T \subseteq V$ *the sets* $T-$ *and* $T+$ *are defined by* $T- := \bigcup_{v \in T} v- = \{k \in K \mid k+ \in T\}$ *and* $T+ := \bigcup_{v \in T} v+ = \{k \in K \mid k- \in T\}$.
- *a subset of vertices* $E \subseteq V$, *which elements are called* external *nodes. The other vertices are called* internal *nodes. The* input edges *are denoted by* $I = E+ = \{k \in K \mid k- \in E\}$. *The* output edges *are denoted by* $O = E- = \{k \in K \mid k+ \in E\}$. *All other edges are called* internal *edges.*
- *a subset of the input edges* $C \subseteq I$. *The elements of* $C$ *are called* control *edges. The other input edges* $U := I \setminus C$ *are called* user *edges.*
- *a finite set* $A$, *which is used as alphabet for communication between the nodes.*
- *a family* $a = (a_v)_{v \in V}$ *of stochastic automatons* $a_v = (A^{v-}, A^{v+}, S_v, P_v, d_v)$, *where*
  - $A^{v-} = \{(x_k)_{k \in v-} \mid x_k \in A, k \in v-\}$ *are the* local input values,
  - $A^{v+} = \{(x_k)_{k \in v+} \mid x_k \in A, k \in v+\}$ *are the* local output values,
  - $S_v$ *is the set of states,*
  - $P_v : S_v \times A^{v-} \times S_v \times A^{v+} \to [0, 1]$ *is a function, such that* $P(q, x, \cdot, \cdot) : S_v \times A^{v+} \to [0, 1]$ *is a probability distribution on* $S_v \times A^{v+}$ *for each* $q \in S_v$ *and* $x \in A^{v-}$. *The value* $P(q, x, q', y)$ *is the probability, that the automaton moves from state* $q \in S_v$ *into the new state* $q' \in S_v$ *and gives the local output* $y \in A^{v+}$ *when it receives the local input* $x \in A^{v-}$.
  - $d_v : S_v \to \mathbb{R}^+$ *is a map, where* $d_v(q)$ *describes the delay between two pulses of the clock when the automaton is in state* $q \in S_v$ *(there is no global clock, but each automaton has its own clock).*

These definitions allow us to model complex systems of the real world: Assume that we would like to analyze a system, e.g. a computer network. Then each node of the network corresponds to a vertex of the multigraph. If one node of the network is able to communicate with another node, then we draw an edge between the vertices in the graph. The behavior of each node is modeled by a stochastic automaton, which describes, how the internal state changes for each input, which it gets from the other nodes. Note that the clock pulses of the automatons are deterministic: For each node $v \in V$ and each state $q \in S_v$ the value $d_v(q)$ is the amount of time, the automaton stays in the state $q$. We could

---

[1] See also [13].

also use nondeterminism for the clock pulses (e.g. with a probability distribution on a discrete set of values for the duration), but this is usually not needed, because this can be modeled by using the nondeterminism for the state transition into different states $q \in S_v$ with different values $d_v(q)$.

If we consider the global view on the system at a point of time, then we see a current local state inside each automaton and a current value on each edge, which is transmitted from one node to another node. Such a global view is a snapshot of the system. It is formally defined in the following definition:

**Definition 2.** *Let $\mathcal{S}$ be a system. A* configuration $c = (c_V, c_K, D)$ *consists of*

- *a tuple $c_V \in \prod_{v \in V} S_v$ of states, which defines the current states of the automatons,*
- *a map $c_K : K \to A$, which defines the current symbols on the edges,*
- *a map $D : V \to \mathbb{R}^+$, where $D(v) \leq d_v(c_V(v))$ describes the length of the time interval between the current point of time and the next pulse of the clock in the automaton $a_v$.*

*The set of all configuration is denoted as* Conf. *For a configuration $c = (c_V, c_K, D)$ and a set $T \subseteq K$ of edges the assignment $c_K|_T : T \to A$ of the edges in $T$ is also denoted by $c|_T$. The restriction of $c$ to the external nodes is defined by $c|_{ext} = (c_V|_E, c_K|_{E+}, D|_E)$. The restriction of $c$ to the internal nodes is defined by $c|_{int} = (c_V|_{V \setminus E}, c_K|_{(V \setminus E)+}, D|_{V \setminus E})$. An initialization of $\mathcal{S}$ is a pair $(\Gamma, P_\Gamma)$, where $\Gamma$ is a set of configurations and $P_\Gamma : \Gamma \to [0, 1]$ is a probability distribution on $\Gamma$, which describes, with which probability the system starts in a certain configuration $c \in \Gamma$. For a configuration $c = (c_V, c_K, D)$ the duration of $c$ is defined by $d_c = \min\{D(v) \mid v \in V\}$. Let $N_c = \{v \in V \mid D(v) = d_c\}$. Then the elements of $N_c$ are the nodes with the soonest clock pulse after the current point of time. A configuration $c' = (c'_V, c'_K, D')$ is a* successor configuration *of $c = (c_V, c_K, D)$ with probability $p \in [0, 1]$ (notation: $P(c \to c') = p$) if*

- $c'_V(v) = c_V(v)$ *for $v \in V \setminus N_c$,*
- $c'_K(k) = c_K(k)$ *for $k \in (V \setminus N_c)+$,*
- $D'(v) = D(v) - d_c$ *for $v \in V \setminus N_c$,*
- $D'(v) = d_v(c'_V(v))$ *for $v \in N_c$,*
- $p = \prod_{v \in N_c} P_v(c_V(v), (c_K(k))_{k \in v-}, c'_V(v), (c'_K(k))_{k \in v+})$.

*A (finite or infinite) tuple $s = (c_0, c_1, c_2, \ldots)$ of configurations is called* configuration sequence *if for each $j \geq 0$ we have $P(c_j \to c_{j+1}) > 0$. For a configuration $c$ let $succ(c)$ be a random variable with the probability distribution $P(succ(c) = c') = P(c \to c')$ for each successor configuration $c'$ of $c$. This concept of successor can be extended in a canonical way to arbitrary sequences $(c_0, c_1, c_2, \ldots, c_j)$ of configurations to get the probability $P(c \to^* c')$, that $c'$ is reached from the configuration $c$, where the steps are considered as independent. The duration of a sequence $s = (c_0, c_1, c_2, \ldots, c_j)$ of configurations $c_i \in$ Conf is $d_s = \sum_{i=0}^{j} d_{c_i}$. For a given duration $t \geq 0$ let $P(c \to^t c')$ be the probability, that $c'$ is active $t$ time* *units after the time of c, i.e. we consider the probabilities of the sequences $s = (c_0, c_1, c_2, \ldots, c_j)$ of configurations with $c_0 = c$ and $c_j = c'$ with $d_s - d_{c'} \leq t < d_s$. For $t \geq 0$ let $P(\Gamma \to^t c)$ be the probability, that $c$ is active at time $t$. Define*

$$\Gamma_t = \{c \mid P(\Gamma \to^t c) > 0\},$$

*i.e. $\Gamma_t$ is the set of all configurations $c$ that may be active at time $t$, where we assume that the initialization of the system is at time $t_0 = 0$. Let $\mathrm{Conf}_t$ be the random variable taking values in $\Gamma_t$ with the probability distribution $P(\mathrm{Conf}_t = c) = P(\Gamma \to^t c)$ for $c \in \Gamma_t$.*

To analyze the behavior of a system, we initialize it at time $t_0 = 0$ by choosing a start configuration $c_0 \in \Gamma$ and then the automatons produce a sequence $c_0 \to c_1 \to c_2 \to \ldots$ of configurations during the run of the system. When we do a snapshot of the system at time $t \geq 0$, we see a current configuration $c \in \Gamma_t$. Since the automaton and the initialization are not deterministic, the sequence $c_0 \to c_1 \to c_2 \to \ldots$ is not uniquely determined by the system, but it depends on random events. So for each time $t \geq 0$, we have a random variable $\mathrm{Conf}_t$, which describes, with which probability $P(\mathrm{Conf}_t = c)$ the system is in a given configuration $c$ at time $t$.

A single node $v$ in the system has not the global view, it only sees its local input and output values and its internal state. This concept is given in the following definition.

**Definition 3.** *Let $S$ be a system, $v \in V$ and $c = (c_V, c_K, D)$ be a configuration. The* local configuration *of $c$ in $v$ is defined by*

$$Loc_v(c) = (c_V(v), (c_K(k))_{k \in v- \cup v+}, D(v))$$

*Let $Conf_{t,v} = Loc_v(Conf_t)$ be the random variable for the local configuration of $v$ at time $t \in \mathbb{R}_0^+$.*

For measuring the information in a system we use the statistical entropy:

**Definition 4.** *For a discrete random variable $X$ taking values from a set $W$ the entropy $H(X)$ of $X$ is defined by [16]*

$$H(X) = -\sum_{w \in W} P(X = w) \log_2 P(X = w).$$

*For discrete random variables $X, Y$ the* conditional entropy *$H(X|Y)$ is defined by*

$$
\begin{aligned}
H(X|Y) =\, & H(X,Y) - H(Y) \\
=\, & -\sum_{w,w' \in W} P(X = w, Y = w') \log_2 P(X = w, Y = w') \\
& + \sum_{w \in W} P(Y = w) \log_2 P(Y = w)
\end{aligned}
$$

The entropy measures, how many bits are needed to encode the outcome of the random variable in an optimal way. In Section 6 we use this concept to define quantitatively the homogeneity of a system. Another concept, that we need for the quantitative definitions in Sections 4-7, is the average value of a function:

**Definition 5.** *Let* $f : \mathbb{R}_0^+ \to \mathbb{R}$ *be a real function, which is integrable on every finite interval. For points of time* $s > r \geq 0$ *the* average value *of* $f$ *in the interval* $[r, s]$ *is defined by* $\mathrm{Avg}_{[r,s]}(f) = \frac{1}{s-r} \int_r^s f(t)dt$. *The* average value *of* $f$ *is defined by* $\mathrm{Avg}(f) = \liminf\limits_{t \to \infty} \mathrm{Avg}_{[0,t]}(f)$.

## 4    Target Orientation

Before a new system is designed, we have the goal of the system in our mind: The system should fulfill a given purpose. The behavior of each node is defined in such a way, that this goal is reached, so the design of a system needs a *target orientation*, which is specified in the following definition.

**Definition 6.** *Let* $S$ *be a system and* $(\Gamma, P_\Gamma)$ *be an initialization. Let* $b : \mathrm{Conf} \to [0, 1]$ *be a valuation map for the configurations. For a point of time* $t \geq 0$ *the* level of target orientation *of* $S$ *at time* $t$ *is defined by* $\mathrm{TO}_t(S, \Gamma) = E(b(\mathrm{Conf}_t))$, *where* $E$ *is the mean value of the random variable. The* level of target orientation *of the system* $S$ *is defined by* $\mathrm{TO}(S, \Gamma) = \mathrm{Avg}(t \mapsto \mathrm{TO}_t(S, \Gamma))$. *The system* $S$ *is called* target oriented *with respect to* $b$ *if* $\mathrm{TO}(S, \Gamma) = 1$.

For the target orientation, the valuation map $b$ describes which configurations are "good": A high value $b(c) \approx 1$ means that the configuration $c$ is a part of our goal which we had in mind during the design of the system. The level of target orientation measures the valuations $b(c)$ of the configurations during the whole run of a system: A high level of target orientation $(\mathrm{TO}(S, \Gamma) \approx 1)$ means, that the mean valuation of the configurations during a run of the system often is nearly 1.

In [17], a practical application of target orientation is given in the theory of cellular automatons. In this example, the target goal is the classification of the initial state. This is the reason, why [17] uses a different definition of target orientation: A system is target oriented, if the nodes are able to classify the initial state with respect to a given equivalence relation, which describes the classes of interest.

## 5    Resilience

For computer networks, there are different forms of resilience:

- resilience with respect to malfunctioned nodes
- resilience with respect to attacks by an intruder, which is inside the network

- resilience with respect to attacks by an intruder, which is outside the network
- resilience with respect to natural disasters or other external influence, which might cause a breakdown of some nodes

The following definition can be used to measure these different forms of resilience:

**Definition 7.** *Let $\mathcal{S}$ be a system and $(\Gamma, P_\Gamma)$ be an initialization. Let $\Theta$ be a set and $p_\Theta : \Theta \to [0,1]$ be a probability distribution. Let $Z = (Z_{\theta,v})_{\theta \in \Theta, v \in V}$ be a family of stochastic automatons*

$$Z_{\theta,v} = (A^{v-}, A^{v+}, ZS_{\theta,v}, ZP_{\theta,v}, Zd_{\theta,v})$$

*For $\theta \in \Theta$ let $\mathcal{S}^\theta$ be the system $\mathcal{S}$ after replacing $a_v$ by $Z_{\theta,v}$ for all $v \in V$. Let $(\Gamma^{\mathcal{S}^\theta}, P_{\Gamma^{\mathcal{S}^\theta}})$ be an initialization of $\mathcal{S}^\theta$. Let $\mathrm{Conf}^\theta$ be the set of the configurations of $\mathcal{S}^\theta$. Let $b = (b_\theta)_{\theta \in \Theta}$ be a family of valuation maps $b_\theta : \mathrm{Conf}^\theta \to [0,1]$ for the configurations. For a point of time $t \geq 0$ let $\mathrm{Conf}_t^\Theta$ be the random variable, which applies the random variable $\mathrm{Conf}_t$ in the system $\mathcal{S}^\theta$ after choosing $\theta \in \Theta$ randomly according to the probability $p_\Theta$. The level of resilience of $\mathcal{S}$ at time $t$ is defined by $\mathrm{Res}_t(\mathcal{S}, \Gamma) = E(b(\mathrm{Conf}_t^\Theta))$, where $E$ is the mean value of the random variable. The level of resilience of the system $\mathcal{S}$ is defined by $\mathrm{Res}(\mathcal{S}, \Gamma) = \mathrm{Avg}(t \mapsto \mathrm{Res}_t(\mathcal{S}, \Gamma))$. The system $\mathcal{S}$ is called resilient with respect to $b$ if $\mathrm{Res}(\mathcal{S}, \Gamma) = 1$.*

In this definition the automaton $Z_{\theta,v}$ can be used to describe the malfunctioned behavior of a node $v$. In a computer network, this behavior could be caused by hardware failure, it could be the behavior of an intruder inside the network ($v \in V \setminus E$) or outside of the network ($v \in E$) or it does not send data to its successor nodes due to a breakdown. The system is resilient if despite the malfunctioned nodes the system still runs through many "good" configurations.

If there are only few malfunctional nodes, then we can use $Z_v = a_v$ for the other nodes. If the behavior of a malfunctional node $v$ depends on the original behavior $a_v$, then the automaton $Z_v$ can be a modification of the original automaton $a_v$ to describe the malfunctional behavior of $v$.

## 6 Homogeneity

The following definition describes how homogenous a system is.

**Definition 8.** *Let $\mathcal{S}$ be a system and $(\Gamma, P_\Gamma)$ be an initialization. Let $<$ be a strict linear order on $V$. For a point of time $t \geq 0$ the level of homogeneity of $\mathcal{S}$ at time $t$ is defined by $\mathrm{Ho}_t(\mathcal{S}, \Gamma) = 1 - \dfrac{\sum\limits_{v,w \in V, v < w} |H(\mathrm{Conf}_{t,v}) - H(\mathrm{Conf}_{t,w})|}{\sum\limits_{v,w \in V, v < w} \max(H(\mathrm{Conf}_{t,v}), H(\mathrm{Conf}_{t,w}))}$. The level of homogeneity of the system $\mathcal{S}$ is defined by $\mathrm{Ho}(\mathcal{S}, \Gamma) = \mathrm{Avg}(t \mapsto \mathrm{Ho}_t(\mathcal{S}, \Gamma))$. The system $\mathcal{S}$ is called homogeneous, if $\mathrm{Ho}(\mathcal{S}, \Gamma) = 1$.*

The level of homogeneity measures the similarity of the local configurations of different nodes: A high level of homogeneity means, that the local configurations have a similar entropy, while a low level indicates, that the entropies of the local configurations in different nodes differ very much.

## 7   Adaptivity

The following definition measures the adaptivity of a system.

**Definition 9.** *Let $S$ be a system and $(\Gamma, P_\Gamma)$ be an initialization. Let $(\Theta, p_\Theta)$ be a set with a probability distribution $p_\Theta : \Theta \to [0,1]$. Let $Z = (Z_{\theta,v})_{\theta \in \Theta, v \in C-}$ be a family of stochastic automatons*

$$Z_{\theta,v} = (A^{v-}, A^{v+}, ZS_{\theta,v}, ZP_{\theta,v}, Zd_{\theta,v})$$

*For $\theta \in \Theta$ let $S^\theta$ be the system $S$ after replacing $a_v$ by $Z_{\theta,v}$ for all $v \in C-$. Let $\Gamma^{S^\theta}$ be an initialization of $S^\theta$. Let $\mathrm{Conf}^\theta$ be the set of the configurations of $S^\theta$. Let $b = (b_\theta)_{\theta \in \Theta}$ where $b_\theta : \mathrm{Conf}^\theta |_{int} \to [0,1]$ is a valuation map for the configurations of the internal nodes. For a point of time $t \geq 0$ let $\mathrm{Conf}_t^\Theta$ be the random variable, which applies the random variable $\mathrm{Conf}_t$ in the system $S^\theta$ after choosing $\theta \in \Theta$ randomly according to the probability $p_\Theta$. The level of adaptivity of $S$ at time $t$ is defined by $\mathrm{Ad}_t(S, \Gamma) = E(b(\mathrm{Conf}_t^\Theta |_{int}))$, where $E$ is the mean value of the random variable. The level of adaptivity of the system $S$ is defined by $\mathrm{Ad}(S, \Gamma) = \mathrm{Avg}(t \mapsto \mathrm{Ad}_t(S, \Gamma))$. The system $S$ is called adaptive with respect to $b$ if $\mathrm{Ad}(S, \Gamma) = 1$.*

The level of adaptivity measures the influence of the change of control data: A high value of $\mathrm{Ad}(S, \Gamma)$ means that the mean valuation of the configurations during each run of the system with the new control data is nearly 1, so many "good" configurations are reached.

If the system has no external nodes ($E = \emptyset$), then no automaton in replaced. In this case, the concept of target orientation can be seen as a special case of the concept of adaptivity: By choosing a one element set $\Theta = \{\theta\}$ we get $\mathrm{TO}(S, \Gamma) = \mathrm{Ad}(S, \Gamma)$. For $E = \emptyset$ with $|\Theta| > 1$ the level $\mathrm{Ad}(S, \Gamma)$ is the weighted mean level of target orientation $\mathrm{TO}(S, \Gamma)$ with respect to $\theta \in \Theta$:

$$\mathrm{Ad}(S, \Gamma) = \sum_{\theta \in \Theta} p_\Theta(\theta) \cdot \mathrm{TO}_\theta(S, \Gamma)$$

where $\mathrm{TO}_\theta(S, \Gamma)$ is the level of target orientation with respect to $b_\theta$.

## 8   Slot Synchronization in Wireless Networks

In this section we apply the definitions of the previous sections to the process of self-organized slot-synchronization in wireless networks [18]. We consider a set of nodes, each node being able to communicate with some of the other nodes. The access on the shared medium is organized in time slots. Since there is no central clock, which defines when a slot begins, the nodes perform a slot synchronization in a completely distributed manner. An algorithm for this purpose is proposed by Tyrrell, Auer, and Bettstetter in [18]. It is based on the model of pulse-coupled oscillators by Mirollo and Strogatz [19].

In the latter synchronization model, the clock is described by a phase function $\phi$ which starts at time instant 0 and increases over time until it reaches a threshold value $\phi_{th} = 1$. The node then sends a "firing pulse" to its neighbors for synchronization. Each time a node receives such a pulse from a neighbor, it adjusts its own phase function by adding $\Delta\phi := (\alpha - 1)\phi + \beta$ to $\phi$, where $\alpha > 1$ and $\beta > 0$ are constants.

In [18] the pulse-coupled oscillator synchronization model is adapted to wireless systems, where also delays (e.g., transmission delay, decoding delay) are considered. The duration of an uncoupled period is now $2T$ with $T > 0$. This period is divided into four states (see Figure 1). Let $\gamma \in [0, 2T]$ be a time instant. Then the node is in a

- waiting state, if $\gamma \in [0, T_{wait}) =: I_{wait}$,
- transmission state, if $\gamma \in [T_{wait}, T_{wait} + T_{Tx}) =: I_{Tx}$,
- refractory state, if $\gamma \in [T_{wait} + T_{Tx}, T_{wait} + T_{Tx} + T_{refr}) =: I_{refr}$,
- listening state, if $\gamma \in [T_{wait} + T_{Tx} + T_{refr}, 2T) =: I_{Rx}$,

where the constants are defined as follows:

- $T_{Tx}$: Delay for transmitting a value,
- $T_{wait}$: Waiting delay after the phase function reached the threshold. The transmission of the firing pulse begins after this delay. The waiting delay is calculated by $T_{wait} = T - (T_{Tx} + T_{dec})$, where $T_{dec}$ is the delay for decoding the received value.
- $T_{refr}$: Refractory delay after the transmission of the firing pulse to avoid an unstable behavior.

Let $T_{Rx} = 2T - T_{wait} - T_{Tx} - T_{refr}$ be the duration of an uncoupled listening state. We assume that each of these durations $T_{wait}, T_{Tx}, T_{refr}, T_{Rx}$ is less than $T$. The listening state is the only state, in which firing pulses from the neighbors can be received and decoded, and the phase function is changed only during the listening state.

This network can be modeled as a discrete system $\mathcal{S} = (G, E, C, A, a)$, where the internal state of an automaton consists of the current value of the phase

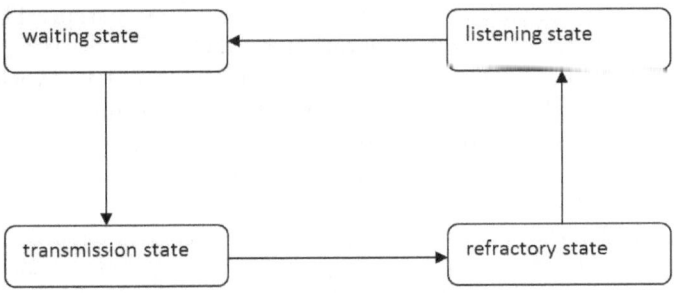

**Fig. 1.** State diagram

function $\phi \in [0, 1]$, the current position $\gamma \in [0, 2T]$ in the cycle in Figure 1 and some information about the decoding delays. In [13] the complete system and the algorithm is modeled in full details, so we omit the details here. Simulation results in [18] show that during the run of the system, groups of synchronizations are built, i.e. inside each group we have good synchronization (each node of the group fires at nearly the same time like the other nodes of the group), and if we wait long enough, then there are only two groups left firing $T$ time units apart from each other. If we use $T$ as the size of the slots, where the beginning of each slot is at $\gamma = 0$ or $\gamma = T$, then both groups have the same slot allocation, so the slots of all nodes are synchronized.

We now apply our definitions of the previous sections.

Concerning the level of target orientation of $\mathcal{S}$, the good configurations are those, where nearly all nodes work synchronously. Since the beginning of a slot is at $\gamma = 0$ or $\gamma = T$, we define the slot distance of two nodes $v, w \in V$ in a configuration $c \in \text{Conf}$ by $dist_c(v, w) = dist(\gamma_v + T\mathbb{Z}, \gamma_w + T\mathbb{Z})$, where $\gamma_v$ is the current value of $\gamma$ at node $v$ and $dist(X, Y) = \inf\{|x - y| \; : \; x \in X, y \in Y\}$ is the usual distance between sets of numbers. The slot distance $dist_c(v, w) \in [0, \frac{T}{2}]$ is the amount of time elapsed between the beginning of the slot of one node and the beginning of the slot of the other node. For a good configuration the slot distances should be low: $b(c) = 1 - \dfrac{\sum\limits_{v,w \in V} dist_c(v,w)}{|V^2| \cdot T/2}$. In this example, the level of target orientation does not depend on the initialization $(\Gamma, P_\Gamma)$, so the mean value $E(b_\theta(\text{Conf}_t))$ can be calculated by using an arbitrary start configuration as a deterministic initialization. We have calculated the level of resilience for a complete graph $G$ with the parameters, which had also been used for the analysis in [18]:

- $|V| = 30$
- $T = 100$
- $T_{dec} = 15$
- $T_{Tx} = 45$
- $T_{refr} = 35$
- $T_{wait} = 40$
- $\alpha = 1.2$
- $\beta = 0.01$

We have calculated $\text{TO}_t(\mathcal{S}, \Gamma) = E(b(\text{Conf}_t))$ for $t \leq 10^6$ steps. Then we can approximate the level of target orientation by $\text{TO}(\mathcal{S}, \Gamma) \approx \text{Avg}_{[10^5, 10^6]}(t \mapsto \text{TO}_t(\mathcal{S}, \Gamma))$. The result is $\text{TO}(\mathcal{S}, \Gamma) \approx 0,997$, so the system has a very high level of target orientation. This can also be derived from the results of [18]: After the groups of synchronizations are built, the slot distances are zero for almost every pair of nodes, so $\text{TO}_t(\mathcal{S}, \Gamma) \approx 1$ and therefore the system is target orientated: $\text{TO}(\mathcal{S}, \Gamma) \approx 1$.

Now we consider the level of resilience. Assume that a set $\theta \subseteq V$ of nodes have a breakdown in the wireless network. Then the parameter set $\Theta$ consists of all subsets of $V$, i.e. $\Theta = \mathfrak{P}(V)$ and $p_\Theta(\theta) \in [0, 1]$ is the probability that the nodes of $\theta$ brake down and the other nodes are still working. After the breakdown we have a system $\mathcal{S}^\theta$, where the automaton $Z_{\theta,v}$ does not send any pulses for $v \in \theta$ and $Z_{\theta,v} = a_v$ works like before for $v \in V \setminus \theta$. The valuation map $b_\theta$ describes the good configurations, i.e. all working nodes should still be able to synchronize: $b_\theta(c) = 1 - \frac{\sum\limits_{v,w \in V \setminus \theta} dist_c(v,w)}{|(V \setminus \theta)^2| \cdot T/2}$. If the working nodes still form a strongly connected graph, they are able to synchronize. If $G$ is the complete graph, where each node is connected to each other node, then the system is resilient: $\text{Res}(\mathcal{S}, \Gamma) \approx 1$. If $G$ is not the complete graph then the working nodes might not be strongly connected anymore, so the synchronization only take place inside each connection component. In this case we have $\text{Res}(\mathcal{S}, \Gamma) < 1$, where the exact value depends on the graph $G$ and on the probabilities $p_\Theta$: If $p_\Theta(\theta)$ is small for large sets $\theta$, then with high probability, the working nodes still form a connected graph, so $\text{Res}(\mathcal{S}, \Gamma)$ is near 1.

Instead of a breakdown of nodes, we also can model the resilience with respect to an intruder at a node $v_0 \in V$, who wants to disturb the communication. In this case, the parameter set $\Theta$ can be used to describe the behavior of the intruder. Here we use $\Theta$ as a discrete subset of $\mathbb{R}^+$, where $\theta \in \Theta$ is the duration between two consecutive pulses, that the intruder sends periodically to the neighbors. The system $\mathcal{S}^\theta$ is the system $\mathcal{S}$ after replacing the automaton $a_{v_0}$ by $Z_{v_0}$ and leave all other automatons as they are: $Z_v = a_v$ for $v \neq v_0$. The good configurations are those, where all other nodes are synchronized: $b_\theta(c) = 1 - \frac{\sum\limits_{v,w \in V \setminus \{v_0\}} dist_c(v,w)}{|(V \setminus \{v_0\})^2| \cdot T/2}$. If the graph without the node $v_0$ is not connected anymore, then the two connection components are not able to synchronize anymore, so we get $\text{Res}(\mathcal{S}, \Gamma) < 1$. For the complete graph with the parameters, which have already been used above for the target orientation, we calculated the level of resilience (with the approximation $\text{Res}(\mathcal{S}, \Gamma) \approx \text{Avg}_{[10^5, 10^6]}(t \mapsto \text{Res}_t(\mathcal{S}, \Gamma))$). The results for different sets $\Theta$ with the uniform distribution $p_\Theta : \Theta \to [0,1]$ are given in Table 1.

**Table 1.** Level of Resilience

| $\Theta$ | {45} | {70} | {100} | {120} | {150} | { 45, 70, 100, 120, 150 } |
|---|---|---|---|---|---|---|
| $\text{Res}(\mathcal{S}, \Gamma)$ | 0.987 | 0.985 | 0.996 | 0.991 | 0.996 | 0.991 |

Therefore the system in this model has a high level of resilience with respect to an intruder, which periodically sends pulses.

Now let us consider the level of homogeneity. If the in-degree $|v - |$ of all nodes are the same (e.g. in a complete graph), then the local configuration $\text{Conf}_{t,v}$ has the same entropy for all nodes $v$ since all automatons are the identical and the initialization is uniform. Therefore the have $\text{Ho}_t(\mathcal{S}, \Gamma) = 1 - \frac{\sum\limits_{v,w \in V, v < w} |H(\text{Conf}_{t,v}) - H(\text{Conf}_{t,w})|}{\sum\limits_{v,w \in V, v < w} \max(H(\text{Conf}_{t,v}), H(\text{Conf}_{t,w}))} = 1$ and the system is homogeneous. If the in-degrees of the nodes differ, then the entropies $H(\text{Conf}_{t,v})$ and $H(\text{Conf}_{t,w})$ need not be the same, but after the two groups of synchronization have been built, we have $H(\text{Conf}_{t,v}) \approx H(\text{Conf}_{t,w})$, because the only difference of these entropies is the information which predecessor $q \in v-$ belongs to which group during the transmission state. So in this case the system is nearly homogeneous.

For the level of adaptivity, we note that the system has no external nodes. For the valuation map, which we already used for the level of target orientation, we get $\mathrm{Ad}(\mathcal{S}, \Gamma) = \mathrm{TO}(\mathcal{S}, \Gamma) = 1$.

# 9    Discussion of the Results

The main result of this paper is, that we get a formalism to analyze complex systems with respect to self-organizing properties. For the adaptivity, target orientation, homogeneity and resilience the quantitative definitions given in the previous sections can be useful for the analysis of complex systems.

When we have a real world system and we would like to analyze this system with respect to self-organizing properties, we first create the model with the definitions of Section 3. The connections between the objects are described by the multigraph and the behavior of the objects are described stochastic automatons. In this model we can apply the definitions of sections 4-7 to calculate

- the level of target orientation,
- the level of adaptivity,
- the level of resilience,
- the level of homogeneity.

The level of target orientation shows, how good the system satisfies the goal, for which the system is designed for. The adaptivity and the resilience are two of the main properties of self-organizing systems, so together with the level of emergence and the level of autonomy of [13] they are also indicators, how self-organizing the system is. Finally, the level of homogeneity shows how different or equal the nodes behave.

One major problem with the definitions of these levels is the complexity: For a very complex system, it might be difficult to compute the exact values for the levels. But the examples in this paper show, that even if the system is too large to compute the exact levels, the definitions can still be useful to get an approximation of these levels. For example, if we consider the level of homogeneity for the slot synchronization in Section 8, then we can derive $\mathrm{Ho}(\mathcal{S}, \Gamma) \approx 1$ from Definition 8 without computing the exact value.

# 10    Conclusion

In this paper we described how self-organzing properties of complex systems can be measured quantitatively. While [13] and [14] have already given definitions for the emergence and autonomy, the novel contribution of this paper are the quantitative definitions for some other properties of self-organizing systems:

- adaptivity,
- target orientation,
- homogeneity,
- resilience.

These definitions may help for the analysis of existing systems and for the design of new systems with self-organizing properties.

# References

1. De Meer, H., Koppen, C.: Characterization of self-organization. In: Steinmetz, R., Wehrle, K. (eds.) Peer-to-Peer Systems and Applications. LNCS, vol. 3485, pp. 227–246. Springer, Heidelberg (2005)
2. Heylighen, F.P.: The science of self-organization and adaptivity. In: Kiel, L.D. (ed.) Knowledge Management, Organizational Intelligence and Learning, and Complexity, The Encyclopedia of Life Support Systems. EOLSS Publishers (2003)
3. Nicolis, G., Prigogine, I.: Self-Organization in Non-Equilibrium Systems: From Dissipative Structures to Order Through Fluctuations. Wiley, Chichester (1977)
4. Shalizi, C.R.: Causal Architecture, Complexity and Self-Organization in Time Series and Cellular Automata. PhD thesis, University of Wisconsin-Madison (2001)
5. von Foerster, H.: Self-Organizing Systems. In: ch. On Self-Organizing Systems and their Environments, pp. 31–50. Pergamon, Oxford (1960)
6. Ashby, W.R.: Principles of Self-organization. In: ch. Principles of the Self-organizing System, pp. 255–278. Pergamon, Oxford (1962)
7. Heylighen, F., Joslyn, C.: Cybernetics and second order cybernetics. Encyclopedia of Physical Science & Technology 4, 155–170 (2001)
8. Haken, H.: Self-organizing Systems: An Interdisciplinary Approach. In: ch. Synergetics and the Problem of Selforganization, pp. 9–13. Campus (1981)
9. Gershenson, C.: Design and Control of Self-organizing Systems. PhD thesis, Vrije Universiteit Brussel, Brussels, Belgium (May 2007)
10. Boccara, N.: Modeling Complex Systems. Springer, Heidelberg (2004)
11. Di Marzo Serugendo, G., Foukia, N., Hassas, S., Karageorgos, A., Mostéfaoui, S.K., Rana, O.F., Ulieru, M., Valckenaers, P., Van Aart, C.: Self-organisation: Paradigms and applications. In: Di Marzo Serugendo, G., Karageorgos, A., Rana, O.F., Zambonelli, F. (eds.) ESOA 2003. LNCS (LNAI), vol. 2977, pp. 1–19. Springer, Heidelberg (2004)
12. Holzer, R., de Meer, H.: On modeling of self-organizing systems. In: Autonomics 2008 (2008)
13. Holzer, R., de Meer, H., Bettstetter, C.: On autonomy and emergence in self-organizing systems. In: Hummel, K.A., Sterbenz, J.P.G. (eds.) IWSOS 2008. LNCS, vol. 5343, pp. 157–169. Springer, Heidelberg (2008)
14. Mnif, M., Mueller-Schloer, C.: The quantitative emergence. In: Proc. of the 2006 IEEE Mountain Workshop on Adaptive and Learning Systems (SMCals 2006), pp. 78–84. IEEE, Los Alamitos (2006)
15. Auer, C., Wuechner, P., de Meer, H.: The degree of global-state awareness in self-organizing systems. In: IWSOS 2009, Springer, Heidelberg (2009)
16. Cover, T.M., Thomas, J.A.: Elements of Information Theory, 2nd edn. Wiley, Chichester (2006)
17. Auer, C., Wuechner, P., de Meer, H.: Target-oriented self-structuring in classifying cellular automata. In: Automata 2009 (2009)
18. Tyrrell, A., Auer, G., Bettstetter, C.: Biologically inspired synchronization for wireless networks. In: Dressler, F., Carreras, I. (eds.) Advances in Biologically Inspired Information Systems: Models, Methods, and Tools. Studies in Computational Intelligence, vol. 69, pp. 47–62. Springer, Heidelberg (2007)
19. Mirollo, R., Strogatz, S.: Synchronization of pulse-coupled biological oscillators. SIAM Journal of Applied Mathematics 50, 1645–1662 (1990)

# Resolving the Noxious Effect of Churn on Internet Coordinate Systems

Bamba Gueye[*] and Guy Leduc

University of Liege, Belgium
{cabgueye,guy.leduc}@ulg.ac.be

**Abstract.** Internet Coordinate Systems (ICS) provide easy and practical latency predictions in the Internet. However, peer dynamics (*i.e*, churn), which is an inherent property of peer-to-peer (P2P) systems, affects the accuracy of such systems. This paper addresses the problem of churn in an ICS without landmarks, like Vivaldi. We propose a framework to assess the robustness of such an ICS in the presence of churn, and evaluate two models for handling churn. The key idea is to reactively recover lost neighbours, either by picking new nodes at random, or by selecting a new one among the node's two-hop neighbours, while maintaining high reliability and low communication overhead. We then show by simulations that our models mitigate the impact of churn, and lead to a good accuracy compared to an instance of an ICS running without churn.

**Keywords:** ICS, Node Churn, Accuracy, Clustering.

## 1 Introduction

Nowadays, a new class of large-scale globally-distributed network services and applications such as distributed overlay network multicast, content addressable overlay networks, and peer-to-peer file sharing (*e.g.*, Gnutella BitTorrent, etc.) have emerged. To achieve network topology-awareness, most, if not all, of these overlays rely on the notion of proximity, usually defined in terms of network delays or round-trip times (RTTs), for optimal neighbor selection during overlay construction and maintenance.

It is important for the new applications presented above to limit the resources consumption and particularly the number of on-demand measurements. In such a context, Internet Coordinate Systems (ICS) [1,2,3] have been proposed to allow hosts to estimate delays without performing direct measurements and thus reduce the consumption of network resources. The key idea of an ICS is to model the Internet as a geometric space and characterize any node in the Internet by a position (i.e., a *coordinate*) in this space. The network distance between any two nodes is then predicted as the geometric distance between their coordinates. Explicit measurements are, therefore, not required any longer.

Generally, an ICS follows a three step procedure. The first step is neighbor selection. In this step, each node in the system chooses a constant number of neighbors. In the

---

[*] This work has been partially supported by the EU under projects FP6-FET ANA (FP6-IST-27489).

T. Spyropoulos and K.A. Hummel (Eds.): IWSOS 2009, LNCS 5918, pp. 162–173, 2009.

second step, each node measures the delays to its neighbors. After collecting the delay measurement, all the nodes use an optimization algorithm to compute the coordinates based on these delays.

In a large-scale P2P system, peer dynamics (i.e., churn) is a prevalent phenomenon, which makes maintenance a challenging task [4]. Almost every distributed system has to deal with churn: *i.e.*, the continuous process of node arrival and departure due to various reasons, *e.g.*, link outage, graceful leaves, failure, etc. In an ICS, in the presence of churn, nodes often did not have time to settle into a stable position before they exited the system. In such case, some nodes will update their coordinates according to neighbors that have not stabilized their coordinates, leading to skewed coordinates [5,6]. As consequence, they had a deleterious effect not only on themselves, but on the overall system convergence. Since churn is inherent to P2P system, it is mandatory to build an ICS that should predict latency with accuracy under churn situation. The remainder of this paper is dedicated to determining whether an ICS can be built so that it continues to perform well under churn.

We study how to reduce the harmful effect of churn on Vivaldi by intelligently remplacing reactively node's neighbors which have left the system. In so doing, we considered two potential solutions to the problem of sustaining a coordinate system under high churn rates. The first one, the *Random Replacement* (RR) replaces a failed neighbor with a randomly chosen node. The second strategy, the *Two-hop Neighbors Replacement* (TNR), considers the two-hop neighbors as a preference list, and thus picks randomly in this list the set of nodes that will be used to replace the failed ones. Note that, the set of two-hop neighbors is formed by the union of the direct neighbors, *i.e.*, the set of peer nodes that are used as neighbors in the ICS for the purpose of coordinate computation, and the neighbors' neighbors. We then provide a comparison of the performance of a range of different node selection strategies in three real-word traces. One of our contribution is to show that Vivaldi can in fact handle churn following an approach that reactively recovers lost neighbors. Our second contribution is an examination of churn in a Self-Organized network, according to a Two-Tier approach of Vivaldi, where nodes cluster themselves based on their network distance [7].

Our main results can be summarized as follows: (*i*) Coordinate systems that experience churn have trouble converging; (*ii*) The strategies of node replacement perform well compared to the case where no recovery mechanism is settled; (*iii*) The Two-tier approach, where we apply the TNR and the RR techniques for addressing churn, is more accurate under high node churn rate compared to a flat Vivaldi.

In this paper, we begin by giving an overview on Vivaldi and a few proposed works that do provide a mechanism for handling churn in Vivaldi, but do not rely on the "original" Vivaldi (Sec 2). Then, we describe different churn scenarios and analyze the behaviour of the RR and TNR strategies in the presence of churn (Sec. 3). We also test as case study the RR and the TNR strategies in a Two-Tier Vivaldi environment (Sec. 4). We finally conclude this paper by reminding its main contributions (Sec. 5).

## 2 · Background on Vivaldi

Vivaldi [2] is probably the most successful Internet coordinates system that has been proposed so far. It does not require any fixed network infrastructure and makes no distinctions between nodes. In fact, Vivaldi [2] is based on a simulation of springs, where the position of the nodes that minimizes the potential energy of the spring also minimizes the embedding error. A Vivaldi node collects distance information towards other nodes (its neighbors) and computes its new coordinates with the collected measurements (sample). The sample used by each node $A$ is based on measurement to a node, $B$, its coordinates $x_B$ and the estimated error $e_B$ being reported by $B$. A relative error of this sample is then computed with respect to $d_{AB}$ and $\hat{d}_{AB}$. Note that $d_{AB}$ and $\hat{d}_{AB}$ represent respectively the measured distance and the estimated distance. The node then computes the sample weight, balancing local and remote error. The local (resp. remote) error represents node $A$'s (resp. node $B$'s) confidence in its own coordinate. Thus, the coordinates are updated by moving a small step toward the perfect position that best reflects the RTT measured. The Vivaldi algorithm quickly converges towards a solution when latencies satisfy the triangle inequality. Vivaldi considers a few possible coordinate spaces that might better capture the underlying structure of the Internet Euclidean spaces, spherical coordinates, etc. For the present study, we use a 2D Euclidean space and each node computes its coordinates by doing measurements with 32 neighbors.

It is worth noticing that Dabek et al. [2] have only studied Vivaldi under stable environment. In such case, nodes will not leave the system after their join. This assumption is not realistic due to the prevalence of peer dynamics in P2P systems. Therefore, no recovery mechanism is proposed in [2] when nodes have lost their neighbors. To overcome this limitation, Ledlie et al. in [5] propose a simple technique by considering a *"full Vivaldi embedding"* that runs over Azureus [8] (now called *Vuze*) which is a popular clients for BitTorrent, a file sharing protocol. Note that, in Azureus metadata are stored in a Distributed Hash Table (DHT) which enables Vivaldi to choose node's neighbors. In this approach, *i.e.*, full Vivaldi embedding, Vivaldi nodes have no dedicated set of neighbors as designed in [2]. Therefore, the information necessary for a coordinate update is piggybacked in the other application level messages, such as routing table heartbeats [5]. In such case, nodes have no control over the selection of which nodes we gossiped with, and nodes communicate only with a limited set of nodes that was much smaller than the number of nodes with which it actually communicated.

In the same way, Chen et al. in [6] propose also *Myth* a Landmark-based NCS which used partially a full Vivaldi embedding as in [5] running on Bamboo DHT. Myth has a initial coordinates prediction scheme that is used before the Vivaldi algorithm. This scheme like GNP [1] uses the nodes already in the system as landmark to compute its coordinates. Requiring that some nodes be designated as landmarks may be a burden on symmetric systems (such as P2P systems). To obtain a neighbor, a given node randomly generates a global unique identifier and queries the DHT for it, and then the DHT returns the node that owns this identifier which will be used as neighbor. Assuming that nodes did not have a fixed set of neighbors, as in [5,6], is an easy way to let down the problem of how a new neighbor will take the place of the left one. It is worth noticing that a fixed set of neighbors is very important because nodes could expect regular exchanges for updating their coordinates. In contrast to previous works, our approaches for handling churn is not based on a structured P2P overlay network or DHT. As a consequence, we will rely on the "original" Vivaldi, where each node has a fixed set of neighbors.

# 3    Handling Node Churn in Vivaldi

In this section we formally model peer churn, describe and evaluate our two replacement strategies for reducing the harmful effect of churn on Vivaldi.

## 3.1    Churn Scenario: Node Arrival and Departure Rates

Churn can be modelled by two kinds of peer-level characterization. Firstly, the session length distribution, which is one of the most basic properties of churn, captures how long peers remain in the system each time they appear. Secondly, the downtime can be defined as the interval between the moment a peer departs and its next arrival. The characterization of churn has been relatively well addressed in the literature. One issue is whether a good mathematical distribution exists to model network churn. In fact, previous works [9,10,11] have shown that both distributions in typical P2P systems are exponential, though other studies claim that the distribution of session length follows a Pareto distribution [5,12,13]. In contrast, the results in [14] show that the distribution of session lengths does not exactly follow the exponential distribution or the long-tailed Pareto distribution across different P2P systems, (e.g., Kad, Gnutella, BitTorrent). Thus, there is still no clear answer on how to model the peer behavior appropriately. Nevertheless, as some studies have shown that session lengths are either exponential or Pareto, we model churn in our simulations by testing both distributions. Note that, when the session length is modelled as a Pareto distribution we have considered that nodes sleep for a random period with uniform distribution and rejoin the system as a newcomer. In contrast, when the session length follows an exponential distribution, the downtime is modelled also as exponential [9,10,11].

We study the set of strategies described above and we believe that they are all relevant in practice. Since this paper focuses more on handling churn in network coordinate systems, finding a good distribution that fits well churn in P2P system is left for future work.

**Modeling peer churn:** In the previous section we showed that the session length and the downtime can be modelled by different kind of distributions. We concentrate primarily on the use of the quantile function for the formulation of distributions. In fact, the quantile function can be used as the basis for a range of approaches to the construct of models of populations. After the probability density function and the cumulative density function, the *Quantile Function*, $QF$, denoted by $QF(p)$, provides a third way of defining a distribution. By definition, the $QF$ of a probability distribution is the inverse of its cumulative distribution function. Formally, we have

$$x_p = \text{the value of } x \text{ for which Probability}(X \le x_p) = p$$

For instance, if $\mathcal{F}$ is a probability distribution function, the $QF$ may be used to "construct" a random variable having $\mathcal{F}$ as its distribution function. This fact serves as the basis of a method of simulating the churn from an arbitrary distribution with the aid of a random number generator. In the following, we present a detailed system model based on the above observation.

In our simulations, the individual peers have different arrival rates for the join/leave events. Theses events can be scheduled as follows according to a fixed distribution. The quantile function for the exponential distribution at time $t$ can be computed as:

$$QF(p) = -\frac{\ln(1-p)}{\lambda} \qquad \text{for } 0 \le p < 1 \tag{1}$$

where $\lambda$ is the parameter of the distribution.

To model the peers behavior following a Pareto distribution, the quantile function is obtained by:

$$QF(p) = \frac{\beta}{(1-p)^{\frac{1}{\alpha}}} \qquad \text{for } 0 \le p < 1 \tag{2}$$

where $\alpha$ is a shape parameter that determines how skewed the distribution is, and $\beta$ is a location parameter that determines where the distribution starts.

Finally, the quantile function of the uniform distribution, which defines an equal probability over a given range, is expressed as follows:

$$QF(p) = (x_{min} + x_{max} \times p) \times mean \qquad \text{for } 0 \le p < 1. \tag{3}$$

## 3.2 Approaches for Handling Churn

We focus only on agnostic strategies, *i.e.*, where we ignore past uptime or availability of individual node because we do not explicitly try to minimize churn, but rather to deal with its presence. Therefore, we study two set of strategies that we believe are both relevant in practice: (*i*) the *Random Replacement* (RR) where each node replaces a failed neighbor reactively with a uniform-random available node; (*ii*) the *Two-Hop Neighbors Replacement* (TNR) where each node replaces lost neighbors by one of its neighbors' neighbors (*i.e.*, node's *Two-hop neighbors*). It should be noted that the list of neighbors' neighbors can be obtained in the network by simply piggybacking the information in the messages exchanged by the ICS system.

We allow our selection algorithm to react immediately after each change in node's neighbors state. We feed the sequence of events into the P2psim simulator [15] following the different distributions of churn characteristics described in Sec. 3.1. Events are nodes joins and failures. The obtained results are shown in the next section.

## 3.3 Experimental Results

In this section we present the results of an extensive simulation study of Vivaldi under churn using the P2Psim discrete-event simulator [15]. Each Vivaldi node has 32 neighbors and results are obtained for a 2-dimensional Euclidean space.

We performed a set of simulations using three datasets: the P2psim data (1740 nodes) [15], the Meridian data ( 2500 nodes) [16], and the PlanetLab data which we collected using *ping* measurements between 180 PlanetLab nodes [17]. Note that, the King and Meridian data sets are obtained following the *King* measurement technique [18] which is similar to ping in the sense that it estimates the latency between arbitrary end hosts by using recursive DNS queries. Based on these delay matrices, we study through the basic *Absolute Estimation Error* (AEE) and *Relative Estimation Error* (REE) metrics the accuracy of Vivaldi under churn. For a given link between two nodes $A$ and $B$ we have the following definitions:

$$AEE(AB) = |EST(A, B) - RTT(A, B)|$$
$$REE(AB) = \frac{AEE(AB)}{RTT(A, B)}$$

where $RTT(X, Y)$ is the measured RTT between the nodes $X$ and $Y$, and $EST(X, Y)$ is the estimated RTT obtained with the coordinates of the nodes $X$ and $Y$.

During our simulations, in some cases, churn is temporary as the departed peers may rejoin the system; churn can also be permanent as peers may depart the system forever. In particular, we set the Pareto distribution parameters in Eq. 2 as: $\alpha = 1.03$ and $\beta = 300s$. The choice of the value of $\alpha$ is guided by Wang *et al.* in [19] using traces of *PPLive*, which is a popular P2P live streaming system. They have shown that the node's stay duration of PPLive is well approximated by a Pareto distribution of $\alpha = 1.03$. The value of $\beta$ represents the time when the churn starts in the system. After a node leaves the network, it sleeps (*i.e.*, downtime) during a time which is uniformly distributed between $0.1 \times mean$ and $1.9 \times mean$ (see Eq. 3), where $mean = 100s$, and rejoins the system (if the downtime is not beyond the simulation time) as a newcomer to stay another Pareto distribution. In the second approach of characterizing churn, events for each node is exponentially distributed with a mean of $100s$ (Eq. 1). Indeed, peers within the network are assigned exponentially distributed session lengths. When a peer reaches the end of its session length, it leaves the network and waits an exponentially distributed time for another potential join. It should be noted that the choice of mean session time is consistent with past studies of P2P networks [20].

Figures 1, 2, and 3 illustrate the general behavior of nodes for the King, PlanetLab, and Meridian data sets during our simulations. Note that *No Recovery Nodes* (NRN) means an instance of Vivaldi under churn without replacement of lost nodes. When the churn is intensive it remains at least 80% of the nodes in the system following the Pareto distribution, whereas for the exponential distribution the average of available

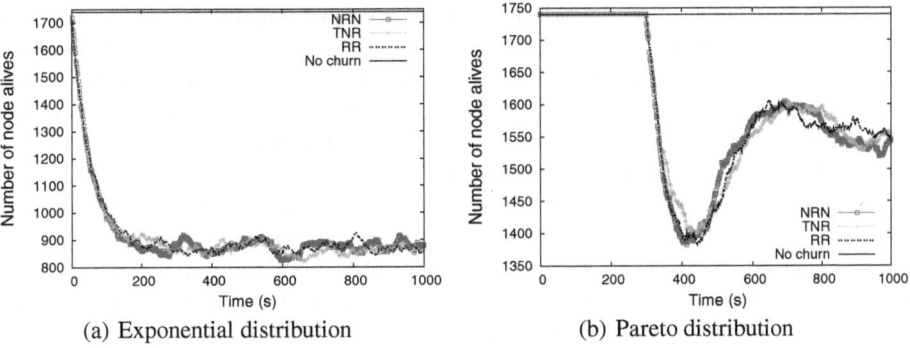

(a) Exponential distribution          (b) Pareto distribution

**Fig. 1.** King dataset: nodes alive as function of time

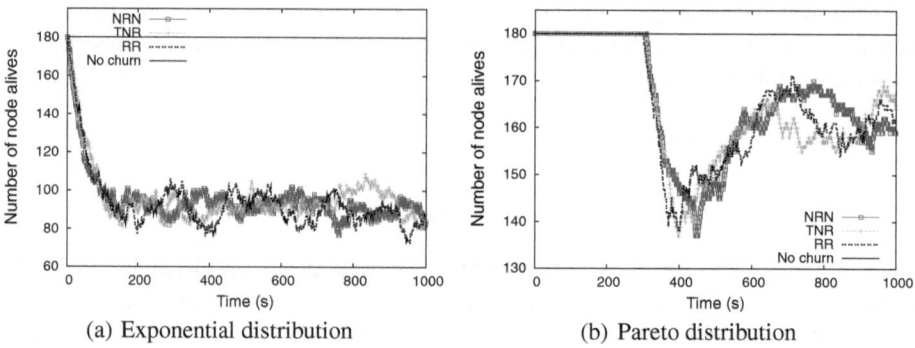

(a) Exponential distribution          (b) Pareto distribution

**Fig. 2.** PlanetLab dataset: nodes alive as function of time

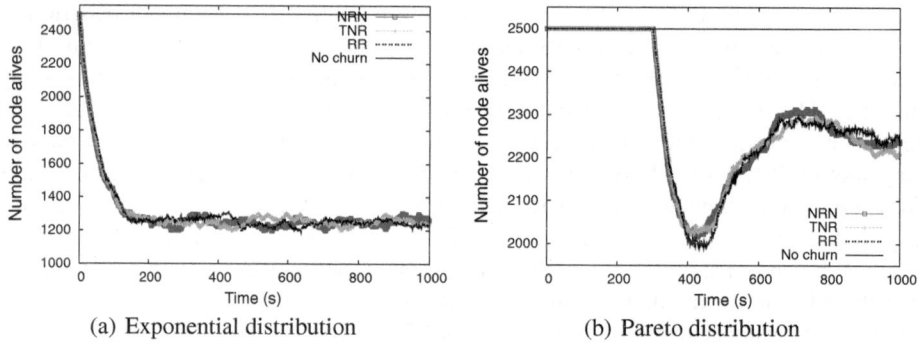

(a) Exponential distribution          (b) Pareto distribution

**Fig. 3.** Meridian dataset: nodes alive as function of time

nodes is roughly $52\%$. In summary, churn is more intensive following the exponential distribution.

We ran Vivaldi on King, Meridian, and PlanetLab respectively and recorded the co-ordinates of the nodes every 10 ticks as well the corresponding time to this tick in order to plot the evolution of coordinates as function of time. Note that a tick represents an update coordinates once. Furthermore, every 10 ticks we compute the percentile AEE and ERR overall links and we considered the 5th, 10th, 25th, 50th, 75th, 90th, and 95th percentile error. Each simulation runs for $1000s$ of simulated time. Unless otherwise noted, all figures are for simulations done in the churn environment.

Figures 4, 5, and 6 illustrate the general behavior of Vivaldi under different churn scenarios as function of time according to King, PlanetLab and Meridian data set. As expected, the curve labelled NRN (No Recovery Nodes) in Figures 4, 6, 5 which shows an instance of Vivaldi under churn without replacement of lost nodes, has always the worst accuracy with respect to AEE metric: the performance of Vivaldi degrades in node churn scenario and Vivaldi never converges to a steady state. The main observation one can notice is that it exists a benefit of doing neighbors recovery according to

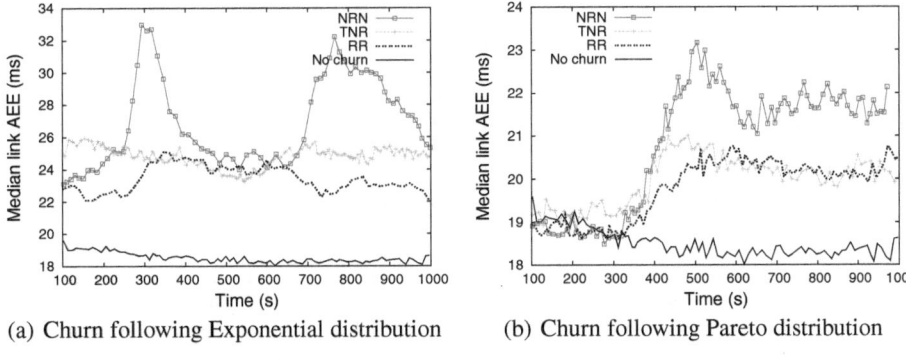

(a) Churn following Exponential distribution     (b) Churn following Pareto distribution

**Fig. 4.** King dataset: Median absolute error as function of time

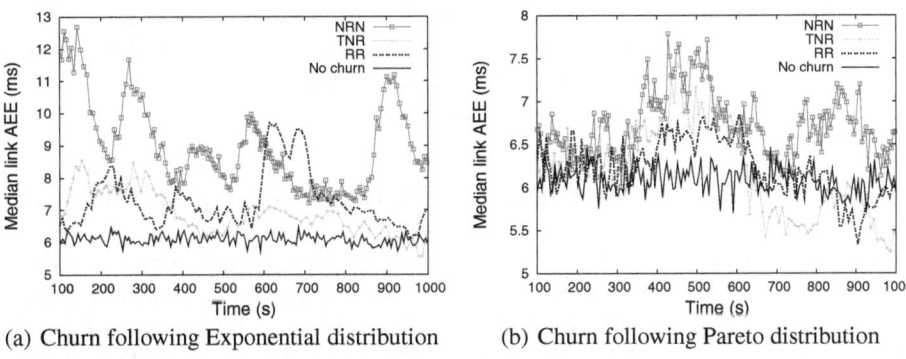

(a) Churn following Exponential distribution     (b) Churn following Pareto distribution

**Fig. 5.** PlanetLab dataset: Median absolute error as function of time

our strategies. For instance, Figure 4(b) shows the AEE as a function of time in the case where the session length follows a Pareto distribution. It illustrates that the AEE increases suddenly (roughly at $300s$), as soon as the churn starts in the network. As consequence, one can see the abrupt jump of the NRN curve as well the TNR and RR curves but in less proportion. The same trend is observed in Fig. 5(b) and Fig. 6(b).

We can see also that the accuracy when the churn follows a Pareto distribution is better than the accuracy obtained with an exponential distribution. This is due to the fact that the churn intensive workload is more important in the exponential distribution (Fig.1(a) and Fig. 3(a)). Nevertheless, with our replacement strategies, TNR and RR, one can see in Figure 5(a), 6(a) that the obtained accuracy if quite similar to a stable Vivaldi and then the RR strategy outperforms the stable Vivaldi.

Additionally the RR strategy performs better than the TNR strategy with respect to the Meridian dataset (Fig. 6). Nevertheless, for other datasets the difference is less important. It is worth noticing that in the RR strategy a node can select out of all nodes available in the system to replace its failed neighbors whereas in the TNR strategy a node chooses only out of node's two-neighborhood which is in fact less important.

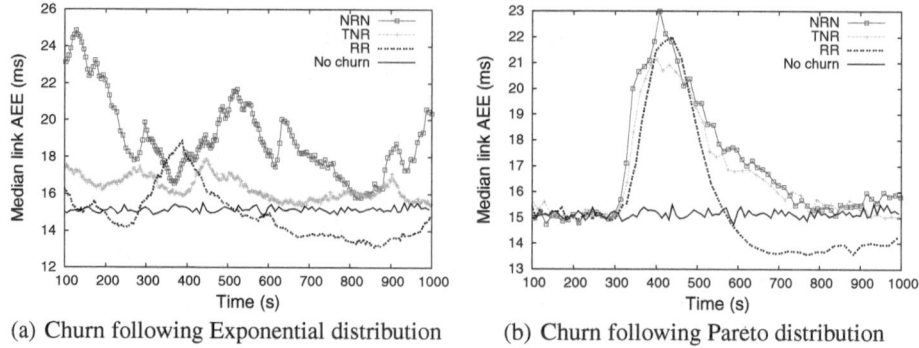

(a) Churn following Exponential distribution    (b) Churn following Pareto distribution

**Fig. 6.** Meridian dataset: Median absolute error as function of time

It should be noted that in terms of communication overhead the cost of retrieving the list of all nodes that are in the system is more important compared to the TNR strategy.

By lack of space we have shown only the median AEE. Note that we found similar trend for the other percentiles and for the REE metric.

## 4    Case Study: Handling Node Churn in a Two-Tier Architecture

Internet latencies, due to routing policies or path inflation [21], do sometimes violate the triangle inequalities which must hold in a metric space. Such Triangle Inequality Violations (TIV) could be a major barrier for the accuracy of Internet coordinate systems. Kaafar et *al.* have shown that longer edges cause more severe TIVs, and thus proposed a Two-Tier architecture opposed to a flat structure of Vivaldi, based on the clustering of nodes [22]. In fact, coordinates computed at the lower (resp. higher) level of clusters are called local coordinates (resp. global coordinates). Within their cluster, nodes use more accurate local coordinates to predict intra-cluster distances, and keep using global coordinates when predicting longer distances towards nodes belonging to foreign clusters. In this section, according to this Two-tier approach and the self-organizing clustering scheme proposed in [7], we test the strategies studied, in Sec. 3, in a peer dynamics system.

To construct clusters in a self-organizing fashion, each node relies on coordinates provided by their two-hop neighbors, but also on measurements towards a potential existing cluster. For instance, if a node has 32 neighbors in order to estimate its coordinates, its Two-hop neighbors will be formed by at most 1024 nodes. Therefore, nodes do not need global knowledge of nodes in the network, nor distances between these nodes, nor a common landmark/anchor infrastructure. In general, the clustering forming phase can be described as follows: each time a node joins a network, it gets the list of cluster heads from its Two-hop neighbors set and verifies if the measurement towards the cluster heads satisfies the cluster diameter. If the constraint diameter is satisfied, the node joins the cluster owned by these cluster head. Nevertheless if none of the distances to existing cluster heads satisfies the clustering criterion, the node starts the clustering algorithm on the basis of the coordinates of its Two-hop neighbors set [7]. For more details about the clustering algorithm we suggest the reader to refer to [7]. Next, we describe how we set up the clusters we experimented with to illustrate our results on the Two-tier Vivaldi approach.

For handling churn in a Two-Tier architecture, we have first based our clusters recognition on the coordinates as observed by running a flat Vivaldi over the P2psim data. Our second step has then consisted of using the algorithm proposed in [7] to self organize nodes into clusters. Following this cluster selection method, we run an extensive simulation either without churn, or under node churn scenario without recovery mechanism, or churn recovery with random replacement (RR), or churn recovery with Two-hop neighbors replacement (TNR). Note that, if a node belongs to a given cluster, it takes halfot its neighbors inside its own cluster and the remaining out of the available nodes in the network. We use the absolute error as our main performance indicator. Again, we compute the AEE over all links to represent the accuracy of the overall system. Nevertheless, if two nodes belong to the same cluster, we used their local coordinates to compute the AEE. Otherwise, we consider their global coordinates in order to estimate the AEE.

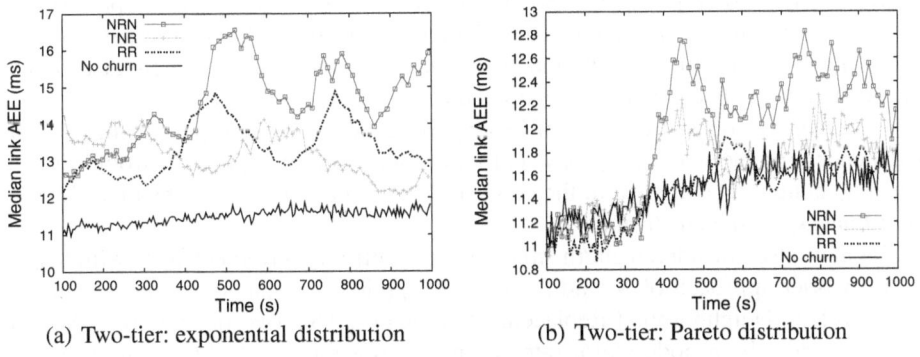

(a) Two-tier: exponential distribution          (b) Two-tier: Pareto distribution

**Fig. 7.** King dataset: Median absolute error as function of time

Figures 7(a) and 7(b) represent the median Absolute Estimation Error (AEE) belonging to our Two-tier Vivaldi according to the P2psim data. We see that the same trend is observed with respect to an instance of a "flat Vivaldi" (Fig. 4). In other words, the churn NRN still has the worst accuracy compared to an instance of Vivaldi where we replace nodes leaving the system. Furthermore, Fig. 8 clearly illustrates that the AEE computed based on Two-tier architecture are much less than errors as computed using the flat Vivaldi. More generally, improvements inside these clusters is explained by the fact that intra cluster nodes, when computing their local coordinates select only close by nodes as their neighbors. This constraints the node-to-neighbors edge lengths and thus reduces the selection of severe TIVs likelihood.

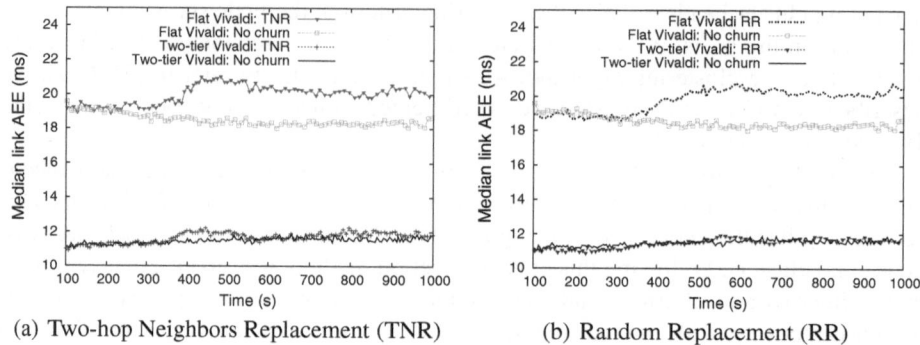

(a) Two-hop Neighbors Replacement (TNR)    (b) Random Replacement (RR)

**Fig. 8.** King dataset: Comparison of median absolute error between Two-tier and Flat Vivaldi

## 5    Conclusion

We have shown that the performance of Vivaldi degrades in churn scenario. We designed and evaluated two strategies for reducing the harmful effect of churn in Vivaldi. These strategies were deployed under different churn distribution characteristics. The experimental results according to three data sets show that the Random Replacement (RR) and the Two-Hop Neighborhood Replacement (TNR) improve the precision of Vivaldi under churn environment. The RR outperforms the TNR with respect to churn following an exponential distribution. Nevertheless, in the case of a Pareto distribution, the gap noticed between both strategies is much smaller.

Although we focused on Vivaldi for experimentations, the proposed strategies are independent of the embedding protocol used. Our proposed approach would then be general enough to be applied in the context of coordinates computed by other Internet coordinate system than Vivaldi.

We have also considered churn situations in a Self-Organized network, with respect to a Two-tier approach of Vivaldi, where we applied our Random Replacement and the Two-Hop Neighborhood Replacement strategies. The Two-tier approach is more accurate under high node churn rate compared to a flat Vivaldi. Our findings show that the performance of the Two-tier Vivaldi exceed that of flat Vivaldi a lot in churn scenario as well as in a scenario without churn. Our future work would then consist on the deployment of the Two-tier approach on Internet (*e.g.*, PlanetLab).

Even though this paper does not address the problem of comparing the Pareto and the exponential distribution, we note that the obtained accuracy, when the session length is modelled as a Pareto distribution, is lower than when it follows an exponential distribution. We are pursuing further study for more general conclusion, considering different values for the parameters of both distributions.

## References

1. Ng, T.S.E., Zhang, H.: Predicting Internet network distance with coordinates-based approaches. In: Proc. IEEE INFOCOM, New York, NY, USA (June 2002)
2. Dabek, F., Cox, R., Kaashoek, F., Morris, R.: Vivaldi: A decentralized network coordinate system. In: Proc. ACM SIGCOMM, Portland, OR, USA (August 2004)

3. Donnet, B., Gueye, B., Kaafar, M.A.: A survey on network coordinates systems, design, and security. IEEE Communications Surveys & Tutorials (to appear)

4. Godfrey, P.B., Shenker, S., Stoica, I.: Minimizing churn in distributed systems. SIGCOMM Comput. Commun. Rev. 36(4), 147–158 (2006)

5. Ledlie, J., Gardner, P., Seltzer, M.I.: Network coordinates in the wild. In: Proc. NSDI, Cambridge (April 2007)

6. Chen, Y., Zhao, G., Li, A., Deng, B., Li, X.: Myth: An accurate and scalable network coordinate system under high node churn rate. In: IEEE International Conference on Networks ICON', Adelaide, Australia (November 2007)

7. Cantin, F., Gueye, B., Kaafar, M.A., Leduc, G.: A self-organized clustering scheme for overlay networks. In: Hummel, K.A., Sterbenz, J.P.G. (eds.) IWSOS 2008. LNCS, vol. 5343, pp. 59–70. Springer, Heidelberg (2008)

8. Azureus bittorent client, http://azureus.sourceforge.net/index.php

9. Liben-Nowell, D., Balakrishnan, H., Karger, D.: Analysis of the evolution of peer-to-peer systems. In: Principles of Distributed Computing, pp. 233–242 (2002)

10. Rhea, S., Geels, D., Roscoe, T., Kubiatowicz, J.: Handling churn in a dht. In: USENIX Conference (2004)

11. Li, J., Stribling, J., Morris, R., Kaashoek, M.F., Gil, T.M.: A performance vs. cost framework for evaluating dht design tradeoffs under churn. In: INFOCOM, pp. 225–236 (2005)

12. Bustamante, F., Qiao, Y.: Friendships that last: Peer lifespan and its role in p2p protocols. Web Content Caching and Distribution, 233–246 (2004)

13. Liang, J., Kumar, R., Ross, K.W.: The kazaa overlay: A measurement study. Computer Networks Journal, Elsevier (2005)

14. Stutzbach, D., Rejaie, R.: Understanding churn in peer-to-peer networks. In: IMC, Rio de Janeriro, Brazil, pp. 189–202 (2006)

15. A simulator for peer-to-peer protocols, http://www.pdos.lcs.mit.edu/p2psim/index.html

16. Wong, B., Slivkins, A., Sirer, E.: Meridian: A lightweight network location service without virtual coordinates. In: Proc. the ACM SIGCOMM (August 2005)

17. PlanetLab: An open platform for developing, deploying, and accessing planetary-scale services (2002), http://www.planet-lab.org

18. Gummadi, K.P., Saroiu, S., Gribble, S.D.: King: Estimating latency between arbitrary Internet end hosts. In: Proc. the SIGCOMM Internet Measurement Workshop, Marseille, France (November 2002)

19. Wang, F., Xiong, Y.Q., Liu, J.C.: mtreebone: A hybrid tree/mesh overlay for application-layer live video multicast. In: ICDCS (June 2007)

20. Saroiu, S., Gummadi, P.K., Gribble, S.D.: A measurement study of peer-to-peer file sharing systems, vol. 9, pp. 170–184 (August 2003)

21. Zheng, H., Lua, E.K., Pias, M., Griffin, T.G.: Internet routing policies and round-trip-times. In: Dovrolis, C. (ed.) PAM 2005. LNCS, vol. 3431, pp. 236–250. Springer, Heidelberg (2005)

22. Kaafar, M.A., Gueye, B., Cantin, F., Leduc, G., Mathy, L.: Towards a two-tier internet coordinate system to mitigate the impact of triangle inequality violations. In: Das, A., Pung, H.K., Lee, F.B.S., Wong, L.W.C. (eds.) NETWORKING 2008. LNCS, vol. 4982, pp. 397–408. Springer, Heidelberg (2008)

# Tuning Vivaldi:
# Achieving Increased Accuracy and Stability

Benedikt Elser[1], Andreas Förschler[2], and Thomas Fuhrmann[1]

[1] Computer Science Department, Technical University of Munich, Germany
{elser,fuhrmann}@so.in.tum.de
[2] Computer Science Department, University of Karlsruhe, Germany
foerschler@so.in.tum.de

**Abstract.** Network Coordinates are a basic building block for most peer-to-peer applications nowadays. They optimize the peer selection process by allowing the nodes to preferably attach to peers to whom they then experience a low round trip time. Albeit there has been substantial research effort in this topic over the last years, the optimization of the various network coordinate algorithms has not been pursued systematically yet. Analyzing the well-known Vivaldi algorithm and its proposed optimizations with several sets of extensive Internet traffic traces, we found that in face of current Internet data most of the parameters that have been recommended in the original papers are a magnitude too high. Based on this insight, we recommend modified parameters that improve the algorithms' performance significantly.

## 1 Introduction

Self-organizing peer-to-peer systems cannot rely on a well-designed network topology. They rather form their overlay network topology on their own. For many such applications it is crucial that the resulting topology reflects the underlying topology of the Internet. Matching both topologies can, for example, reduce the latency which important not only for telephony applications. It can also increase the throughput, for example, when a file sharing application is able to find peers in the same organization.

Ideally, a peer would know which potential new neighbors provide a low latency. It is important that, to this end, we cannot first measure the latency and decide afterwards, because sampling the network in such a way has an extremely high overhead. It is thus necessary to predict the latencies.

A popular solution to this problem is the use of network coordinates. They assign positions in an euclidean space to the peers, so that their metric distance reflects the respective latencies. There are several such systems, but Vivaldi [1] has obtained the most attention. Its simple spring model is easy to understand and led to a practical implementation in the BitTorrent [2] client *Azureus* (now called Vuze).

The inclusion of the Vivaldi algorithm in Azureus sparked even greater interest in network coordinates. Ledlie et al. [3], e.g., proposed an optimization that changed the force computation in Vivaldi's spring model into a round based approach. Both this optimization of Vivaldi and the original algorithm contain several parameters. Their complex interaction offers various possibilities to fine tune the algorithm.

T. Spyropoulos and K.A. Hummel (Eds.): IWSOS 2009, LNCS 5918, pp. 174–184, 2009.
© IFIP International Federation for Information Processing 2009

To the best of our knowledge, this fine tuning has not been studied systematically yet. In this paper we report on an in-depth analysis of Vivaldi and its optimizations. For our analysis we collected a number of latency matrices that complement other such data sets. We created a packet level simulator that uses these latency matrices to create a overlay between peers running the Vivaldi algorithm. Besides static latency matrices we also used dynamic trace-based input data for the simulator. Based on our analysis we then propose parameter choices that can improve the algorithms' performance significantly.

The rest of this paper is structured as follows: Section 2 gives an overview of Vivaldi and its proposed optimizations. Section 3 describes the data sets and simulation method that we used for our analysis. Section 4 discusses the most interesting results from our analysis. Section 5 gives a brief overview of related work. Section 6 concludes with an outlook to future work.

## 2   Overview of Vivaldi

The Vivaldi algorithm models a spring network: The peers are the endpoints of springs whose length is set to the actually measured round trip time (RTT) between the peers. The underlying metric combines the Euclidean coordinate distance with an additional 'height' displacement. Based on continued RTT measurements, the peers update their coordinates according to their displacement error estimations. Two constants $c_c$ and $c_e$ control these updates. Algorithm 1 gives a summary of this Vivaldi NE algorithm in pseudo code.

Studies of the Azureus client revealed problems with the algorithm in real live situations [3]. First of all, a node does not contact its peers equally often. The resulting imbalance has a negative impact on the global optimization process. Secondly, real world influences inevitably create spikes in the RTT measurements, which distort the coordinates of an otherwise stable system massively.

Ledlie et al. [3] proposed two improvements to the Vivaldi NE algorithm that compensate for these problems: A low pass RTT filer based on median values includes only plausible RTT values. A *neighbor decay formula* addresses the imbalanced measurement frequencies. To this end, each node keeps a list of its most recently used peers (*recent neighbor set*). It contains those peers with whom a node ran the Vivaldi algorithm at most a time $e_t$ ago. The force vector $F$, which changes the peers' coordinates, is then modified as follows

$$\tilde{F} = \sum_{j=1}^{N} F_j \cdot \frac{a_{max} - a_j}{\sum_{i=0}^{n-1} a_{max} - a_i} \tag{1}$$

where $a_{max} \leq e_t$ is the maximum age of an entry in the recent neighbor set, $a_k$ is the age of the $k$th entry, and $F_k$ is the force pushing in the direction of that entry. Algorithm 2 summarizes this Vivaldi ND variant in pseudo code.

The most important contribution of this revised algorithm is that it does not only take the most recent measurement into account, but that many measurements jointly contribute to the coordinate adjustment. This greatly reduces spikes and other fluctuations.

**Algorithm 1.** Original Vivaldi Neighbor Error Algorithm [1]

**Input:**

$x_i$: local coordinate of node $i$

$x_j$: remote coordinate of node $j$

$RTT$: RTT to node $j$

$e_i$: error estimation of node $i$

$e_j$: error estimation of node $j$

$c_c, c_e$: constants

**Output:**

$x_i$: updated coordinate of node $i$

$e_i$: updated error estimation of node $i$

1 **function vivaldiNE**$(RTT, x_i, x_j, e_i, e_j, c_e, c_c)$
2 **begin**
3 $\quad\left| \quad w \leftarrow \frac{e_i}{e_i + e_j} \right.$
4 $\quad\left| \quad e_s \leftarrow \frac{|\|x_i - x_j\| - RTT|}{RTT} \right.$
5 $\quad\left| \quad e_i \leftarrow e_s \times c_e \times w + e_i \times (1 - c_e \times w) \right.$
6 $\quad\left| \quad \delta \leftarrow c_c \times w \right.$
7 $\quad\left| \quad x \leftarrow x + \delta \times (rtt - \|x_i - x_j\|) \times u(x_i - x_j) \right.$
8 $\quad\left| \quad \textbf{return } (x_i, e_i) \right.$
9 **end**

In order to judge this and other improvements quantitatively, we use the following three indicators:

1. A node $i$'s error is the median over all absolute errors between all node pairs $(i, j)$. The *median error* of the system is the median of all node errors [4].

2. The *relative application-level penalty* (RALP) [5]

$$RALP \equiv \frac{1}{n} \cdot \sum \frac{v_i - p_i}{p_i}, \tag{2}$$

describes the penalty that a node experiences when choosing a peer based on network coordinates, as compared to the perfect choice that an omniscient oracle could recommend based on 'real' RTTs. Here, $p$ a sorted list of *measured* RTTs between a peer and its neighbors, and $v$ is a sorted list of *predicted* RTTs between the peer and its neighbors.

3. We define the *degree of stability* that the embedding reaches according to [3] as

$$stability \equiv \frac{\sum_i \Delta x_i}{\Delta t} \tag{3}$$

where $\Delta x_i$ as the drift of $x_i$ in the time period $\Delta t$.

## 3   Data Sets and Simulator

In our analysis we used four data sets:

– **Azureus-to-PlanetLab** is the trace from Ledile et al. [3]. It yields a 249x249 RTT matrix.

**Algorithm 2.** The Neighbor Decay Optimization of the Vivaldi Algorithm [3]

> **Input**:
>> $x$: locale coordinate of node
>> $Y = \{y_1, y_2, \ldots, y_k\}$: coordinate of nodes from the $k$ sized neighbor set
>> $R = \{rtt_1, rtt_2, \ldots, rtt_k\}$: RTT to nodes in the neighbor set
>> $A = \{a_1, a_2, \ldots, a_k\}$: time of last contact with all nodes in the neighbor set
>> $t$: constant damping adjustment of local coordinate, similar to VivaldiNE's $c_c$
>
> **Output**:
>> $x$: updated coordinate of local node

```
1  function VivaldiND(x, Y, R, A, t)
2  begin
3  |     s ← 0
4  |     F ← 0
5  |     a_max ← max {a_1, a_2, …, a_k}
6  |     for i = 1 to k do
7  |     └   s ← s + a_max − a_k
8  |     for i = 1 to k do
9  |     |   e ← rtt_i − ‖x − y_k‖
10 |     |   F_i ← e × u(x, y_k)
11 |     └   F ← F + F_i × (a_max − a_i)/s
12 |     return x ← x + t × F
13 end
```

- **MITKing** is the data set of Dabek et al. used to derive the original Vivaldi algorithm [1]. It is based on measurements with the King technique among 1740 DNS servers.
- **KingBlog** is a dataset similar to the MIT King data. It was extracted from 2500 DNS servers [6].
- **Dynamic PlanetLab Dataset** is a dynamic trace of 13,4 million single measurements between 83 fully interconnected PlanetLab nodes, which we collected between March 6 and 9, 2009.

In order to evaluate the different Vivaldi variants, we built a simulator that takes either the static RTT matrices or the dynamic RTT traces as input. The method of deriving a Vivaldi simulation from a static matrix of pairwise RTTs was introduced by Cox et al. [1]. We extended their idea with a dynamic trace-based simulation.

On startup, the simulator randomly chooses a neighborhood of 32 peers for each node. In the static case, the simulator also determines a sequence of RTT measurements. Here, making a measurement means to look up the RTT from the RTT matrix. In the dynamic case, the sequence of measurements in the trace is predefined. A node obtains the RTT values for all the peers in its respective neighborhood and calculates an estimated RTT according to the describe median filter.

There are two variants of how to determine the sequence of RTT measurements in the static case: In [1] a node starts a new measurement immediately after the previous measurement has completed (*continuous adjustment*, CA). This leads to an imbalance, because nodes with low RTT conduct more rounds of the Vivaldi algorithm. Another drawback is the huge traffic overhead that those continuous measurements cause. In order to avoid this imbalance and reduce the overhead, we propose to use a predefined average time interval for the measurement rounds (*uniform adjustment*, UA).

## 4    Results

In our extensive study [7] we analyzed all the proposed algorithm variants with the four datasets and with various parameter settings. Here we briefly discuss the most interesting results. All use the described Vivaldi variants with four dimensions and a height component. In case of uniform adjustment (UA), we chose an interval of 10 seconds.

Fig. 1 shows the error, figure 2 the stability of the original Vivaldi Neighbor Error algorithm with continuous adjustment (CA) and different values for $c_c$ and $c_e$ using the Azureus-to-PlanetLab dataset. In accordance to Cox et al. [1] we find that smaller values for $c_c$ and $c_e$ lead to more stable coordinates. But as our analysis shows, this

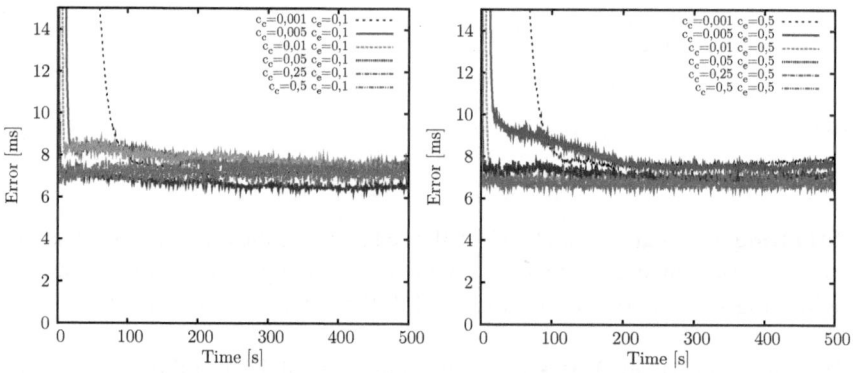

**Fig. 1.** Embedding error (Vivaldi NE CA, Azureus dataset)

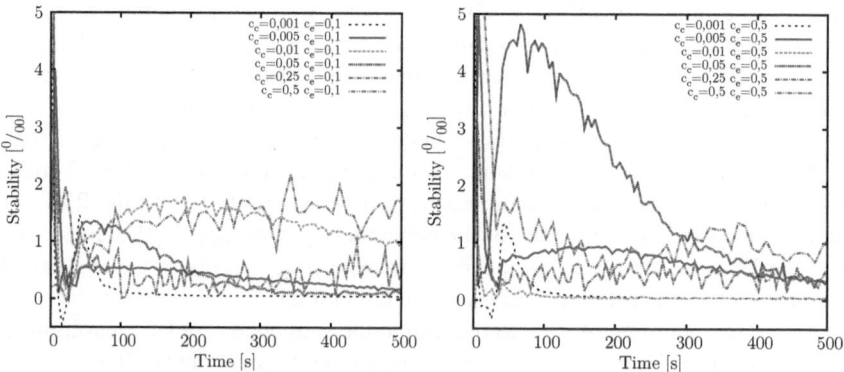

**Fig. 2.** Stability (Vivaldi NE CA, Azureus dataset)

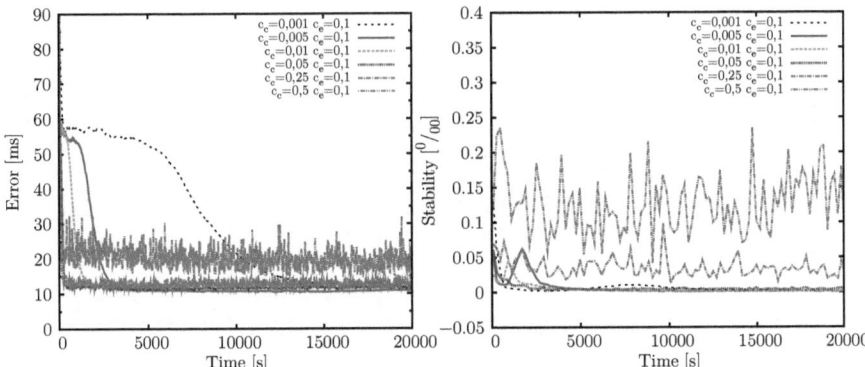

**Fig. 3.** Embedding error and stability (Vivaldi NE UA, Azureus dataset)

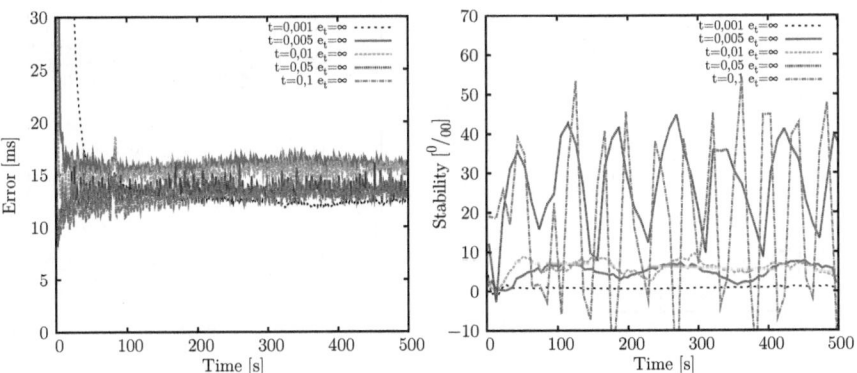

**Fig. 4.** Vivaldi ND with Azureus dataset (CA)

improved stability is only relevant in the initial phase: After 500 seconds, the stability of the embedding is similar for all parameter combinations. Moreover, we find through the course of our simulations that $c_e$ has only little influence on the embedding error and the stability. A conclusion that is also supported from both figures.

Fig. 3 shows the same scenario for the uniform adjustment (UA) variant. Due to the much lower measurement frequency the shown time scale covers a larger range. In all settings we see that both stability and median error never reach the accuracy of the CA variant. Furthermore, we see that small $c_c$ values delay the convergence enormously, whereas large values lead to great instabilities. For large $c_c$ values the low stability also leads to a high, fluctuating embedding error. As a result we find an optimal parameter choice at $c_c = 0,005$. (We do not show different $c_e$ values, because they have only little effect.)

Fig. 4 and 5 show the Neighbor Decay optimization of Vivaldi for different RTT probing intensities. In the CA variant in figure 4 we show the results for $e_t = \infty$, This parameter choice causes the peers to adjust their position relative to all the peers that

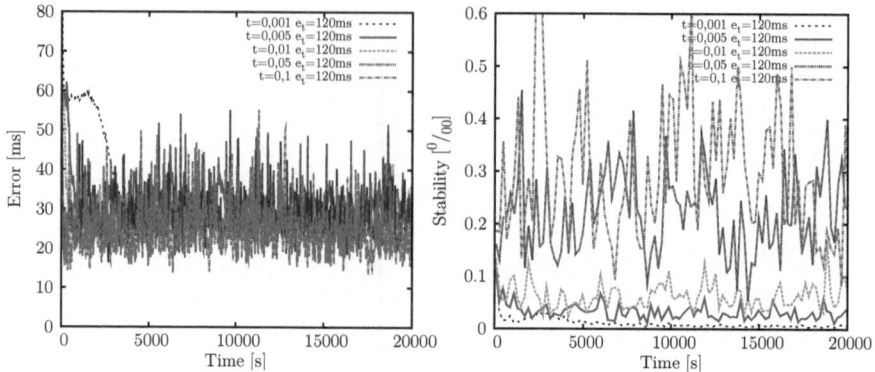

**Fig. 5.** Vivaldi ND with Azureus dataset (UA)

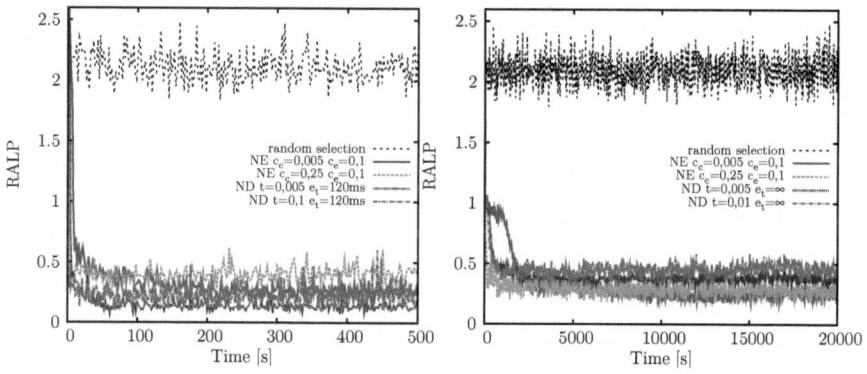

**Fig. 6.** RALP with Azureus data set, using CA (left) and UA (right)

they ever contacted. In the UA variant in figure 5 we use $e_t = 120\,\text{ms}$, where the 120 ms correspond to the median value of the RTTs. This choice causes the peers to adjust their position only relative to active peers, i.e. those that have just send their RTT information. Our results demonstrate that the UA variant produces the results from Ledlie et al. within a factor of two, while requiring only sparse RTT probing (10 sec versus 120 ms). However the results also show a rather poor performance of the algorithm in terms of the median embedding error. Furthermore we learn again that there is an optimal choice for the algorithm's parameter, namely $t = 0,005$.

In order to complement our analysis with an application-level metric, we further analyzed the different variants using the RALP quality measure. Fig. 6 shows the results for continuous adjustment (left) and uniform adjustment (right) using the Azureus dataset, while figure 7 illustrates the same for the KING Blog dataset. Clearly, all variants improve over the random peer selection case, which does not use network coordinates at all.

Similarly to almost all of our measurements, the difference between the two adjustment variants depends on the dataset. Simulations with the KING dataset behave

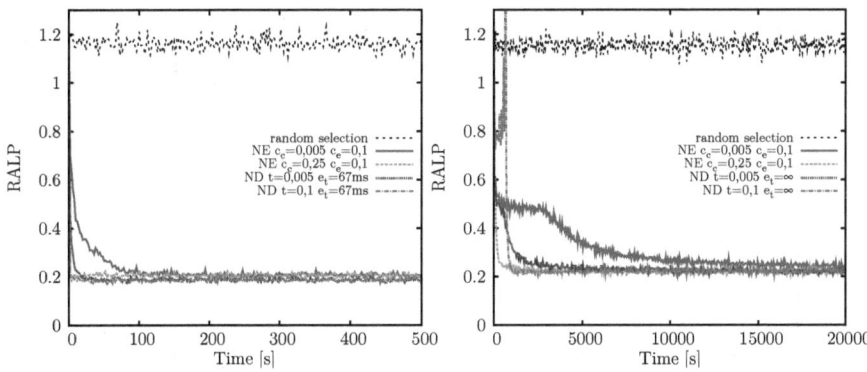

**Fig. 7.** RALP with King data set, using CA (left) and UA (right)

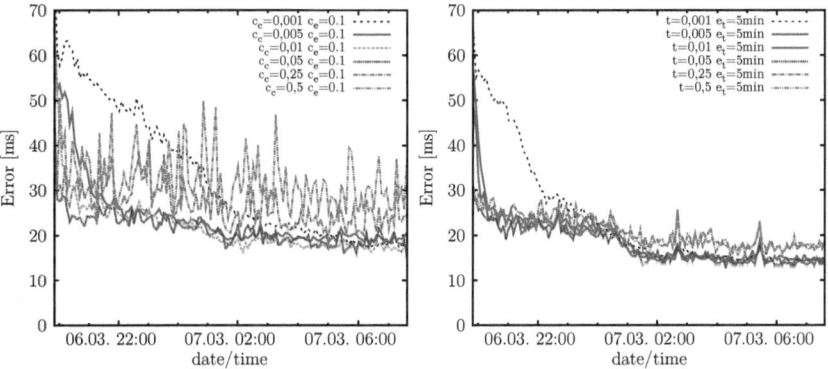

**Fig. 8.** Vivaldi NE (left) vs. ND (right) with median filter

comparably for all variants (NE, ND, w/ and w/o CA, UA) of the Vivaldi algorithm, whereas the Azureus dataset exhibits a significant dependence on the parameter choices. In particular, when using the uniform adjustment variant, the original neighbor error algorithm outperforms the neighbor decay optimization. This indicates that the claimed improvement of ND [3] might not carry over to a general application 'in the wild', despite the suggestive title of the respective publication.

In order to better understand highly dynamic systems and their differences to static systems, we analyzed the PlanetLab RTT trace. Fig. 8 shows the embedding error for the original NE algorithm and the ND optimization with the median filter proposed by Ledile et al. [3], figure 9 shows the same data without this enhancement. First of all, we do not observe any significant effect of the median filter. Furthermore, the advantage of the neighbor decay optimization manifests most, when it is compared to a high $c_c$ parameter in the original neighbor error algorithm. When $c_c = 0,01$ NE and ND behave almost identically.

Fig. 10 illustrates the same setup stability wise. All measurements show large instabilities for large values of $c_c$ and $t$ respectively. Again we conclude that the choice of

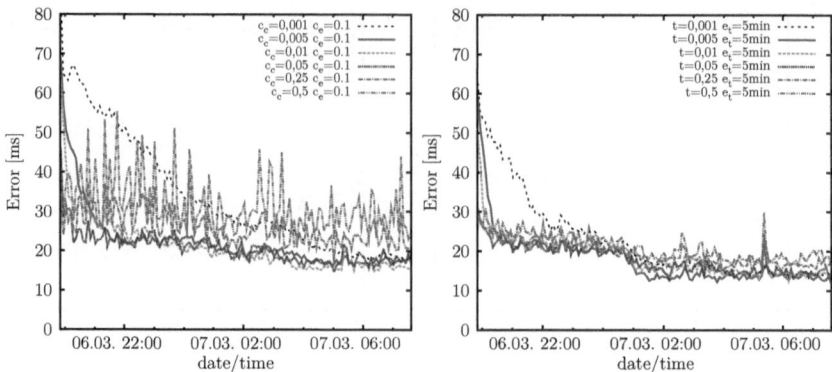

**Fig. 9.** Vivaldi NE (left) vs. ND (right) without median filter

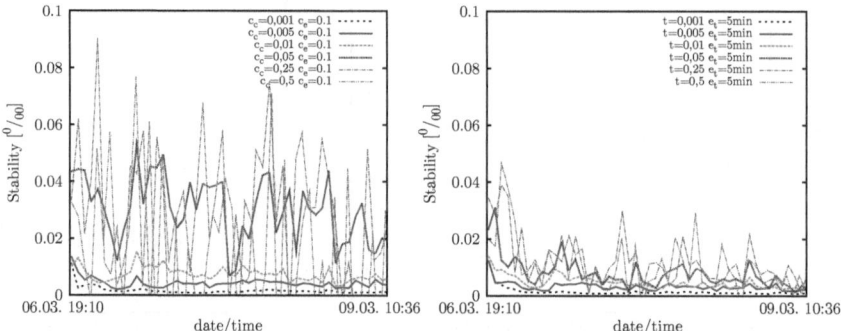

**Fig. 10.** Vivaldi NE (left) and ND (right) with median filter

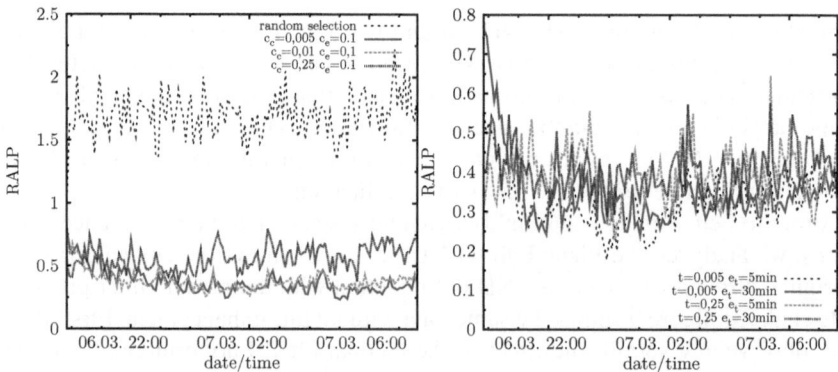

**Fig. 11.** RALP with NE (left) and ND (right) for the dynamic data trace

large values for $c_c$ and $t$ does not only lead to faster convergence (as stated in literature), but also to an increased instability that in the end leads to a worse quality of embedding. As a result, we conclude that even though the neighbor decay optimization can yield a better stability in principle, it does not improve the embedding for the moderate values of $t$ and $c_c$ that overall yield the best results.

Finally we also checked our finding in the dynamic setting using RALP (cf. fig. 11). Here we find that the neighbor decay optimization shows a better performance than the original Vivaldi NE algorithm. Again, $c_c = 0.005$ and $c_e = 0.1$ outperform all other parameter combinations in the NE case. For the ND case $t = 0.005$ is the best choice.

## 5   Related Work

Finding neighbors with low RTT in an overlay network is an important issue. Several other works besides Vivaldi address the same problem. Global network positioning (GNP) [8] proposed to use RTT measurements to landmark servers as components for network coordinates. Vivaldi adopted the idea of network coordinates, but replaced the fixed set of landmarks with its spring approximation model. This makes Vivaldi a self-organizing system in the sense that is independent from a landmark infrastructure.

Meridian [9] is a gossip based system. It sorts a node's peers into exponentially growing buckets according to the respectively measured RTTs. When a node queries one of its peers for recommended further peers, it obtains answers from the bucket that matches the querier's RTT. This algorithm induces an iterative process that allows the peers to find proximate peers. Comparing Medidian to Vivaldi, we see that the need for direct measurements between the peers leads to increased costs that are not present in network coordinate systems [10].

Ono [11] derives peer proximity from querying a content distribution network (CDN) to resolve a vector of DNS names. It assumes that peers that receive similar results from the CDN reside in the same AS and wile likely have a low mutual RTT. This is similar to GNP, but replaces the RTT measurements to the landmark servers with the CDN queries. Aggarwal et al. introduce in [12] the idea of an oracle that pursues a similar goal as Ono. But instead of living on a CDN the oracle is an ISP-operated recommendation server that allows peers to obtain lists of proximate peers. Contrary to Ono and the oracle we aim at improving fully decentralized solutions that do not require the ISP to support the overlay network.

## 6   Conclusion

In this paper we studied Vivaldi and its proposed optimizations both with static RTT matrices and dynamic RTT traces. We found that in the original works, most constants are chosen too high to produce stable coordinates in dynamic real world settings. Seemingly the respective authors decided to sacrifice the stability of the resulting embedding to obtain a faster convergence. We believe this to be a bad trade, especially because most parameters produce similar relative embedding errors.

As a result of our analysis we recommend the following values: $c_e = 0.1$, $c_c = 0.005$, $t = 0.005$, and $e_t = 30\,\text{min}$. In our analysis we also spotted rare cases of especially poor performance of the Vivaldi algorithm family. We therefore recommend using conservatively low parameters to also safeguard against such cases.

Especially, we could not reproduce the claimed drastic superiority of the Neighbor Decay optimization method proposed by Ledlie et al. [3]. The only effect we could reproduce was a faster convergence, which we consider not as important as stability.

What struck us at most in our study were the large differences between the datasets. Even when comparing static data among each other, we found significantly different reactions of the different algorithm variants. The differences between the results from the static RTT matrices and the dynamic RTT traces were even greater. Ledlie et al. [3] argue that this depends on the magnitude of RTTs that the algorithm receives. We could not confirm this hypothesis and therefore recommend further research in that area, to better understand the cause of the large deviations.

# References

1. Cox, R., Dabek, F., Kaashoek, F., Li, J., Morris, R.: Practical, distributed network coordinates. In: Proceedings of the Second Workshop on Hot Topics in Networks (HotNets-II), Cambridge, Massachusetts (2003)
2. Cohen, B.: Incentives build robustness in bittorrent. In: Proceedings of the first Workshop on Economics of peer-to-peer systems, Berkley, California (2003)
3. Ledlie, J., Gardner, P., Seltzer, M.: Network coordinates in the wild. In: Proceedings of USENIX NSDI 2007, Cambridge, Massachusetts (2007)
4. Dabek, F., Cox, R., Kaashoek, F., Morris, R.: Vivaldi: A decentralized network coordinate system. In: Proceedings of the ACM SIGCOMM 2004 Conference, Portland, Oregon (2004)
5. Choffnes, D.R., Bustamante, F.B.: What's Wrong with Network Positioning and Where Do We Go From Here? Technical Report NW-EECS-09-03 (2009), http://www.eecs.northwestern.edu/research/tech_reports/number
6. Syrah: King-Blog-Dataset (2006), http://www.eecs.harvard.edu/~syrah/nc (accessed December 30, 2008); Variant with at least 8 neighbors was picked
7. Förschler, A.: Einbettung von Peer-to-Peer-Netzwerkgraphen zur Vorhersage von Paketumlaufzeiten, Diplomarbeit, Universität Karlsruhe (2009)
8. Ng, T.S.E., Zhang, H.: Predicting Internet Network Distance with Coordinates-Based Approaches. In: Proceedings of the Twenty-First Annual Joint Conference of the IEEE Computer and Communications Societies, INFOCOM 2002 (2002)
9. Wong, B., Slivkins, A., Sirer, E.G.: Meridian: a lightweight network location service without virtual coordinates. In: Proceedings of the ACM SIGCOMM 2005 Conference, Philadelphia, Pennsylvania (2005)
10. Pietzuch, P., Ledlie, J., Seltzer, M.: Supporting network coordinates on PlanetLab. In: WORLDS 2005: Proceedings of the 2nd conference on Real, Large Distributed Systems, Berkeley, California (2005)
11. Choffnes, D.R., Bustamante, F.E.: Taming the torrent: a practical approach to reducing cross-isp traffic in peer-to-peer systems. Comput. Commun. Rev. 38(4), 363–374 (2008)
12. Aggarwal, V., Feldmann, A., Scheideler, C.: Can ISPS and P2P users cooperate for improved performance?. Comput. Commun. Rev. 37(3), 29–40 (2007)

# A Stable Random-Contact Algorithm for Peer-to-Peer File Sharing

Hannu Reittu

VTT Technical Research Center of Finland
Hannu.Reittu@vtt.fi

**Abstract.** We consider a BitTorrent type file sharing algorithm with randomized chunk copying process. The system functions in completely distributed way without any 'Tracker', just relying on randomness. In such case the stability becomes an issue. It may happen, say, that some chunk becomes rare. This problem can persist and cause accumulation of peers in the system, resulting in unstable system. The considered algorithms result in processes similar to urn-processes. The rare chunk phenomenon corresponds to Polya-urn type process, where common chunks are favored. However, some urn-processes like the Friedman-urn can provide good balance by favoring rare chunks in copying process. Recently, we showed that an algorithm based on Friedman-urn is efficient in two chunk case. We generalize this algorithm for the more realistic case of many chunks. It shows good performance in terms of balance of chunks in an open system with constant flow of incoming peers. Further, the system is able to cope with instances like 'flash crowd', with large burst of incoming peers. The open system can also quickly reach equilibrium after an initial imbalance, when the system starts from a state with one rare chunk. We constructed a simplified model, assuming a good balance of chunks, and get results surprisingly close to simulations for Friedman-urn based random process.

**Keywords:** file-sharing, urn-models, randomized algorithms.

## 1 Introduction

File sharing has been one of the first and the most popular application of peer-to-peer systems. The early applications like 'Napster' and 'Gnutella' were replaced by more advanced algorithms like BitTorrent, [1]. Such systems have shown capacity for large scale file distribution, [2]. These protocols are thus highly untrivial in performance and scalability, and are good motivation for interesting models to reveal what is their 'secret' of success. Further, such abstract models, if good enough, should also indicate ways to improve the protocol design. Here we report first results in this direction.

We consider a BitTorrent type system, which, however, does not contain any centralized elements like the 'Tracker' in real BitTorrent that controls who contacts who. Rather, our algorithm relies on randomness and is distributed. However, the

T. Spyropoulos and K.A. Hummel (Eds.): IWSOS 2009, LNCS 5918, pp. 185–192, 2009.

simplest randomized algorithm leads to system that is similar to Polya-urn like system, with natural instability in open system setting with constant flow of new peers. In BitTorren the main innovation is that the file is divided into large number of smaller pieces or 'chunks'. Such chunks are copied from peer to peer. Since the chunks are small they are copied swiftly, which improves the performance, see also [3] for performance limits. In a randomized setting, it can easily happen that some chunks become rare, thus forming a bottleneck of performance. This happens because the most common chunks are easier to find, and, if no measures are taken, are favored in copying process. This is exactly what happens if we assume the simplest, Polya-type algorithms: peers make uniformly random contacts and copy what they find and after collecting all chunks departure.

Recently, we examined such a problem in a systems with just two chunks and the above problem of instability, [4, 5] was pinpointed. However, the two chunk case is unrealistic, because one should have many chunks to speed up copying. This is the issue of the current short paper. First we see that the same type of instability arises in the many chunk case as well.

The problem of stability in BitTorrent type systems has been studied also in [6] mostly with quite similar assumptions. However, the setting is different since the authors assume that the inflowing peers receive one uniformly random chunk upon arriving. The peers could obtain this chunk from the seed node that has all chunks. However, such a seed node becomes a server-like centralized element and possibly a bottleneck of performance. Then the system shows provable stability in fluid limit. We avoid the assumption of first random chunk, the peers arrive with no chunks and obtain every chunk from the network in a distributed manner. The untrivial result seems to be that the instability problem arises and can be, probably, avoided by a specific yet simple design.

We noticed, [5], that by modifying the random contact procedure to one that imitates the so called Friedman-urn (see e.g. [7]), the two chunk sharing process becomes remarkably stable and efficient. In this scheme, a peer that arrives does not have neither of the chunks, called chunk 0 and 1. The peers that have both chunks immediately leave the system. As a result there are three types of peers in the system: peers without any chunks, and peers with chunk 0 or 1. There is also one peer called the seed, a permanent node with both chunks,which , acts as it has a random chunk, chosen independently for each time it is contacted. Peers arrive with constant rate $\lambda$.

The Friedman-urn process in two chunk case would mean that peers make uniformly random contacts to acquire missing chunks, if the target has chunk 0, it downloads the chunk 1. However, in our setting this is impossible since this chunk was not found. Our solution was ([5]) that the peer makes three simultaneous uniformly random contacts and downloads the chunk 0 if it sees the configuration $\{0, 1, 1\}$ and chunk 1 in case $\{0, 0, 1\}$, in other cases it does not download anything and makes another three contact trial until it does find such a configuration. The idea is that under the condition that a node succeeds to obtain a chunk, the probability that it downloads a particular chunk equals to that for the Friedman-urn process, in which the rarest chunk is favored. Such a system is able to cope with substantial imbalance, say, when the system starts in a state with one chunk very rare with respect to the other. Then, with constant rate of arrivals, the system quickly relaxes toward the equilibrium. Such a system in equilibrium shows also a good performance, the peers go through the system quickly.

## 2   Open System with Random Contacts

The two chunk case is not realistic, because the very idea of BitTorrent is to use many small chunks. So we have natural question: how to obtain stable system in such a case? If we have $m$ chunks then we have $2^m - 1$, possible states of a peer. This is of course very large number, say, for $m = 100$, and the system is very complicated. In this situation, we generalized the algorithm as simply as possible. We use a slightly modified three random contact procedure, described in the Introduction.

The main points of our algorithm are: (i) All peers run the same procedure (ii)-(iii) independently of each other. (ii) A peer makes 3 simultaneous and uniformly random contacts with peers in the system. The peer that makes the contacts, learns which chunks those 3 contacts posses. Those chunks that only one contact has, are called the 'minority chunks'. The peer makes a list of those minority chunks found, that it self does not have. If this list is not empty, the peer downloads one of such minority chunk from the list, chosen randomly if there is more than one options. If the list is empty, the peer proceeds (iii). (iii) Repeat (ii) until all chunks are collected, then leave the system.

Quite surprisingly, such approach seems to produce a stable and efficient system. Although not proven rigorously, simulations and simplified models seems to support this conjecture.

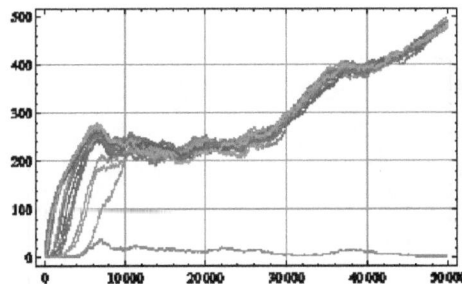

**Fig. 1.** A result of a computer simulation with a simple random contact system, without favoring rare chunks. Each line represents the size of a particular chunk population, number of nodes in the system with given chunk, for system starting from a system with one 'seed' as a function of time (a sample path). Number of chunks $m = 20$ and $\lambda = 10$ times the contact rate. The populations of nodes that have certain chunks blows up, expect for one chunk that remains 'rare'.

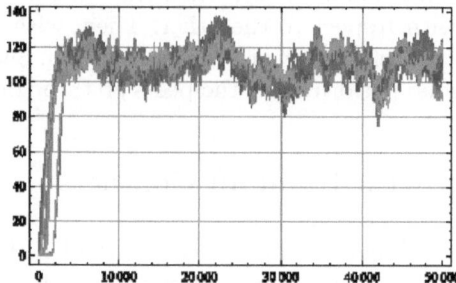

**Fig. 2.** A computer experiment with random contact system with Forced-Friedman random contacts, $m = 20$, $\lambda = 100$. The sizes of chunk populations are shown starting from system with only the seed node. A good balance seem to prevail and a good performance, since no accumulation of peers is not seen although there is a constant flow of incoming peers into the system.

First we consider Polya-urn like, 'greedy' algorithm, in which a peer makes uniformly random contacts and downloads a missing chunk if it sees one. Then it repeats until it collects all chunks and departures. This seems to result in an inefficient and unstable system. One chunk becomes rare, and the number of peers keeps growing. This means that it takes longer and longer time for a peer to complete. This case is shown in Fig. 1.

This problem is persistent from case to case and is similar to the two chunk case, [5]. The other algorithm described above we call 'Forced-Friedman-algorithm'. As we can see, the results of simulation in Fig. 2 are promising. Indeed, the system shows very good performance and balance, peers go through the system almost with maximal possible rate, almost every contact is productive, the peer can find something to copy and moves ahead. This picture is also persistent from case to case.

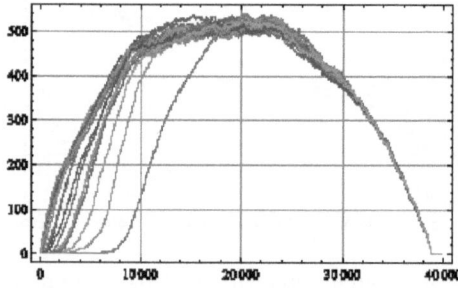

**Fig. 3.** An unstationar scenario with Forced-Friedman algorithm. The systems starts from the empty state and with constant rate of incoming peers, then after a while the flow of peers stops. The case with 50 chunks, populations of nodes having particular chunks. The system manages to complete, all peers complete without any long tail of delay.

Furthermore, it seems to have other good properties as well. Indeed, such a system seems to be able to cope with unstationar scenarios, a kind of 'flash crowds'. By this we mean that first there is a constant flow of incoming peers, but after some time this flow completely shuts down. If the system is unstable with poor balance of chunk populations, the system would not be able to complete, there would be a left-over, see also [8, 4]. The left-over is situation when some peers would not be able to complete (in system without seed) or would be forced to complete slowly by obtaining the last chunk from the seed. Our system seems to be able to avoid such difficulties as shown in Fig 3. Another good feature is the systems ability to cope with large initial unbalance. Even if there is a initially extremely rare chunk, the system quickly relaxes to steady state, as shown in Fig. 4.

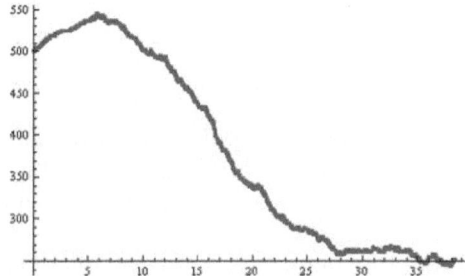

**Fig. 4.** Relaxation of the system with Forced-Friedman algorithm. The system starts from a state with 500 nodes missing the same chunk (a rare chunk), however, the system quickly relaxes to steady state. A case with 20 chunks, number of peers in the system is shown.

## 3  A Simple Analytical Model

The state-space of our system is enormous, so it seems to be impossible to create a useful model for this system. However, something can be done in this direction. Obviously some simplified assumptions must be done. We observed that the performance of the system is very close to ideal. By postulating this kind of behavior, a surprisingly accurate model can be found.

More precisely, we assume that the system is in an ideal state, meaning that all chunks are equally likely to be found in system. A peer that enters the system makes tree random contacts, and uses the Friedman-type logic to decide which chunk it can copy. If it founds at least one such chunk, it moves to state where it has one chunk, and so on. From this assumption, we deduce that probability that a particular chunk can be copied under the Friedman constrain is $3\frac{1}{2}\frac{1}{2}\frac{1}{2} = \frac{3}{8}$, let denote by $p = 1 - \frac{3}{8} = \frac{5}{8}$, the probability of the complement event. Then probability that a node with $k$ missing chunks can copy a chunk in its current contact is $1 - p^k$. Thus it is plausible to describe system by magnitudes $n_i, i = 1, 2, \cdots, m - 1$, where $n_i$ is number of nodes with $i$ chunks. In the fluid limit one can assume the system of differential equations:

$$\frac{d}{dt}n_1 = \lambda - (1 - p^{m-1})n_1$$

$$\frac{d}{dt}n_2 = (1 - p^{m-1})n_1 - (1 - p^{m-2})n_2$$

$$\frac{d}{dt}n_3 = (1 - p^{m-2})n_2 - (1 - p^{m-3})n_3$$

$$\cdots$$

$$\frac{d}{dt}n_{m-1} = (1 - p^2)n_{m-2} - \frac{1}{2}n_{m-1}$$

$$p = \frac{5}{8}$$

They have the stationary solutions:

$$n_1 = \frac{\lambda}{1-p^{m-1}}, n_2 = \frac{\lambda}{1-p^{m-2}}, n_3 = \frac{\lambda}{1-p^{m-3}}, \cdots$$

$$\cdots, n_{m-2} = \frac{\lambda}{1-p^2}, n_{m-1} = 2\lambda.$$

The last relations mean that all populations have different sizes although they have an accumulation point $= \lambda$, as $m$ grows, see Fig 5. These stationary solutions seems to be those that the real simulated system with Forced-Friedman algorithm yields, as shown in Fig. 6.

We made some computer experiments to see whether the empirical expectation is close to solutions of the simplified systems behavior, see Fig. 7. Based on those we conjecture that such means have damping oscillations around the curves of simplified model, and with very close to stationary level of population size. However, it can also be due some inaccuracy of the differential equations. Ideed, in the stochastic model the peer can point to itself and thus fail to download.

These simulations indicate also that the performance is almost ideal. This is because in the steady state, there is no 'bottleneck' chunks that are hard to find. That is why, there are few such contacts that do not lead to a download of a chunk.

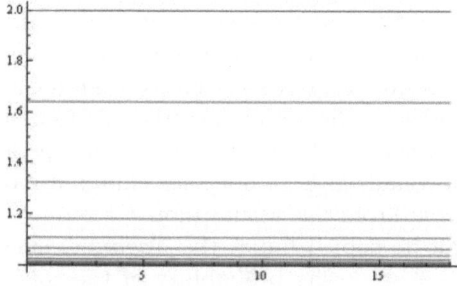

**Fig. 5.** Levels of steady state population sizes of nodes with $1, 2, \cdots, m - 1$ chunks in units of $\lambda$, the accumulation point of lines equals to 1.

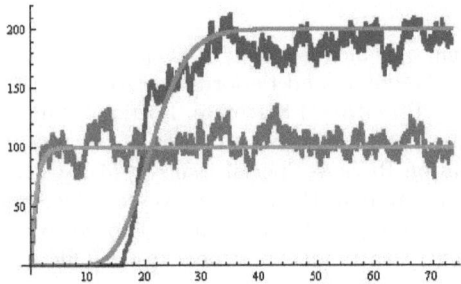

**Fig. 6.** A simulation for 20 chunk-system with Forced-Friedman algorithm, $\lambda = 100$. The rugged lines are simulated processes for $n_1$ and $n_{19}$, while the smooth lines are solutions of the differential equations for the simplified model. Other components have similar behavior.

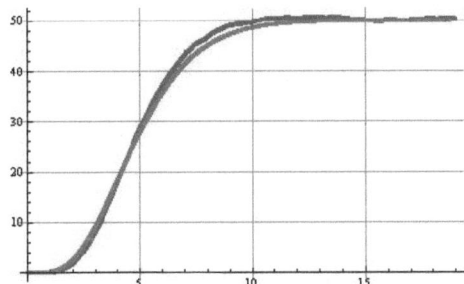

**Fig. 7.** A simulation for 20 chunk-system with Forced-Friedman algorithm, $\lambda = 50$. A bit rugged line is simulated processes for $n_5$, empirical average over 1000 experiments, the smooth line is solution of the differential equations for the simplified model. It seems, that the average of the steady state is converging to the one for the simplified process. However, the transient state is slightly deviating from it, possibly having damping oscillations around the simplified system curve.

## 4  Conclusions

In this short paper we describe preliminary results on chunk copying system that relies entirely on randomness. Previously we studied two chunk case. It was shown that the system with Friedman-urn like algorithm is efficient and stable. In the case of many chunks we imitate this algorithm as far as possible. The resulting system shows stability and good performance under dynamical conditions. Unlike the two chunk case, the simple proof based on Friedman-urn is not usable. The main challenge is to find rigorous foundation for this algorithm.

# References

[1] Cohen, B.: BitTorrent specification (2006), http://www.bittorrent.org
[2] Qiu, D., Srikant, R.: Modeling and Performance Analysis of BitTorrent-Like Peer-to-Peer Networks. In: Proc. ACM Sigcomm, Portland, OR (2004)
[3] Mundinger, J., Weber, R., Weiss, G.: Analysis of peer-to-peer file dissemination. Performance Evaluation Review, Special Issue on MAMA 2006 (2006)
[4] Norros, I., Prabhu, B., Reittu, H.: Flash crowd in a file sharing system based on random encounters. In: Inter-Perf, Pisa, Italy (2006), http://www.inter-perf.org
[5] Reittu, H., Norros, I.: Urn models and peer-to-peer file sharing. In: Proc. IEEE PHYSCOMNET 2008, Berlin (2008)
[6] Massoulie, L., Vojnovic, M.: Coupon Replication Systems. In: Proc. ACM SIGMETRICS, Banff, Canada (2005)
[7] Pemantle, R.: A survey of random processes with reinforcement. Probability Surveys 4, 1–79 (2007)
[8] Reittu, H., Norros, I.: Toward moldeling of a single file broadcasting in a closed network. In: Proceedings of IEEE SPASWIN 2007, Limassol, Cyprus (2007)

# A Decentralized Architecture for Distributed Neighborhood Based Search

Pascal Katzenbach, Yann Lorion, Tjorben Bogon, and Ingo J. Timm

Institute of Computer Science
Goethe University Frankfurt
{katzenb,lorion,tbogon,timm}@cs.uni-frankfurt.de

**Abstract.** We present a decentralized self-X architecture for distributed neighborhood based search problems using an overlay network based on random graphs. This approach provides a scalable and robust architecture with low requirements for bandwidth and computational power as well as an adequate neighborhood topology, e.g. for several instances of parallel local search and distributed learning. Together with an adapted load balancing schema our architecture is self-organizing, self-healing and self-optimizing.

## 1 Introduction

As the number of computational resources increases faster than the computational power of single resources, in almost every discipline of computer science flexible and scalable parallel solutions are needed in order to reduce computation time. Usually, effective parallel and distributed algorithms are custom-made for each problem and no general solutions exist for bigger class of problems.

In this paper we present an architecture for the class of distributed neighborhood based search problems. The problems that we are interested in vary from parallel local search to distributed learning – basically, every search problem over a measurable search-space that can be solved efficiently by distributed search entities that are only allowed to communicate to a small given neighborhood. One possible scenario is the distribution of population based metaheuristics using island structures, i.e. encapsulated sub-populations, that are often used in evolutionary algorithms or particle swarm optimization [1].

The architecture should be self-organizing, self-healing and self-optimizing, since it should be able to deal with heterogeneous, dynamic and unreliable large-scale environments. During the last decade, peer-to-peer networks, i.e. a communication structure in which individuals interact directly without going through a centralized system or hierarchy, has proven to be an adequate solution for such requirement. In order to achieve self-organization, overlay networks are placed on top of the physical networks. Since in contrast to well-known P2P applications like file sharing we have no requirement for lookup-operations to locate explicit peers or data in the network, there is no need for structured overlay topologies, e.g. based on distributed hash tables. Instead, in order to minimize the overhead

T. Spyropoulos and K.A. Hummel (Eds.): IWSOS 2009, LNCS 5918, pp. 193–200, 2009.

for construction and maintenance, we design a new randomized overlay network that meets all requirements.

## 2   Fundamentals

Since the architecture, in particular the overlay network, presented in this paper is based on a construction schema for random graphs with low mixing times, we first introduce some basic definitions for graphs, markov chains, mixing times and random graphs.

An *undirected graph* $G$ is a pair $G = (V, E)$ of a set $V$ of $n$ vertices or nodes and a set $E \subseteq V \times V$ of edges. The degree $\deg(v)$ of a node $v$ is the number of edges starting (or ending) in $v$. A Graph $G$ is *d-regular*, iff all nodes have degree $d$. A *path* is a sequence $(v_1, \ldots, v_m)$ of nodes where each successive pair is connected by an edge in $E$. A graph $G$ is *connected*, iff $G$ contains a path between each pair of different nodes.

For the construction schema used in this paper, we will need to draw nodes nearly uniformly distributed out of the set of all nodes $V$ without knowing the entire set. For that matter a random walk on the graph comes in handy. Given a pair $(\Omega, P)$ of the finite denumerable state space $\Omega$ and the stochastic $(\Omega \times \Omega)$-matrix $P = (p_{xy})$. A sequence $X_1, X_2, \ldots$ of random variables is called *Markov chain*, iff it satisfies the *Markov property*

$$\Pr[X_{n+1} = y | X_n = x_n, \ldots, X_1 = x_1] = \Pr[X_{n+1} = y | X_n = x_n] = p_{x_n y},$$

for all $n \geq 0$ and $x_0, \ldots x_n, y \in \Omega$. By interpreting nodes as states and edges as transitions a graph can be seen as a Markov chain where $\Omega = V$ and the stochastic matrix $P$ is denoted by $p_{ij} := 1/\deg(i)$ iff there is an edge from node $i$ to node $j$. Starting with the probability vector $\mu^{(0)}$, where $\mu_i^{(0)} := \Pr[X_0 = x_i]$, the distribution at time step $t+1$ can be calculated from time step $t$ with $\mu^{(t+1)} = \mu^{(t)} P$, hence $\mu^{(t)} = \mu^{(0)} P^t$. For $t \to \infty$ on a non-bipartite undirected graph the corresponding Markov chain has a unique stationary distribution $\pi = P\pi$, where $\pi_i = \deg(i)/2|E|$ [2]. Hence when starting at an arbitrary node $x$ and performing a random walk on a $d$-regular graph the distribution $P^t(x, .)$ of the node visited at time $t$ will converge towards a uniform distribution over all nodes in $V$.

The *mixing time* $\tau(\epsilon)$ is the minimum number of time steps needed to almost reach the stationary distribution, i.e. to have a *maximal total variation distance* between both distributions of less than $\epsilon$ (for details see [3]). A markov chain is considered *rapid mixing* if $\tau(\epsilon) \in \mathcal{O}(\text{poly}(\log(n/\epsilon)))$ holds. It can be shown, that the mixing time $\tau(\epsilon)$ is bounded by

$$\left(\frac{1}{\gamma(P)} - 1\right) \ln \frac{1}{2\epsilon} \leq \tau(\epsilon) \leq \frac{1}{\gamma(P)} \ln \left(\frac{1}{\epsilon \pi_{min}}\right) \tag{1}$$

where $\pi_{min} := \min_{x \in \Omega} \pi(x)$ and $\gamma(P)$ is the *spectral gap*, i.e. the distance between the largest and the absolute second largest eigenvalue of the stochastic matrix $P$[3]. Hence having a large spectral gap leads to fast mixing times.

# 3   Designing the Overlay Network

## 3.1   Requirements and Goals

The design purpose of our overlay network is to provide an architecture for distributed neighborhood based search in dynamic and unreliable large-scale environments. Since we focus on self-organizing, self-healing and self-optimization our main requirements for the overlay network are:

**The communication structure should be well suited for neighborhood based search.** We study distributed search entities, that are only allowed to communicate to a small neighborhood, given by the overlay network. In order to ensure fast spreading of important information the shortest way to the furthest entity should be as short as possible. Hence the graph of the overlay network should have a small diameter. A little opposing to the small diameter the number of direct neighbors of each entity, i.e. their degree in the respective graph, should be relatively small. This way even the communication costs of events that have to be spread to the entire neighborhood stay small for each entity.

**The overlay network should allow dynamic joining and leaving and be robust against loss of connections and sub systems (self-organization and self-healing).** One possible consequence of the requirement for robustness is that the corresponding graph should have a high degree, so that with high probability the loss of one connection has no significant effect. Another way is to let each entity have a list of backup-neighbors in case connections to neighbors fail. However losing an edge could still lead to decomposition of a connected graph to sub graphs. Hence as major requirement the graph should have a high bisection width so that a decomposition becomes highly improbable.

**Operations for construction and maintenance need to be scalable to the network size (self-adaptive) and low cost intensive in computational, memory and bandwidth terms.** The construction and maintenance of the overlay network should be at most logarithmic in the number of peers for the above-mentioned terms. Additionally the graph structure should be easily expandable, i.e. construction and maintenance should never necessitate restructuring and hence blocking times for communication.

**Finally, the structure should be well suited for diffusion based load balancing (self-optimization).** The *diffusion schema* is an iterative and local procedure for decentralized load balancing, as it only uses information of a node and its direct neighbors. We deal with a dynamic and heterogeneous environment, where each entity only knows a small neighborhood. Diffusion based load balancing is the only procedure known to the authors, that meets all requirements. In order to have a fast convergence to a balanced state the corresponding graph needs to have a high spectral gap in the diffusion matrix [4].

## 3.2   Model and Algorithmic Description

Some of the requirements given in the previous section result in opposed objectives. While lower boundable degrees are desirable for scalability and resource allocation reasons, they may not lead to dense graphs with small diameters. In order to reach a trade-off between these objectives, an appropriate graph structure has to be chosen.

One example for a good compromise are expander graphs. They are defined over the expansion ratio $h(G) = \min_{1 \le |S| \le \frac{n}{2}} \frac{|\partial S|}{|S|}$, which is a measure for density. A family of expander graphs is an infinite sequence $(G_i)_{i \in \mathbb{N}}$ of $d$-regular graphs with increasing sizes, if there is a $\epsilon > 0$, so that $h(G_i) \ge \epsilon$ for all $i \in \mathbb{N}$. Expander graphs as network models result in very good run-time performance in diffusion load-balancing, e.g. a static expander graph achieves in $O(1/\epsilon)$ steps a $\epsilon$-balanced state [4]. Also, expander graphs have the useful property of logarithmic mixing-times for approximately uniform sampling.

Distributed construction of expander graphs is not a trivial task, especially in unreliable dynamic environments. One possible method are $\mathbb{H}$-Graphs [5]. A $2d$-regular random graph is constructed, which consists of $d$ Hamilton circles. Every node knows its predecessors and successors for each circle. When a new node enters the network, for each Hamilton circle a node is chosen as insertion position via random walk sampling. Afterwards the predecessor/successor-tables of the involved nodes are updated. Hence each node has exactly $2d$ neighbors. It can be shown that these graphs are expander graphs and therefore provide uniform distributed sampling via logarithmic random walks [5]. A negative aspect for our purpose is that amongst others the failure of some connections can lead to a total restructuring process and that small graphs have to be considered separately [5].

In order to avoid these downsides we want to build an expander-similar graph with a lower bounded expansion ratio but not necessarily $d$-regularity and hope to keep the resulting properties of an expander graph like small diameter, a low degree and logarithmic mixing-times. This is achieved by a randomized graph construction schema where each node has the same expected degree with an as low as possible variance:

The network is modeled by a tuple $(G, S, T)$ with an undirected graph $G = (V, E)$, where $v \in V$ are participating peer nodes and $\{v_1, v_2\} \in E$ are direct connections between peer nodes. Variable $S = (S_v)_{v \in V}$ describes the configuration of sample pools with $S_v \subseteq V \setminus v$ from which we assume at this moment, that members of $S_v$ are (approximately) uniformly distributed in $V/\{v\}$. Symbol $T$ is a tuple $T = (\overline{d}, \tau^*, \eta, d_{max})$ which describes global parameters, where $\overline{d}$ indicates the target degree, $\tau^*$ indicates the minimum random walk length, $\eta$ indicates the maximum sample pool size and $d_{max}$ indicates an upper bound for degrees. Because $G$ and $S$ are time mutable, the precise notation would be $(G^{(t)}, S^{(t)}, T)$ with time $t$, but for better readability we leave $(t)$ out.

Each peer $w$ regulates its current degree $|Nb_w|$ to the target degree $\overline{d}$ by periodically adding or removing the corresponding number of neighbors. To add a neighbor, a node $v$ is randomly drawn from $S_w \setminus Nb_w$ and a connection to $v$ is established. To remove a neighbor, a node $v$ is randomly drawn from $Nb_w$ and the connection to $v$ is dropped.

Figure 1 shows the basic algorithms to fill a sample pool and to sample through a random walk. FILLPOOL$_w$, which is called periodically on each peer $w$, is

FillPool$_w$()

1  $k \leftarrow \eta - |S_w|$
2  for $i \leftarrow 1, ..., k$ do
3      $v \leftarrow$ Sample$_w(\tau^*)$
4      if $v \notin S_w \land v \neq w$ then
5        $S_w \leftarrow S_w \cup \{v\}$

Sample$_w(t)$

1  if $t \leq 0$ then  return $w$
2  $i \leftarrow$ uniform i.i.d. over $\{1, ..., d_{max}\}$
3  if $i > |Nb_w|$ then  return Sample$_w(t-1)$
4  $u \leftarrow i$-th node in $Nb_w$
5  return Sample$_u(t-1)$

**Fig. 1.** Pseudocode for sample pool filling and random walk sampling

responsible for filling sample pool $S_w$ with approximately uniform distributed elements over $(V \setminus w)$. The number of added samples is denoted by the difference between target sample pool size $\eta$ and the current pool size. Sampling is done via random walks of length $\tau^*$ through call of Sample$_w$. Pooled sampling brings two major advantages against on-demand sampling: Instant access on samples for faster regulation in case of differing degrees and advanced capability for network reconstruction in case of global failures (when many nodes or connections fail at once), in which on-demand sampling may not be able to reach the entire network.

Sample$_w(t)$ performs a random walk of (remaining) length $t$. Due to possible irregularity of network graph $G$ (which would lead to a non-uniform stationary distribution), a technique called *max degree random walk* is used to simulate a random walk on a undirected regular graph $G' = (V, E')$ with $E' \supseteq E$ : For each node $w \in V$, a number of $(d_{max} - |Nb_w|)$ self-loops are additionally added to $E'$. Parameter $d_{max}$ should be with almost sure probability an upper bound for all occurring node degrees. For $G'$ the resulting transition matrix $P$ has the entries $P_{v,w} = 1/d_{max}$ for $w \in Nb_v$, the diagonal entries $P_{v,v} = 1 - |Nb_v|/d_{max}$ and zero entries elsewhere. This markov chain is simulated by the described local or remote recursive calls depending on a random value $i$ in Sample$_w(t)$. In case of a connected network graph $G$, graph $G'$ will also be connected and because of the added self-loops $G'$ cannot be bipartite. Hence the markov chain on transition matrix $P$ has a unique stationary distribution $\pi = (1/|V|, \ldots, 1/|V|)$ (see section 2). If spectral gap $\gamma(P)$ can be lower bounded, a mixing-time $\tau(\epsilon)$ in $O(\log(|V|))$ is sufficient (see inequation (1)). Resulting spectral gaps of our model are analyzed in section 4.

### 3.3  Load Balancing

In diffusion load balancing for a network $G = (V, E)$ with heterogeneous subsystems $v \in V$ with benchmark factors $c_v$, the update equations of work load $W_v$ and work float $y_{vw}$ between two subsystems $v$ and $w$ at time step $t$ are:

$$W_v^{(t)} = W_v^{(t-1)} - \sum_{w:\{v,w\}\in E} y_{vw}^{(t)} \qquad y_{vw}^{(t)} = \alpha_{vw}\left(\frac{W_v^{(t-1)}}{c_v} - \frac{W_w^{(t-1)}}{c_w}\right).$$

**Table 1.** Spectral gaps of different network sizes $n$ and minimum random-walk lengths $\tau^*$ in 30 simulations

| n | $\tau^* = 8$ | $\tau^* = 12$ | $\tau^* = 16$ | $\tau^* = 20$ | $\tau^* = 24$ |
|---|---|---|---|---|---|
| 500 | 0.0863 | 0.0889 | 0.0897 | 0.0889 | 0.0900 |
| 1000 | 0.0809 | 0.0854 | 0.0862 | 0.0866 | 0.0870 |
| 2000 | 0.0796 | 0.0848 | 0.0855 | 0.0858 | 0.0859 |
| 4000 | 0.0783 | 0.0845 | 0.0850 | 0.0852 | 0.0852 |
| 8000 | 0.0780 | 0.0842 | 0.0847 | 0.0849 | 0.0851 |
| 16000 | 0.0767 | 0.0841 | 0.0846 | 0.0848 | 0.0849 |
| 32000 | 0.0765 | 0.0841 | 0.0845 | 0.0847 | 0.0848 |
| 64000 | 0.0768 | 0.0840 | 0.0845 | 0.0847 | 0.0848 |

(a) Target degree $\bar{d} = 6$, Mean values

| n | $\tau^* = 8$ | $\tau^* = 12$ | $\tau^* = 16$ | $\tau^* = 20$ |
|---|---|---|---|---|
| 500 | 0.0823 | 0.0846 | 0.0844 | 0.0843 | 0.0845 |
| 1000 | 0.0714 | 0.0829 | 0.0838 | 0.0845 | 0.0854 |
| 2000 | 0.0746 | 0.0822 | 0.0838 | 0.0840 | 0.0842 |
| 4000 | 0.0711 | 0.0836 | 0.0843 | 0.0843 | 0.0841 |
| 8000 | 0.0724 | 0.0833 | 0.0841 | 0.0842 | 0.0845 |
| 16000 | 0.0663 | 0.0834 | 0.0842 | 0.0845 | 0.0847 |
| 32000 | 0.0684 | 0.0836 | 0.0843 | 0.0845 | 0.0845 |
| 64000 | 0.0648 | 0.0819 | 0.0842 | 0.0845 | 0.0845 |

(b) Target degree $\bar{d} = 6$, Minimum values

$M = (\alpha_{vw})$ is called diffusion matrix with $0 < \alpha_{vw} < 1$ for $\{v, w\} \in E$, otherwise $\alpha_{vw} = 0$. Rate of convergence (how fast a balanced state is reached) depends on the spectral gap of $M$ [4]. If we choose $\alpha = (1/d_{max})$, the diffusion matrix equals our transition matrix $P$, hence we can use the results for spectral gaps in section 4. Since distributed computation often only allows unnormalized benchmarking factors $b_v$, a locally Euclidean normalization $c_v = b_v/\sqrt{b_v^2 + b_w^2}$ can be applied for each pair of neighbors $(v, w) \in E$. If the distributed search problem requires steady cooperation between entities (e.g. migration in island structured metaheuristics), there is a need of constant information exchange between neighbors. Therefore a minimal float $s_{min}$ is introduced for the computation of each outgoing float $s_{vw}^{(t)}$:

$$s_{vw}^{(t)} = s_{min} + \max\left\{0 \ , \ \alpha_{vw}^{(t)} \sqrt{b_v^2 + b_w^2} \left(\frac{W_v^{(t-1)}}{b_i} - \frac{W_w^{(t-1)}}{b_w}\right)\right\}$$

It must be pointed out that the minimal float interferes with load balancing on the way to the balanced state.

## 4   Evaluation

The main question of our evaluation is whether our construction schema produces graphs with expander-similar properties, i.e. leads to lower boundable expansion ratios? The verification is done by locally simulating several networks based on the described algorithm and analyzing the spectral gaps of their resulting graphs.

During simulation, new nodes are injected with rate $|V| \cdot \alpha_{join}$, existing nodes leave the network with rate $|V| \cdot \alpha_{leave}$ (rates lower than 1 are interpreted as probabilities) and existing connections fail with probability $\alpha_{fail}$. In our simulations $\alpha_{join}$ is set to 0.4, $\alpha_{leave}$ is set to 0.1 and $\alpha_{fail}$ to 0.01. When the networks reaches a designated size, simulation is stopped and the resulting adjacency matrix is saved for eigenvalue and spectral gap computations.

Simulations with network sizes up to 64.000 nodes were performed. Table 1 shows to resulting spectral gaps. As you can see, the mean values of the simulation decrease with expanding network size. The diagonal entries in the tables suggest that spectral gaps can be lower bounded by using logarithmic minimum random-walk lengths.

Compared to structured P2P construction schemas like CHORD [6] the spectral gap of our topology does not appear to decrease monotonously with increasing number of peers. An extensive comparison of different topologies is in work.

## 5   Application

A first application of the presented architecture is based on distributed particle swarm optimization (PSO). PSO is a population-based metaheuristic for which one way of parallelization is the separation of the population into islands of sub-populations. Each island is computed by a different peer node in the described topology. Information exchange between islands is performed by periodic migrations of population members (particles) between adjacent peer nodes. The advantage of our network topology is reflected in an efficient spread of information in the entire network via such local exchange. Compared to other decentralized approaches for PSO (e.g. [7]) with simple topologies (e.g. circles), first results on heterogeneous networks suggests fast global information exchange with a moderate number of messages and a strong convergence to a load balanced state. Further results will be published soon.

## 6   Conclusion and Future Work

In this paper an innovative decentralized self-X architecture for distributed neighborhood based search problems is presented exploiting a P2P-based approach. The overlay network based on random graphs is built using approximate sampling by random walks with local sample pools. This leads on the one hand to a scalable and robust structure with low requirements for bandwidth and computational power and on the other hand provides an adequate neighborhood topology. Diffusion based load balancing ensures efficient computation on heterogeneous systems. First experiment series on a simulation of the overlay network demonstrate the properties of our architecture.

The properties of the overlay network and the corresponding random graphs are shown through simulation, but not theoretically proven yet. It would be interesting to analyze the network more theoretically, e.g. in order to achieve tighter bounds. In Addition, further simulations and evaluations are needed. First experiments with our load balancing schema show promising results but lack precise evaluation. The robustness of the overlay network has to be evaluated and test series with distributed neighborhood based search scenarios are needed. First positive results were achieved for particle swarm optimization and we are currently working on distributed learning scenarios. Another interesting point would be an empirical comparison with other random graphs as well as flooding- and especially DHT-based P2P networks for which we expect expander similar properties. Furthermore, we are currently working on a self-configuration approach for parameters $T = (\bar{d}, \tau^*, \eta, d_{max})$. In particular the choice of a sufficient random walk length $\tau^*$ depends on the network size, for which estimation techniques are needed.

# References

1. Lorion, Y., Bogon, T., Timm, I.J., Drobnik, O.: An agent based parallel particle swarm optimization - APPSO. In: Proc. of Swarm Intelligence Symposium – SIS (2009)
2. Lovász, L.: Random walks on graphs: A survey. Combinatorics, Paul Erdös is Eighty 2 (1996)
3. Levin, D.A., Peres, Y., Wilmer, E.L.: Markov Chains and Mixing Times. American Mathematical Society (2008)
4. Muthukrishnan, S., Ghosh, B., Schultz, M.H.: First- and second-order diffusive methods for rapid, coarse, distributed load balancing. Theory Comput. Syst. 31(4), 331–354 (1998)
5. Law, C., Siu, K.Y.: Distributed construction of random expander networks. In: IEEE INFOCOM 2003., vol. 3, pp. 2133–2143 (2003)
6. Stoica, I., Morris, R., Karger, D., Kaashoek, M.F., Balakrishnan, H.: Chord: A scalable peer-to-peer lookup service for internet applications. In: SIGCOMM 2001: Proceedings of the 2001 conference on Applications, technologies, architectures, and protocols for computer communications, pp. 149–160. ACM Press, New York (2001)
7. Romero, J.F., Cotta, C.: Optimization by island-structured decentralized particle swarms. In: Proceedings of Computational Intelligence, Theory And Applications: International Conference 8th Fuzzy Days in Dortmund, Germany, September 29-October 01, 2004, Springer, Heidelberg (2005)

# Scheduling in P2P Streaming: From Algorithms to Protocols

Luca Abeni and Alberto Montresor[*]

DISI - University of Trento, Trento (IT)
{luca.abeni, alberto.montresor}@unitn.it

**Abstract.** Chunk and peer scheduling is among the main driver of performance in P2P streaming systems. While previous work has analytically proved that optimal scheduling algorithms exist, such strategies are based on a large number of strong assumptions about the knowledge that a single peer has of the rest of the system. This short paper presents a protocol for turning these theoretical results into practical ones, by taking into account practical aspects like the diffusion time of signaling messages and a partial knowledge of the participating peers.

## 1 Introduction

Peer-to-peer streaming systems are becoming increasingly popular [1,2,3] as a way to overcome the scalability limitations of traditional streaming technologies based on a client/server paradigm. In particular, there is an increasing interest in *unstructured* P2P streaming solutions, which distribute a media stream by dividing it in *chunks* that are disseminated by the various peers contributing to the stream diffusion. This approach allows to better exploit the upload bandwidth of all the peers which actively participate in the dissemination, and to reduce the overhead required for maintaining the overlay structure in case of churns. Moreover, unstructured systems are more tolerant to peer failures [4].

Traditionally, unstructured P2P streaming systems tend to distribute the chunks based on *random* decisions, generating a lot of useless network traffic (due to duplicated chunks) and delaying the chunk diffusion. These effects can be avoided by basing the chunks distribution on smarter decisions (this is the scheduling problem), so that every peer can receive the various chunks at the proper time and chunks duplications (the same chunk sent to the same peer multiple times) are reduced or completely eliminated. There has been a good amount of research on scheduling in P2P streaming systems, and some scheduling algorithms providing good mathematical properties have been developed [5,6,7]. Some of such algorithms (namely, Rp/Lb, Rp/LUc, MDp/LUc, LUc/ELp, and Dl/ELp) have been formally analysed, and various optimality properties have

---

[*] This work is supported by the European Commission through the NAPA-WINE Project (Network-Aware P2P-TV Application over Wise Network – www.napa-wine.eu), ICT Call 1 FP7-ICT-2007-1, 1.5 Networked Media, grant No. 214412.

T. Spyropoulos and K.A. Hummel (Eds.): IWSOS 2009, LNCS 5918, pp. 201–206, 2009.

been proved (see the papers cited above for details on the algorithms and on their properties).

An unstructured P2P streaming system is modelled as a set of nodes in which a special one, called the *source*, generates chunks at a fixed rate from an encoded media stream. Chunks are distributed to the peers present in the system through a *push* strategy: each peer contributing to the streaming selects a chunk to be sent and the target peer, and sends (pushes) the chunk to the target. The goal is to select the chunk to be sent and the target peer so that duplicated chunks are reduced and the streaming performance is improved. These two decisions are taken by two *schedulers*, called *chunk* and *peer scheduler*. The peer scheduler can select a target peer from a set of peers named *neighbourhood* (basically, each peer knows the existence of a limited subset of peers).

The works cited above focused on the scheduling algorithms, assuming that each scheduler has some knowledge of the peer's neighbourhood (and knows the chunks that have already been received by all the neighbours). Since such algorithms are often based on a set of assumptions regarding the overlay management or the diffusion of the information about the chunks that have been received by each peer, the creation and maintenance of the neighbourhood (overlay management) and the signalling mechanisms between different peers (chunk buffer state management) have not been considered. Hence, the theoretical properties proved in the original papers have to be verified in more realistic situations.

This paper presents the design of PUSHSTREAM, a protocol that can be used to implement various peers and chunks scheduling algorithms, and evaluates the performance of such protocol through a set of simulations which take in account the overlay construction, the signalling delay, and churn.

## 2    The PushStream Protocol

As explained, to implement an effective scheduler the scheduling algorithm must be complemented with a protocol which takes care of the chunk buffer state management and overlay management issues.

Some overlay management protocols which are well known in the P2P community (such as NEWSCAST [4]) can be used to construct the neighbourhoods, and chunk buffer state management can be performed by sending signalling messages when a peer receives a chunk. However, the overlay management algorithms generally create unidirectional links (that is, if peer $P_i$ is in $P_j$'s neighbourhood, it is not guaranteed that $P_j$ is in $P_i$'s neighbourhood), while the solution mentioned above requires overlays based on bidirectional graphs (because when a peer $P_i$ receives a chunk, it should notify all the peers that can potentially send chunks to $P_i$ - that is, all the peers that have $P_i$ in their neighbourhoods). In the following, this problem will be referred as the *bidirectional neighbourhood* problem.

Another problem is caused by the one-way-delay: if peer $P_i$ receives a chunk $C_k$ at time $t$ and sends a message to peer $P_j$ to notify it about the chunk reception, $P_j$ will receive such notification only at time $t' = t + \delta$, where $\delta$ is the one-way-delay. Hence, if $P_j$'s scheduler runs between $t$ and $t'$ then it might decide to send again $C_k$ to $P_i$. In the following, this problem will be referred as the *delayed notification* problem.

The protocol proposed in this paper, PUSHSTREAM, addresses the bidirectional neighbourhood problem by dynamically constructing an *input neighbourhood*, which is used to send notification messages: when peer $P_i$ receives a chunk from peer $P_j$, $P_j$ is added to $P_i$'s input neighbourhood, and a notification message is sent to all the peers in the input neighbourhood (all but $P_j$, clearly).

The delayed notification problem is particularly dangerous when using "more deterministic" schedulers such as ELp [7] which try to take optimal decisions to reduce the chunks diffusion times as much as possible. In fact, such algorithms increase the probability to have multiple peers sending the same chunk to the same target simultaneously (because of the "deterministic" behaviour of the scheduler, many peers will take the same decision as they are unaware of what the other peers are doing). This problem can be addressed by exploiting some topological properties to reduce the collisions probability: basically, the scheduler does not select the target from the whole neighbourhood provided by the overlay management protocol, but from an *output neighbourhood*, which is a subset of such neighbourhood. In particular, the output neighbourhood of $P_i$ is composed by $P_i$'s neighbours based on some desirable property (in this paper, the minimum topological distance from $P_i$ has been used).

Hence, each peer simultaneously runs an overlay management protocol and PUSHSTREAM, which is a cyclic protocol (with time cycle $T$) and works as follows:

- When a chunk $C_k$ is received from a peer $P_j$
  - Add $C_k$ to the chunk buffer
  - Add $P_j$ to the input neighbourhood $N^{in}$, if not already present
  - Otherwise, reset $P_j$'s timeout to the maximum value $T^{out}$
  - For each peer $P_h \in N^{in}$, send a notification about $C_k$ reception to $P_h$, and decrease $P_h$'s timeout
  - If the timeout of $P_h$ is 0, remove $P_h$ from $N^{in}$
- When a notification message is received from peer $P_j$
  - Update the information about the chunks received by $P_j$
- At every cycle (with period $T$)
  - Invoke the scheduler to select a target peer in the output neighbourhood $N^{out}$ and a chunk to be sent. The scheduler uses the information provided by the previous notification messages
  - Send the selected chunk to the target
- Every $K$ cycles
  - Update $N^{out}$ by using the peers known by the gossipping protocol having the smallest distance from the current peer.

It is worth noting that PUSHSTREAM does not require any notion or global time, nor it assumes any kind of special synchronisation between the peers.

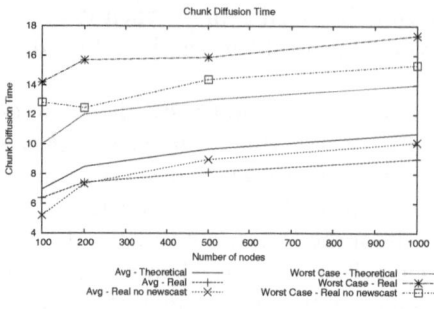

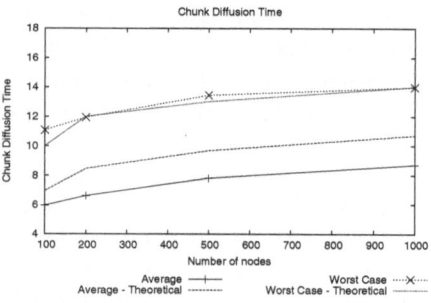

**Fig. 1.** Chunk diffusion time in the ideal case and using PushStream

**Fig. 2.** Chunk diffusion time in realistic situations using PushStream

## 3 Preliminary Results

The PushStream protocol has been implemented in the Peersim[1] P2P simulator, which also emulates the chunk transmission time and the signalling delay, and its performance have been evaluated and compared with the "theoretical results".

In these PushStream tests, Dl/ELp [7] has been selected as a scheduling algorithm (because of its optimality properties), and the Newscast algorithm has been used for overlay management. The simulations have been run assuming a chunk size of $1s$ (each chunk contains 1 second of encoded media), an output neighbourhood size equal to $\log_2(N)$, where $N$ is the number of peers, and a Newscast neighbourhood size equal to $3\log_2(N)$.

**Correctness of the Implementation:** The correctness of the implementation has been verified by simulating an "almost ideal situation" (all the chunk transmission times and the signalling delays are set to the minimum possible value, which is $1ms$) and comparing the results with the theoretical results from [7][2]. The streaming performance have been evaluated by considering the chunk diffusion time (the time needed by a chunk to be diffused to all the peers): in particular, Figure 1 displays the worst case and average values for the chunk diffusion times. The average values obtained by using PushStream are smaller than theoretical ones because each peer receives about 93% of the chunks, and the statistics are computed on the received chunks only (in the theoretical case, no chunk is lost). Such lost chunks are mainly due to Newscast dynamically changing the overlay, and to the initial startup time needed to setup the input neighbourhood (during this time, a lot of duplicated chunks are received). To verify this, the simulations have been repeated disabling the periodic update of

---

[1] http://peersim.sf.net

[2] Such reference values have been obtained by simulating the ideal system (in which every peer knows the exact state of all its neighbours, the overlay is static, and the neighbourhood size is assumed to be equal to $3\log_2(N)$).

the output neighbourhood (in this way, PUSHSTREAM ends up using a static overlay, and NEWSCAST does not affect the streaming performance); as a result, each pears receives more than 98.5% of the chunks and the worst case diffusion times are a little bit larger than the theoretical bound as shown in Figure 1. In any case, the measured values are quite close to the theoretical ones, showing that the protocol is correctly implemented.

**Realistic Setup:** In the previous simulations, the size $T$ of the periodic cycle of each peer has been set to $1s$ because the Dl/ELp algorithm is known to be able to diffuse a media stream when all the peers forward the stream at its bitrate (hence, since each chunk is $1s$ large the system can work if each peer outputs one chunk per second). Since in a real system there will be some duplicate chunks, the output bitrate requested to each peer is larger than 1; hence, in the next set of simulations the cycle size of each peer but the source will be set to $T = 1/1.2$.

In a next batch of simulations, PUSHSTREAM has been simulated in more realistic situations, assuming that a chunk needs a time randomly distributed in $[100ms, 300ms]$ to be transmitted, and the one-way-delay for the signalling messages is randomly distributed in $[50ms, 200ms]$. The results are reported in Figure 2, which again displays the average and the worst case diffusion times, compared to the theoretical values from [7]. By looking at the figure, it is possible to see that the algorithm performance are still near to the theoretical bounds (again, the average diffusion time is smaller than the theoretical one because of the lost packets, and because $T < 1$). In these simulations, the average percentage of chunks not received by a peer is always less than 6%.

**Overhead Measurements:** A potential problem with PUSHSTREAM is that the size of the input neighbourhood can grow too much, forcing a peer to send a lot of signalling messages each time that it receives a new chunk. Hence, the size of the input neighbourhood for the previous simulations has been measured, and it turned out that the maximum value is 20 (forced by the timeout used to remove peers from the input neighbourhood) and the average value is less than 16. Hence, the overhead introduced by the signalling protocol is not too high.

**Dynamic Overlay Management:** To evaluate the effect of the dynamic overlay construction, the simulations have been repeated disabling the periodic update of the output neighbourhood. As a result, it has been observed that the simulation results became heavily affected by the overlay topology: in some runs, the chunk diffusion delays and the fraction of lost chunks became very similar to the theoretical expectations, while in other runs the performance became poor. When, instead, NEWSCAST is used to dynamically update the output neighbourhood, the results become very consistent from run to run.

**Additional Results:** To understand the effect of the output bitrate on the protocol performance, the number of peers has been fixed to $N = 500$ (so, the neighbourhood size is 27 and the output neighbourhood size is 9), and the cycle size has been set to $T = 1.0/(1.0 + \rho)$, (where $\rho$ is called *surplus bandwidth*).

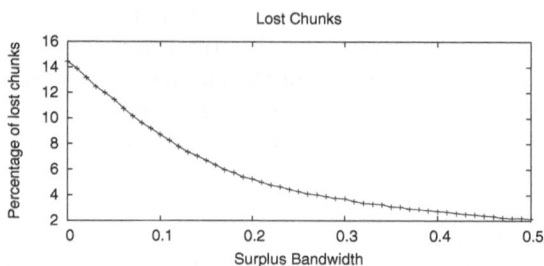

**Fig. 3.** Chunk loss for 500 peers as a function of $\rho$, where $T = 1.0/(1.0 + \rho)$

The results obtained increasing $\rho$ from 0 to 0.5 are shown in Figure 3: note that if peers can forward the chunks at a bitrate that is 130% of the stream bitrate, then the average amount of lost chunks is less than 4%.

Finally, some simulations have been run to verify how the system copes with peers that leave the system without any kind of notification. The experiments showed that the dynamic update of the input and output neighbourhoods allow to tolerate such "leaving peers" without affecting the performance too much: for example, if 1% of the peers leave the system every $500s$, in a system composed by 1000 peers each peer receives about 92% of the chunks (without churn, around 95% of the chunks were received).

## References

1. Hefeeda, M., Habib, A., Xu, D., Bhargava, B., Botev, B.: Collectcast: A peer-to-peer service for media streaming. ACM Multimedia 2003 11, 68–81 (2003)
2. Liu, Y.: On the minimum delay peer-to-peer video streaming: how realtime can it be? In: MULTIMEDIA 2007: Proceedings of the 15th international conference on Multimedia, Augsburg, Germany, September 2007, pp. 127–136. ACM Press, New York (2007)
3. Couto da Silva, A., Leonardi, E., Mellia, M., Meo, M.: A bandwidth-aware scheduling strategy for p2p-tv systems. In: Proceedings of the 8th International Conference on Peer-to-Peer Computing 2008 (P2P 2008), Aachen (September 2008)
4. Jelasity, M., Voulgaris, S., Guerraoui, R., Kermarrec, A.-M., Steen, M.v.: Gossip-based peer sampling. ACM Trans. Comput. Syst. 25(3), 8 (2007)
5. Massoulie, L., Twigg, A., Gkantsidis, C., Rodriguez, P.: Randomized decentralized broadcasting algorithms. In: 26th IEEE International Conference on Computer Communications (INFOCOM 2007) (May 2007)
6. Bonald, T., Massoulié, L., Mathieu, F., Perino, D., Twigg, A.: Epidemic live streaming: optimal performance trade-offs. In: Liu, Z., Misra, V., Shenoy, P.J. (eds.) SIGMETRICS, Annapolis, Maryland, USA, June 2008, pp. 325–336. ACM Press, New York (2008)
7. Abeni, L., Kiraly, C., Cigno, R.L.: On the optimal scheduling of streaming applications in unstructured meshes. In: Fratta, L., et al. (eds.) NETWORKING 2009. LNCS, vol. 5550, pp. 117–130. Springer, Heidelberg (2009)

# Network Heterogeneity and Cascading Failures – An Evaluation for the Case of BGP Vulnerability

Christian Doerr[1], Paul Smith[2], and David Hutchison[2]

[1] Department of Telecommunication, TU Delft, The Netherlands
`c.doerr@tudelft.nl`
[2] Computing Department, Lancaster University, UK
`{p.smith,dh}@comp.lancs.ac.uk`

**Abstract.** Large-scale outages of computer networks, particularly the Internet, can have a significant impact on their users and society in general. There have been a number of theoretical studies of complex network structures that suggest that heterogeneous networks, in terms of node connectivity and load, are more vulnerable to cascading failures than those which are more homogeneous. In this paper, we describe early research into an investigation of whether this thesis holds true for vulnerabilities in the Internet's inter-domain routing protocol – BGP – in light of different network structures. Specifically, we are investigating the effects of BGP routers creating blackholes – observed phenomena in the Internet in recent years. We describe our evaluation setup, which includes a bespoke topology generator that can fluidly create any topology configuration from the current scale-free AS-level to the investigated homogeneous graphs. We find that network homogeneity as suggested by theory does not protect the overall network from failures in practice, but instead may even be harmful to network operations.

## 1 Introduction

Recently, a multitude of studies have been conducted that investigate under which conditions complex network structures are most vulnerable. These studies conclude that scale-free networks are relatively robust against random failure and intentional attacks [1,2]. The failure or breakdown of a network component and its effects may not remain self-contained, but depending on the network characteristics and model, might spread across the infrastructure, resulting in cascading failures in a large part of the system [3]. Many of the studies investigating cascading failures suggest that high levels of heterogeneity, i.e., large differences in node connectivity and traffic share among the network nodes, will increase the vulnerability and worsen the outcome of a failure cascade, and that structural homogeneity will dampen the effects.

While collapses of the Internet (or parts thereof) due to traffic overload, as described in [3], have not been observed in reality, cascading failures and subsequent inoperability of some parts of the Internet have frequently been documented for the case of BGP, which is vulnerable to unintentional misconfiguration and could theoretically also be exploited in intentional attacks [4].

T. Spyropoulos and K.A. Hummel (Eds.): IWSOS 2009, LNCS 5918, pp. 207–212, 2009.

In this paper, we describe early research that aims to merge the theoretical findings on abstract scale-free networks with AS-level structural and BGP behavioral data, and investigate whether structural network heterogeneity is indeed a risk for cascading problems, and may be alleviated when the network would be designed more homogeneous.

## 2   BGP Vulnerabilities

We will focus on two types of BGP vulnerabilities that have been readily observed in the Internet: blackholing via false prefix announcements and interdomain routing instabilities via route flapping. Before we describe these vulnerabilities, we sketch the general functioning of the BGP protocol, and how it enables interdomain routing. For an in-depth discussion of BGP, we refer the reader to [5].

In simple terms, what is most commonly referred to as "the Internet" is a conglomeration of individual networks which have been *inter*linked into a larger network (see Fig. 1). Individual networks are referred to as Autonomous Systems (ASes), since they are maintained by a single administrative entity, such as an Internet Service Provider (ISP). Currently, approximately 27,000 ASes provide global connectivity.

Before this can happen, it is necessary to compute paths along which remote parts of the network may be reached. This is the task of the BGP protocol, through which each AS shares information describing its own connectivity with neighbors and announces currently available routes through its own domain to others. In the example of Fig. 1, AS1 may announce to its neighbors that it has direct connections (1 hop away) to AS 3, 4 and 5. As the other neighboring systems do the same, AS1 can further learn that the IP ranges of AS4 may also be reached through AS3, but through a longer path of length 2, and that AS3 will also provide connectivity to AS2, which in turn can forward data to AS6. While BGP can be configured to prefer specific routes and neighbors over others, it typically chooses the shortest and most specific[1] route to its destination. Routes can be advertised or withdrawn as the network topology changes.

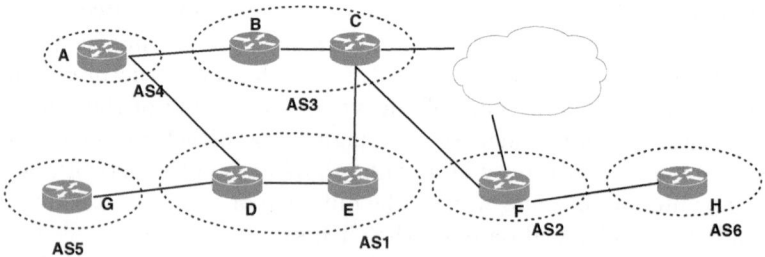

**Fig. 1.** An AS topology vulnerable to blackholing and route flapping

---

[1] The smallest and thereby most specific description of an IP block.

**Blackholing:** While the mechanism of BGP advertisements and withdrawals is designed to be reliable, problems can occur when routers announce wrong routing and reachability information due to accidental or intentional misconfiguration. Consider the situation where AS1 falsely advertises that it has a (new) direct route of length 1 to the IP range of AS 6. This announcement would let AS1's neighbors, AS5 and AS4, send any traffic destined for AS6 to routers D and E in AS1, as this new route of length 2 is shorter than the previously known path of length 4 and 3, respectively. In the area where AS1's announced path is superior to the correct advertisement of AS2, a blackhole is created that disconnects AS6 from this part of the network.

Such situations, where networks falsely announce IP prefixes that they do not own thus leading to blackholes, happen quite frequently in the Internet. In an analysis of historical data, Zhao et al. [6] showed that between 1998 and 2001 about 30 conflicting routes occurred daily. While most of these issues go unnoticed for the vast majority of the Internet, there exist several instances where false advertisements had a global impact on the network and disrupted connectivity on a planetary scale: in 1997, a false advertisement of AS7007 [7] resulted in that network being recognized as the best path to almost the entire Internet. Similar events on a smaller scale were repeated in 1998 and 2001. The most well known recent blackhole event may be the "YouTube incident", where an intentional BGP misconfiguration by Pakistan Telecom to block YouTube in Pakistan was leaked outside its network and disconnected the website from large parts of the Internet [8].

**Route flapping:** Even when each AS only propagates correct routing information, another type of routing problems can occur. Due to the long convergence times of BGP, routers can be configured to employ a mechanism called route dampening to protect the network from routing instabilities. When remote nodes repeatedly announce and withdraw routes in short succession, a large configuration overhead of finding and selecting new routes is introduced throughout the network. Routers will therefore temporarily remove paths from the routing tables when select routes begin to flap, resulting in these nodes becoming unreachable until the dampening expires.

While this mechanism is intended to improve the convergence of BGP and its routes when facing unstable paths, this behavior can be exploited [9]. If router G of AS54 in Fig. 1 would deliberately make repeated path announcements and withdrawals which are then propagated by its neighbor AS1, other upstream networks as such AS2, 3 and 4 may enable route dampening to protect the network, temporarily block these flapping routes, and thereby disconnect AS1 from the network.

Both these error types can create large-scale cascading effects as remote nodes react to the incoming route updates. As discussed earlier, the theoretical analysis of cascading failures in complex networks has resulted in the finding that cascades may be prevented, or at least limited in scope, as the network topology structure becomes more homogeneous. Next, we discuss how we propose to test this claim for the Internet as a concrete complex network.

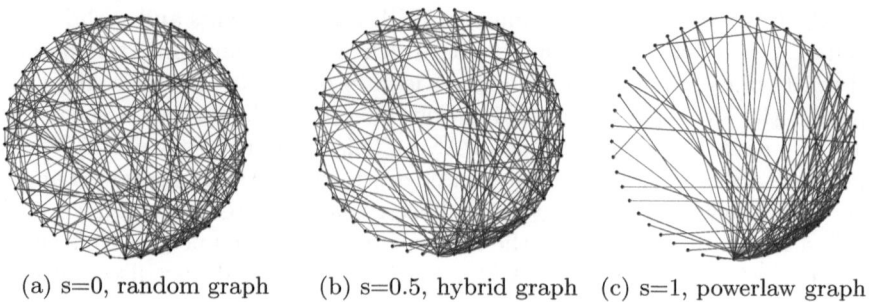

(a) s=0, random graph     (b) s=0.5, hybrid graph   (c) s=1, powerlaw graph

**Fig. 2.** Degree-sorted circular layouts visualize the characteristic degree distributions of random and scale-free graphs

## 3  Evaluation Setup

In order to evaluate whether number and magnitude of cascading failures within the BGP protocol could be reduced, if not avoided, by higher levels of topological homogeneity, we implemented a testing suite that could create, emulate and test a large number of networking scenarios. This testing suite was based upon SSF-Net [10], an open-source discrete event simulator for complex systems with nanosecond-resolution emulation capabilities of all the protocols necessary for this evaluation. This simulator was extended with custom-made modules to mimic the behavior of a malicious agent conducting attacks against random parts of the network.

The simulator was additionally connected to a custom-made topology generator which was used to create customized Internet topologies. Since the goal of our study was to compare current Internet-like topologies with potential alternative, more homogeneous AS topologies, the topology generator was developed in such a way that it could fluidly create any type and mixture of network scenario between these two opposite poles, a scale-free AS-level or a purely random graph. While the procedures of creating good random Erdös-Renyi graphs has been sufficiently explored, there still exists much debate over the best algorithms to create a network topology that can mimic the topology of the Internet. For our evaluation, we implemented a scale-free topology generator from the PLRG class, which has been found to generate network structures with characteristics closely matching those of the Internet at the AS-level [11]. Since the underlying generation procedure of the models is identical, it is possible to overlay and combine the two models into a single generator, where a single parameter (s ∈ [0, 1]) then specifies the level of dominance of one model over the other, where s=0 indicates the full dominance of the random graph and s=1 the full dominance of the scale-free topology. Fig. 2 shows three example topologies with 50 nodes and 200 links for the three parameter values s=0, 0.5, 1, arranged in a degree-sorted circular layout. It is easy to see the powerlaw distribution of node degrees in the case of s=1 and the complete absence of patterns (i.e., randomness) for s=0. The hybrid graph with s=0.5 represents a mixture between the two models, where

50% of the links are created according to a scale-free topology, while the other 50% are added randomly.

## 4   Results

Due to space constraints, this section describes only a selection of results obtained from the simulations, and focuses specifically on the blackholing attack.

As discussed before, the main objective of this work was to evaluate the hypothesis obtained from theoretical modeling that network homogeneity will hinder the spread of failures. To test this prediction, we generated topologies with varying sizes n=100..5000, where the nodes were interconnected according to the scheme discussed in Sect. 3. For each network size, 7 classes of topologies were created, defined by the s-value s=0, 0.1, 0.25, 0.5, 0.75, 0.9, 1. In the SSF-Net simulation environment, each node of the graph was configured to function as an AS, each including a dedicated BGP speaker. Each simulation initialized from zero state and the BGP protocol was given 2,75 hours to reach steady state. After this initial phase, a single node was assumed to become compromised and initiate a blackhole in the network. While in this part of the simulation, the malicious agent was randomly chosen (runs were repeated 100 times), it chose an IP block for the blackhole in such way as it would (in its opinion) maximize the overall impact to the network. After the false route was advertised, we evaluated the state of the BGP tables after an additional 2,75 hour converge time.

In our analysis, we could not find any evidence for the theoretical prediction outlined in previous work. Network homogeneity does not help to reduce the impact of failures, but in fact does hurt the network in two ways: (1) As the topology structure becomes more homogeneous, the importance of the nodes to the network becomes also more homogeneous. This can be shown by the drift between the actual and expected node betweenness metric, which measures the layout of shortest paths, paths that BGP will take, in the network. In other words, as on the way from the current Internet scale-free structure, the well-connected hubs are consecutively replaced by more scale-like structures, the impact of a failure of any random node on the network grows. (2) As the hubs from the scale-free topology are exchanged, the advantage and amount of BGP routing table aggregation also disappears. A more homogeneous structure will correspondingly demand longer BGP forwarding tables across all and especially the backbone routers, which increases the demands on Internet hardware and overall forwarding delays. There exists severe concern about the recent growth of the Internet and trends such as multi-homing, load-balancing and address fragmentation, which increase the BGP routing tables sizes in the Internet by a factor of 10 [12]. Our analysis shows that moving from a pure scale-free topology to a topology in which only 25% of the network follows the homogeneous layout would increase the BGP table on average by an additional 20%. An entirely homogeneous network (s=0) results in a routing table close to the number of total ASes. It may be argued that while a scale-free topology limits the impact that any random node could have on the network, the effect of a well-connected hub node will be dramatically larger. To evaluate this, we conducted a set of experiments where the compromised node was chosen according to its expected impact as measured by a high node betweenness. Our results confirm this assumption and demonstrate wide-spread outages across the network topology, but also indicate that the overall impact of the worst-case attack in a scale-free topology is only 50% worse than the worst-case in a purely homogeneous topology. The overall number of nodes that could successfully initiate such a blackholing attack is, however, very limited, and it may be assumed that these core routers would in practice have special safeguards in place to prevent such corruption.

## Acknowledgements

We would like to thank the anonymous reviewers for the valuable comments. The work presented in this paper is supported by the European Commission, under Grant No. FP7-224619 (the ResumeNet project).

## References

1. Callaway, D.S., Newman, M.E.J., Strogatz, S.H., Watts, D.J.: Network Robustness and Fragility: Percolation on Random Graphs. Physical Review Letters 85, 5468–5471 (2000)
2. Cohen, R., Erez, K., ben-Avraham, D., Havlin, S.: Breakdown of the Internet under intentional attack. Physical Review Letters 86 (2001)
3. Motter, A.E., Lai, Y.-C.: Cascade-based attacks on complex networks. Physical Review E 66 (2002)
4. Nordström, O., Dovrolis, C.: Beware of BGP attacks. SIGCOMM Comput. Commun. Rev. 34(2), 1–8 (2004)
5. Rekhter, Y., Li, T., Hares, S.: RFC4271 - A Border Gateway Protocol 4 (BGP-4) (2006)http://tools.ietf.org/html/rfc4271
6. Zhao, X., Pei, D., Wang, L., Massey, D., Mankin, A., Wu, S.F., Zhang, L.: An Analysis of BGP Multiple Origin AS (MOAS) Conflicts. In: Proceedings of the 1st ACM SIGCOMM Workshop on Internet Measurement (2001)
7. Bono, V.J.: 7007 Explanation and Apology,
   http://www.merit.edu/mail.archives/nanog/1997-04/msg00444.html(1997)
8. YouTube Hijacking: A RIPE NCC RIS case study (2008),
   http://www.ripe.net/news/study-youtube-hijacking.html
9. Mao, Z.M., Govindan, R., Varghese, G., Katz, R.H.: Route Flap Damping Exacerbates Internet Routing Convergence. In: Proceedings of the ACM SIGCOMM Conference (2002)
10. (SSF-NET ), http://www.ssfnet.org/
11. Hernandez, J.M., Kleiberg, T., Wang, H., Mieghem, P.V.: A Qualitative Comparison of Power Law Generators. In: SPECTS (2007)
12. Bu, T., Gao, L., Towsley, D.: On characterizing BGP routing table growth. Computer Networks 45(1), 45–54 (2004)

# A Self-organizing Approach to Activity Recognition with Wireless Sensors*

Clemens Holzmann and Michael Haslgrübler

Institute for Pervasive Computing
Johannes Kepler University Linz, Austria

**Abstract.** In this paper, we describe an approach to activity recognition, which is based on a self-organizing, ad hoc network of body-worn sensors. It makes best use of the available sensors, and autonomously adapts to dynamically varying sensor setups in terms of changing sensor availabilities, characteristics and on-body locations. For a widespread use of activity recognition systems, such an opportunistic approach is better suited than a fixed and application-specific deployment of sensor systems, as it unburdens the user from placing specific sensors at predefined locations on his body. The main contribution of this paper is the presentation of an interaction model for the self-organization of sensor nodes, which enables a cooperative recognition of activities according to the demands of a user's mobile device. We implemented it with an embedded system platform, and conducted an evaluation showing the feasibility and performance of our approach.

## 1 Introduction

The recognition of user activities is an important aspect in context-aware systems and environments, as it enables their adaptation to the user's current situation and hence allows for providing services with reduced human intervention. The recent availability of *body sensor networks* made it possible to recognize activities with wireless sensors which are mounted on different body parts, like for example embedded in wrist bands, belts and clothes, and are able to communicate to each other as well as to a mobile device [1]. A common approach to activity recognition is the use of accelerometers and the classification of acceleration data into a set of output classes using supervised machine learning techniques [2]. However, usually a precise, application-specific deployment of sensors is used, which does not take into account different numbers, displacements and failures of sensors, and therefore limits the widespread use of such context-aware systems. For example, a person which is running in the woods and equipped with a body sensor network for monitoring his activities, may lose sensors, carry more or less sensors, or their on-body locations may vary due to his movement. In the recently started European research project OPPORTUNITY [3], an alternative

---

* This work is supported by the FP7 ICT Future Enabling Technologies programme of the European Commission under grant agreement No 225938 (OPPORTUNITY).

T. Spyropoulos and K.A. Hummel (Eds.): IWSOS 2009, LNCS 5918, pp. 213–219, 2009.

approach is proposed, which is based on an *opportunistic recognition of activities* with sensors that are currently available.

A key issue in this regard is the *self-organization* of the wireless sensors [4] into goal-oriented, cooperative sensing ensembles, in order to recognize the activities which are relevant for a user's mobile device – or, in general, for devices in his environment – from a dynamically varying and a priori unknown sensor configuration. This means, that from a set of available sensors, just those which are capable of providing the relevant information cooperate to achieve the goal. In [3], goal-oriented sensing is achieved based on (i) the formulation of a *recognition goal* which represents the activities to be recognized and (ii) its transformation into a coordinated *sensing mission* which is communicated to the sensor network. The scope of the present paper is the *interaction between body-worn sensors*, which is necessary for their ad-hoc formation according to a received sensing mission. First, in Section 2, we describe our general approach for the opportunistic recognition of activities with a body sensor network. In Section 3, the *self-organization* of sensors to achieve a common recognition goal as well as their *self-adaptation* to changing sensor availabilities, characteristics and on-body locations are explained. Finally, Section 4 presents first results of a performance evaluation using a state-of-the-art hardware platform.

## 2    An Approach for Opportunistic Activity Recognition

Our approach is explained best with an initial application example. Consider a person with a mobile phone in his pocket, and whose clothes are equipped with wireless acceleration sensors; a respective image with our prototype system is shown in Fig. 1(a). These sensors can be used for recognizing his locomotion activity, in order to (i) change the state of the mobile phone accordingly (e.g. to accept or reject phone calls depending on the user's activity) and (ii) notify the caller about the current activity if he is in the user's contact list for example. To become aware of certain activities, the mobile phone *broadcasts* a respective request, which causes the body sensor network to self-organize and provide the needed activity information to the phone. In particular, the phone first formulates a *recognition goal* that basically represents a class of required activities (e.g. locomotion activities such as sitting, standing, walking or running), and automatically translates it into a *sensing mission* describing how the sensors have to cooperate to provide the requested activity. This translation process is conducted by the system and beyond the scope of the present paper.

We define the sensing mission with a *tree data structure* containing those parts of the human body from which sensor data are required to achieve the recognition goal (see Fig. 1(b)). It represents a pre-defined *containment hierarchy* that is given by the human anatomy, with the human as root node and his body parts as child nodes, and by defining which functionality is needed at which node of the tree. As shown in the sensing mission of Fig. 1(c) for example, the sensor

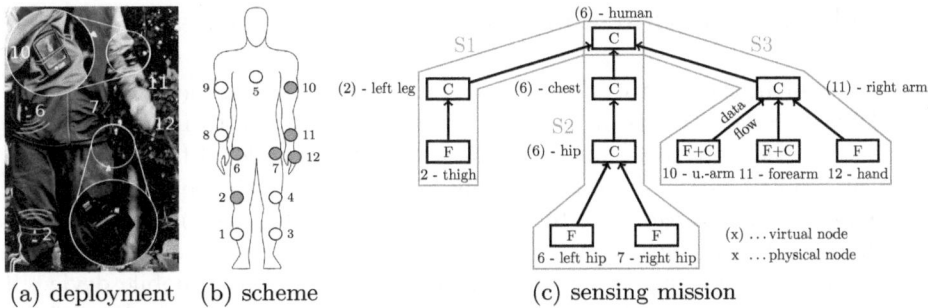

(a) deployment     (b) scheme               (c) sensing mission

**Fig. 1.** Exemplary sensor deployment and sensing mission

mounted on the left thigh has to provide certain features (F), and his parent sensor – which represents the containing left leg – has to do a classification (C) of these features. Which feature extractions and classifications are required at which level of the hierarchy, and how the provided results are linked together, is specified in the tree nodes of the sensing mission. A related approach is described in [5], where so-called "processing layers" are used to structure the information flow in a similar way. We assume that every sensor platform has equal capabilities concerning the extraction of features and their classification, and that a sensor (physical node) can undertake the tasks of multiple tree nodes which do not have a physical representation (virtual nodes); more about physical and virtual nodes is explained in the next section. Note that different configurations of sensors may be able to fulfill a given sensing mission; for example, if acceleration information from the left leg is needed, both a wireless sensor located on the left thigh and on the left shank may be applicable.

Furthermore, it is possible to orchestrate simple activities into complex activities by classifying features and the results of lower-level classifiers to high-level activities. This has also the advantage that the amount of information which has to be delivered bottom-up in the tree decreases drastically, which preserves energy due to the reduced wireless communication, and makes such a system feasible for accurate activity recognition (cf. [6]). Since different activity classes (e.g. locomotion and household activities) may require the same low-level activities, which causes overlapping branches in the trees, they can be reused in the fulfillment of *multiple sensing missions* without requiring additional computations. To the best of our knowledge, this *distributed* recognition of activities described above is a novelty of our approach.

For each sensing mission that is broadcast, the receiving body sensors negotiate their participation in the *cooperative* recognition of the respective activities, including especially the extraction of features from the sensor data as well as their classification to activities; a detailed explanation is given in the next section. We have developed a software framework for this approach, which consist of different services. Among others, it comprises a *communication service* for the abstraction of the underlying communication protocol, as well as an *interaction services* for achieving the self-organizing behavior (see Fig. 2).

# 3   Self-organization and -Adaptation of Wireless Sensors

Every sensor in the network has to know its own body position, which can be pre-defined or detected as proposed in [7], and is thus able to infer a *local tree* representing its body position from this knowledge. The local tree is based on the containment hierarchy of the sensing mission tree, with the same root and the sensor itself as a *leaf node* (e.g. S1 for sensor 2 in Fig. 1(c)). This makes it possible to compare those trees, which is fundamental for our self-organization approach. When a sensor receives a sensing mission from the mobile device, it participates by providing features that are required for the classification, if and only if its local tree covers a leaf node of the broadcast sensing mission. As every participating sensor is a leaf node in the sensing mission, the leaves are called *physical nodes*; all the ancestor nodes are called *virtual nodes*. The virtual nodes, which contain classifications of features belonging to the corresponding body parts, are also executed by one of their physical child nodes. In the tree S3 of Fig. 1(c) for example, the nodes 10 to 12 extract features from their sensors, and some also perform classification functions. The data of all three nodes are input for the classifier located on the virtual node *right arm*. In this example, sensor 11 additionally undertakes the tasks of this virtual node; this means, that the nodes 10 and 12 send the respective data to the physical node 11, which in turn delivers it – together with its own data – to the virtual node *right arm*.

In order to *self-organize*, each sensor which receives a sensing mission compares it with its local tree, and determines the number of nodes in which they differ. Based on this comparison, we distinguish four different ways to proceed, which are also visualized in the sequence diagrams of Fig. 2:

- *No overlap of the local tree with a leaf node of the sensing mission*: The sensor does not have valuable information for the respective sensing mission and simply ignores the request (e.g. sensor 1 in Fig. 1(c)).
- *Partial overlap of the trees (difference > 1)*: The sensor is part of the solution of the sensing mission, and will therefore try to get elected as a *master* to provide the results to the sensor which has broadcast the sensing mission. For the election, each sensor broadcasts the difference number (e.g. difference 7 for the sensors 6 and 7 in Fig. 1(c), as they are both able to provide the functionalities of the nodes *hip*, *chest* and *human*) together with a random number. A sensor elects itself as a master if it has the smallest difference number, or – in the case of equal differences – the highest random number. Instead of the difference and random number, other metrics such as the battery status of the sensor could also be taken into account, but have not been considered yet. The elected master broadcasts new sensing missions for those parts of the tree which it cannot fulfill (e.g. S1, S2 without *left hip*, and S3 for the master sensor 6 in Fig. 1(c)), which are again compared by receiving sensors with their local trees. These sub-requests may lead to the election of sub-masters, like for example sensor 11 which serves as a master for S3. The master responds with the MAC addresses of the sensors assigned to the respective nodes of the received sensing mission tree.

- *Full overlap of the trees (difference = 0)*: The sensor can satisfy the whole sensing mission by itself (e.g. sensor 2 in Fig. 1(c) would satisfy a sensing mission consisting of S1 only), and thus responds with the tree of the sensing mission in which its own MAC address is assigned to all nodes of the tree.
- *Almost full overlap (difference = 1)*: The sensor needs the information of just *one* other node, and directly requests that information by formulating a sensing mission for the missing partial tree by itself (e.g. sensor 6 in Fig. 1(c) would satisfy a sensing mission consisting of S2 only, and request the missing part from sensor 7). Upon receiving a response to this sub-request, the sensor responds with a tree containing its own MAC address and that of the other sensor. This special case has been introduced due to the fact that a master election process would be inefficient for just two nodes.

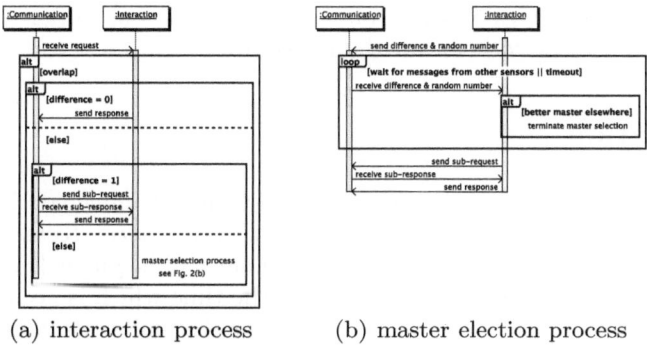

(a) interaction process          (b) master election process

**Fig. 2.** Interaction between sensors depending on their difference to the sensing mission

If multiple sensors respond to the same superior master or to the mobile phone, just the first answer will be used. After the self-organization is complete, the root master reports to the mobile device that the sensing mission can be satisfied. The mobile device acknowledges this by notifying the master that it can start, which in turn relays this notification to its children to start with their feature extraction and classification tasks. As the sensing mission contains the addresses of all participating sensors, the communication between them is done directly by establishing *reliable point-to-point connections*. With regard to related work on the self-organization of body sensor networks (e.g. [8,9]), our approach is novel insofar as it (i) hierarchically organizes the sensors according to the anatomy of the human body, whereby the sensors belonging to a certain body part cooperate in the classification of the respective sub-activities and therewith reduce the overall communication load, and (ii) additionally imposes a data-flow that is optimized for changing sensor configurations and activities.

The tree structure is not only used to self-organize the body sensor network for information processing, it is also essential for its *self-adaptation*. If a node fails, the network will try to recover, whereas three scenarios are possible depending on whether (i) a leaf node, (ii) a sub-master or (iii) the root master is concerned.

For the cases (i) and (ii), the corresponding parent node will detect a time-out, whereupon it has several handling strategies. First, it will try to notify the failed nodes again and continue with sending the data requested by the sensing mission (by reusing previously obtained data). If still no data is received, e.g. due to a loss or failure of the sensor, the corresponding parent will formulate a new sensing mission for the missing part of the tree, and the network will organize itself again (e.g. by switching to a backup sensor with the same capabilities than the lost one); for example, if sensor 2 in Fig. 1(c) fails, sensor 6 (i.e. the root master) would request the tree S1 again. If this also fails, the root node is informed that the activity recognition cannot continue, whereupon it notifies all child nodes participating in the sensing mission to (i) stop working and (ii) also notify their children accordingly. If the children of a failed node receive a transmission error, they also stop working and notify their sub-nodes accordingly. For the case (iii), the application that requested the sensing mission in the first place can also notify the nodes or send out a sensing mission again, or it can request an *alternative sensing mission* for the recognition goal.

## 4   Evaluation

We have conducted a first evaluation of our approach by implementing the framework for the Sun Microsystems *Small Programmable Object Technology (SPOT)* platform [10] and the Openmoko *Neo FreeRunner* mobile phone [11] as the user's mobile device (cf. Fig. 1(a)), and measuring the time which is needed for (i) the self-organization of the sensors for different sensing missions consisting of the trees S1, S2 and S3 of Fig. 1(c) as well as compositions of them, and (ii) for the self-adaptation to failures of a leaf node, a sub-master or the root master, according to the strategies re-notification (strategy 1) and sensing mission re-broadcast (strategy 2) at a time; the results are shown in Fig. 3. For the self-adaptation, the complete sensing mission shown in the Fig. 1(c) has been used. It should be noted that there is a timeout of 300ms if the master does not receive the difference and random numbers of the other sensors as well as a retry after 1500ms if it does not receive a response to a broadcast sub-request (cf. Fig. 2(b)), which

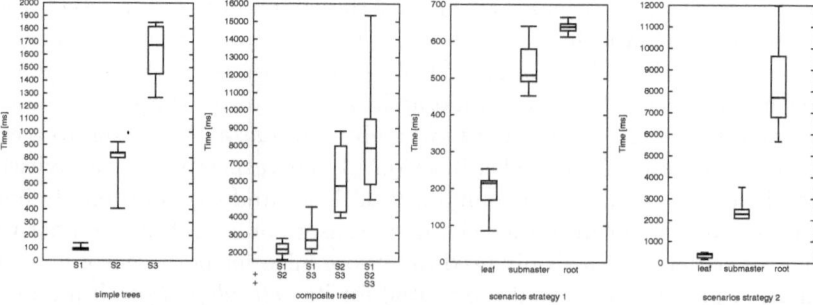

**Fig. 3.** Time required for the self-organization and -adaptation of the network

may lead – due to the unreliability of broadcasts that causes messages to be lost – to high delays for sensing missions with many nodes.

## 5   Conclusions

The presented work is a first step towards building a distributed activity recognition system with opportunistic sensor configurations. We discussed a novel interaction model for the self-organization of a body sensor network according to a given recognition goal and its self-adaptation to changes in the sensor network, and implemented it as a software framework for a wireless sensor platform. The evaluation showed the feasibility of our approach and the time it requires for the self-organization and -adaptation in different real-world scenarios. As for future work, we will first implement feature extraction methods and classification algorithms for recognizing locomotion activities, and evaluate them with respect to their accuracy and the required processing and memory resources; a particular focus will be on the distributed recognition with multiple hierarchically linked classifiers. Another issue of future work is the extension and evaluation of the presented interaction model for its use with multi-hop networks, which would allow for extending it to wireless sensors that are spread in the environment.

## References

1. Farella, E., Pieracci, A., Benini, L., Rocchi, L., Acquaviva, A.: Interfacing human and computer with wireless body area sensor networks: the WiMoCA solution. Multimedia Tools and Applications 38(3) (2008)
2. Ravi, N., Dandekar, N., Mysore, P., Littman, M.L.: Activity recognition from accelerometer data. In: Proc. of IAAI 2005 (2005)
3. Roggen, D., Förster, K., Calatroni, A., Holleczek, T., Fang, Y., Tröster, G., Lukowicz, P., Pirkl, G., Bannach, D., Kunze, K., Ferscha, A., Holzmann, C., Riener, A., Chavarriaga, R., del Millán, R.J.: OPPORTUNITY: Towards opportunistic activity and context recognition systems. In: Proc. of WoWMoM Workshop AOC 2009 (2009)
4. Mills, K.L.: A brief survey of self-organization in wireless sensor networks. Wireless Communications and Mobile Computing 7(7) (2007)
5. Harms, H., Amft, O., Tröster, G., Roggen, D.: Smash: A distributed sensing and processing garment for the classification of upper body postures. In: Proc. of BodyNets 2008 (2008)
6. Lombriser, C., Bharatula, N.B., Roggen, D., Tröster, G.: On-body activity recognition in a dynamic sensor network. In: Proc. of BodyNets 2007 (2007)
7. Kunze, K.S., Lukowicz, P., Junker, H., Tröster, G.: Where am I: Recognizing on-body positions of wearable sensors. In: Strang, T., Linnhoff-Popien, C. (eds.) LoCA 2005. LNCS, vol. 3479, pp. 264–275. Springer, Heidelberg (2005)
8. Watteyne, T., Augé-Blum, I., Dohler, M., Barthel, D.: Anybody: a self-organization protocol for body area networks. In: Proc. of BodyNets 2007 (2007)
9. Osmani, V., Balasubramaniam, S., Botvich, D.: Self-organising object networks using context zones for distributed activity recognition. In: Proc. of BodyNets 2007 (2007)
10. Sun Microsystems Inc.: Sun SPOT World (2009), http://www.sunspotworld.com
11. Openmoko Inc.: Neo FreeRunner (2009), http://www.openmoko.com

# Congestion Control in Wireless Sensor Networks Based on the Bird Flocking Behavior⋆

Pavlos Antoniou[1], Andreas Pitsillides[1],
Andries Engelbrecht[2], Tim Blackwell[3], and Loizos Michael[1]

[1] Department of Computer Science, University of Cyprus, Cyprus
[2] Department of Computer Science, University of Pretoria, South Africa
[3] Department of Computing, Goldsmiths College, University of London, UK

**Abstract.** Recently, performance controlled wireless sensor networks have attracted significant interest with the emergence of mission-critical applications (e.g. health monitoring). Performance control can be carried out by robust congestion control approaches that aim to keep the network operational under varying network conditions. In this study, swarm intelligence is successfully employed to combat congestion by mimicking the collective behavior of bird flocks, having the emerging global behavior of minimum congestion and routing of information flow to the sink, achieved collectively without explicitly programming them into individual nodes. This approach is simple to implement at the individual node, while its emergent collective behavior contributes to the common objectives. Performance evaluations reveal the energy efficiency of the proposed flock-based congestion control (Flock-CC) approach. Also, recent studies showed that Flock-CC is robust and self-adaptable, involving minimal information exchange and computational burden.

**Keywords:** Wireless Sensor Networks (WSNs),Congestion Control (CC).

## 1 Introduction

Typically, WSNs consist of small, cooperative devices (nodes) which may be constrained by computation capability, memory space, communication bandwidth and energy supply. The traffic load injected into the network can be unpredictable (e.g. due to event driven applications) and, in conjunction with variable wireless network conditions, may exceed available capacity at any point of the network resulting in congestion phenomena. Congestion causes energy waste, throughput reduction, increase in collisions and retransmissions at the MAC layer, increase of queueing delays and even information loss, leading to the deterioration of the offered quality and to the decrease of network lifetime.

The focal point of this study is to design a *robust and self-adaptable congestion control (CC) mechanism* for WSNs. Inspiration is drawn from Swarm

---

⋆ This work is supported in part by the GINSENG project funded by the FP7 under Grant No. ICT-224282 and the MiND2C project funded by the Research Promotion Foundation of Cyprus under Grant No. TPE/EPIKOI/0308(BE)/03.

T. Spyropoulos and K.A. Hummel (Eds.): IWSOS 2009, LNCS 5918, pp. 220–225, 2009.

Intelligence (SI) [1], [2] which has been very successful in solving similar types of complex problems [3]. However, the majority of SI-based network approaches like [3] target ad hoc networks which have fundamental differences from WSNs [4]. Our approach mimics the *flocking behavior of birds*, where packets are modeled as birds flying over a topological space, e.g. a sensor network. The main idea is to 'guide' packets to create flocks and 'fly' towards a global attractor (sink), whilst trying to avoid obstacles (congested regions). The direction of motion is influenced by (a) repulsion and attraction forces exercised by neighboring packets, as well as (b) the gravitational force in the direction of the sink. The flock-based congestion control (Flock-CC) approach provides **sink direction discovery**, **congestion detection** and **traffic management**. Flock-CC was initially proposed in [5] and with slight modifications in [6]. Both studies showed that the Flock-CC approach achieves low packet loss resulting in high packet delivery ratio (PDR) and thus reliability, low latency, and fault tolerance. Also it was found to outperform congestion-aware multi-path routing approaches in terms of PDR. This paper shows that Flock-CC also achieves low energy consumption.

The rest of this paper is organized as follows. Section 2 deals with the proposed approach. Section 3 presents performance evaluation results. Section 4 draws conclusions and proposes areas of future work.

## 2  The Flock-Based Congestion Control (Flock-CC)

The main concept of the Flock-CC approach [5], [6] is to *'guide' packets to form groups or flocks, and flow towards a global attractor (sink), whilst trying to avoid obstacles (congestion regions).* In order to make moving packets behave like a flock, each packet interacts with neighboring packets on the basis of attraction and repulsion forces, and experiences a 'gravitational' force in the direction of the sink. These forces are synthesized in the decision making process of a packet, when a new hosting node is about to be chosen. In particular, *packets are (a) attracted to neighboring packets located on nodes experiencing low queue loading, (b) repelled from neighboring packets located on nodes experiencing high wireless channel contention, (c) attracted to the sink by using biased preference to nodes located closer to the sink, whenever a new hosting node is about to be chosen, and (d) experience some perturbation that may help them to pick a random route.*

**Congestion Detection and Sink Direction Discovery**
In wireless environments, packets may be dropped due to queue overflows and link layer failures (e.g. collisions) on each node. Based on these congestion symptoms and the queue state, a *node loading indicator* $p_n(k)$[1] at node $n$ is devised:

$$p_n(k) = \begin{cases} \frac{P_n^{in}(k) + q_n(k-1) - P_n^{out}(k)}{P_n^{in}(k) + q_n(k-1)}, & \text{if } P_n^{in,out}(k), q_n(k-1) \neq 0; \\ 0 & \text{otherwise.} \end{cases} \quad , 0 \leq p_n \leq 1 \quad (1)$$

---

[1] All quantities defined herein, are regularly sampled at discrete time intervals of $T$ seconds, are evaluated at each sensor node, and are broadcasted periodically (every $T$ seconds) to all neighboring nodes, using a dedicated control packet.

where $P_n^{in}(k)$ is the number of incoming packets, $P_n^{out}(k)$ is the number of successful outgoing packets at the end of the $k$-th period, and $q_n(k-1)$ is the queue size at the end of the $k-1$-th period at node $n$. More detailed information is given in [6]. When $p_n(k) \to 0$, both the number of packet drops at node $n$ is close to 0 and the queue is empty or nearly empty. On the other hand, as $p_n(k) \to 1$, node $n$ is considered congested due to either high number of packet drops, or high queue occupancy. Each node estimates the quality of the shared wireless channel (useful link service rate, $r_n(k)$), using information taken from the MAC protocol. This study considers CSMA-like MAC protocols (e.g. IEEE 802.11). The total number of all packet transmission attempts at node $n$ (during sampling period $k$) is denoted by $P_n^{out*}(k)$, where $P_n^{out*}(k) = P_n^{out}(k)$+retransmits within that period. The useful *link service rate* at node $n$ is denoted by:

$$r_n(k) = \begin{cases} 1 & \text{if retransmits= 0 and } P_n^{out}(k) = 0; \\ \frac{P_n^{out}(k)}{P_n^{out*}(k)} & \text{otherwise.} \end{cases} \quad , \quad 0 \le r_n(k) \le 1. \quad (2)$$

When $r_n(k) \to 1$, the channel is not congested and a large percentage of packets are successfully transmitted (few packet retransmissions are observed). As $r_n(k) \to 0$, the channel is congested and a small number of packets are successfully transmitted, often after a large number of retransmissions.

The direction of the sink can be deduced by the hop-distance variable, $h_n(k)$, indicating the number of hops between node $n$ and the sink at the $k$-th sampling period. Nodes located closer to the sink are expected to have smaller hop-distance values and should be chosen with higher probability as next hop hosting nodes.

**Traffic Management and Desirability**
Traffic management is performed on per packet basis in a hop-by-hop manner. Whenever a packet is about to be sent, the decision making process is invoked to determine the next hop node. Consider a network of $N$ nodes that are able to generate packets. Each hosting node evaluates the next hop node for each of its packets based on an $M$-dimensional desirability vector, $\vec{D}(k)$, where $M \le N$, is the number of nodes located within the hosting node's transmission range. Each element, $D_m(k)$, of the vector $\vec{D}(k)$ represents the desirability for each node $m, m \in \{1, .., M\}$. The desirability $D_m(k)$ for every node $m$ is evaluated once in each sampling period $k$ and is used for each packet sent within this period using:

$$D_m(k) = \alpha \cdot r_m(k) + (1 - \alpha) \cdot (1 - p_m(k)), \quad (3)$$

where the parameter $\alpha$, $0 \le \alpha \le 1$, regulates the influence of parameters $r_m(k)$ and $p_m(k)$. In order to address *global attractiveness to the sink*, we allow packets to be forwarded even to nodes that are not closer to the sink, but we place some bias against such a choice, by discounting the desirability of such nodes over the nodes that are closer to the sink, using the discount factor $d_{im}(t)$. In addition, we introduce *randomness* (that allows exploration, and perhaps identification of better solutions) in our model by introducing some noise in the desirability function. This perturbation is achieved by multiplying the desirability of a node by some coefficient drawn randomly from a Gaussian distribution with mean 1 and variance $v$. Let $g$ be a random variable that follows this probability distribution. Thus, we define the *adjusted* desirability of packet $i$ for node $m$ as:

$$D'_{im}(t) = g \cdot d_{im}(t) \cdot D_m(k). \tag{4}$$

Given the modified desirability function, our algorithm is as follows: **For each packet $i$, choose node with the maximum adjusted desirability $m^*$, where:** $m^* = argmax_m\{\overrightarrow{D}'_i(t)\}$. We consider nodes with only an equal or one less hop count to the sink than the hop count of the packet's current location. We set $d_{im}(t)$ equal to $f = 1$ for all nodes $m$ that are closer to the sink, and equal to some constant $e : 0 \leq e \leq 1$ for all nodes $m$ at equal distance from the sink. We also allow nodes of the latter category to be selected if the noise perturbation is sufficiently large to cover the bias against these nodes that is introduced by multiplying their desirability by $e$. The probability with which this bias $f - e$ is covered depends on the standard deviation $\sqrt{v}$ of the Gaussian distribution. It, then, makes sense to define $v$ not entirely independently of $f - e$, but as a linear function of it. We, thus, let $v = c \cdot (f - e)^2$ (and thus $\sqrt{v}$ is linear in $f - e$). Beyond the Gaussian approach the response threshold approach [8] was used in [5] for randomizing the attractiveness to the sink, achieving promising results.

The parameters $\alpha$, $e$, and $c$ are the most important control knobs of our system and are expected to influence its forwarding behavior. The parameter $\alpha$ regulates the influential roles of the node loading indicator $p_m(k)$, and the wireless channel loading $r_m(k)$ on the desirability function of Eq. 4. The parameter $e$ determines the bias against nodes at equal hop-count from the sink. The parameter $c$ determines, indirectly, the variation of the introduced noise.

## 3   Performance Evaluation

This section evaluates the performance of the Flock-CC approach through simulation studies conducted using the ns-2 network simulator. In accordance with [6], the optimal combination of the design parameters values achieving high packet delivery ratio (PDR) and low end-to-end delay (EED) was $a = 0.4, e = 0.6$ and $c = 0.5$. In this paper, the *energy tax metric* is defined to evaluate the performance of the proposed approach. The energy tax calculates the average energy consumption per node per delivered packet. The evaluation topology consists of 200 homogeneous nodes deployed in a 2D lattice. The evaluation scenario involved the activation of 10 nodes located 7 hops away from the sink. Each active node was generating constant bit rate traffic at the rates of $25, 35, 45$pkts/sec. These cases can be considered as slightly congested, congested, and heavily congested, respectively. The sampling period $T$ was set to 1 sec.

Fig. 1 illustrates the impact of $\alpha, e$ and $c$ on energy tax at 35pkts/sec. As can be seen, the lowest energy tax (0.026mJ/delivered packet) was obtained for $\alpha = 0.5, e = 0.6$ and $c = 0$ (dashed line of Fig. 1). Similarly, low energy tax levels were achieved for $0.2 \leq \alpha \leq 0.6$, $0 \leq e \leq 0.7$ and $0 \leq c \leq 0.25$. As stated before, parameter $e$ determines the bias against nodes at equal hop-count from

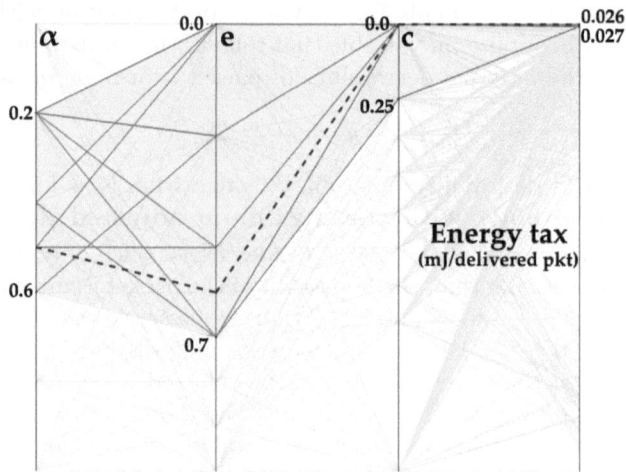

**Fig. 1.** Energy tax at 35pkts/sec

the sink. When $e = 0$ packets can only move towards nodes located closer to sink (high flock coherence). On the other hand, as $e$ approaches to 1, packets are allowed to move towards nodes with equal hop-count with high probability (high traffic spreading). Perturbation is regulated by the parameter $c$. When $e$ and $c$ were close to 0 (i.e. high packet flock coherence and low perturbation), packets were moving on narrow paths to the sink resulting in low energy consumption per node. On the other hand, high packet spreading was observed for $e \to 1$ or $c \to 1$ causing transition loops of packets between uncongested nodes and, as a result, high energy consumption. Also, Fig. 1 shows that high energy consumption was observed for $\alpha > 0.6$. The increase of $\alpha$, amplified the tendency of packets to move towards nodes experiencing low channel utilization (i.e. at the borders of the flock) in order to avoid collision hot spots occurring inside the network. However, packets were caught in loops between nodes experiencing low channel utilization, resulting in high energy consumption.

We implemented (a) a shortest-path congestion-unaware routing mechanism, and (b) a congestion-aware multi-path routing mechanism, according to which each node chooses the least congested next hop node in terms of queue length (among nodes involved in the shortest paths -if many- to the sink) to forward each packet. The Flock-CC approach ($\alpha$=0.4, $e$=0.6, $c$=0.5) was compared with both approaches. This selection of values achieved the highest PDR [6] for the scenario under study, but also corresponds to low energy tax. Fig. 2(a) shows that Flock-CC paid 9.6% less energy tax than the congestion-unaware approach under low loads (25 pkts/sec), as well as 12.5% less under high loads (45 pkts/sec). Also, Flock-CC paid $4-10$% less energy tax compared with the congestion-aware approach. Fig. 2(b) shows that the superiority of Flock-CC is attributed to its ability to deliver larger amount of packets to the sink.

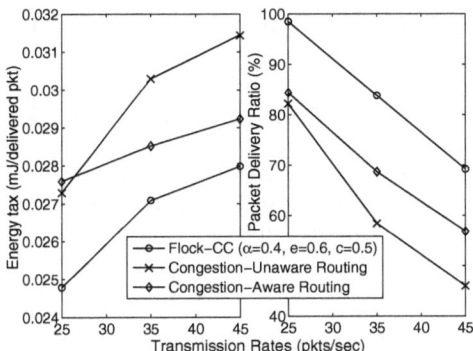

**Fig. 2.** Comparative experiments

## 4 Conclusions and Future Work

The aim of this study was to evaluate the Flock-CC mechanism from the perspective of energy consumption. Performance evaluations showed that under certain parameter values, the Flock-CC mechanism was able to both alleviate congestion [5], [6] and minimize energy tax. Also, Flock-CC outperformed congestion-aware and congestion-unaware routing approaches in terms of energy tax. The future work will provide more extensive comparative evaluations and a more thorough investigation of the influence of the various parameters. Also, the sampling period $T$ and its influence on performance, as well as the communication overhead from exchanging control packets are left for future work.

## References

1. Bonabeau, E., Dorigo, M., Theraulaz, G.: Swarm Intelligence: From Natural to Artificial Systems. In: Proc. of Sciences of Complexity, Santa Fe Institute (1999)
2. Engelbrecht, A.: Fundamentals of Computational Swarm Intelligence. Wiley, Chichester (2005)
3. Di Caro, G., Ducatelle, F., Gambardella, L.M.: AntHocNet: An adaptive nature-inspired algorithm for routing in mobile ad hoc networks. European Trans. on Telecom., Self-organization in Mobile Networking 16, 443–455 (2005)
4. Akyildiz, I.F., Su, W., Sankarasubramaniam, Y., Cayirci, E.: Wireless sensor networks: a survey. Computer Networks 38, 393–422 (2002)
5. Antoniou, P., Pitsillides, A., Blackwell, T., Engelbrecht, A.: Employing the Flocking Behavior of Birds for Controlling Congestion in Autonomous Decentralized Networks. In: Proc. of 2009 IEEE Congress on Evolutionary Computation (2009)
6. Antoniou, P., Pitsillides, A., Engelbrecht, A., Blackwell, T.: A Swarm Intelligence Congestion Control Approach for Autonomous Decentralized Communication Networks. In: Engelbrecht, A., Middendorf, M. (eds.) Applied Swarm Intelligence. SCI. Springer, Heidelberg (to appear)
7. Couzin, I.D., Krause, J., James, R., Ruxton, G., Franksz, N.: Collective Memory and Spatial Sorting in Animal Groups. Jrnl of Theor. Biology 218, 1–11 (2002)
8. Theraulaz, G., Bonabeau, E., Daneubourg, J.: Response Threshold Reinforcement and Division of Labour in Insect Societies, Royal Soc. of London, vol. 265, pp. 327–332.

# A Distributed Range Assignment Protocol

Steffen Wolf, Tom Ansay, and Peter Merz

Distributed Algorithms Group
University of Kaiserslautern
Kaiserslautern, Germany

**Abstract.** We present a new distributed algorithm for creating and maintaining power-efficient topologies in a wireless network. The wireless nodes establish links to neighbouring nodes in a self-organizing fashion. The protocol is designed to first create a connected topology and then iteratively search for better links to reduce the overall power consumption. First results from simulated experiments with various network sizes show that the resulting topologies are close to optimal in respect to the total energy consumption.

## 1 Introduction

Self-organization is an important aspect in wireless ad-hoc networks and wireless sensor networks. In these networks, which work without a wired backbone infrastructure, each node must bring its own energy source. In many scenarios, the wireless nodes are equipped with batteries, which provide them with a limited amount of energy. Since energy is needed for communication, the lifespan of the network is limited as well. In order to increase the lifespan of the wireless network, one could either provide the nodes with better energy sources, or try to use the available energy as efficiently as possible.

In this paper, we demonstrate how the wireless network itself can be used as a self-optimizing system to reduce the energy needed for transmitting data. We will adopt a high level of abstraction from Santi [1], using a graph representation. We assume that the wireless nodes can adjust their transmission power and that a higher transmission power corresponds to a larger coverage area for this node.

Section 2 gives a detailed description of the problem, and Section 3 gives an overview of related work. In Section 4 we present our distributed algorithm, and in Section 5 we provide first results of our experiments. Section 6 concludes the work and gives an outline of future research.

## 2 Problem Formulation

The set of problems of assigning transmission ranges to wireless nodes such that the resulting topology has a pre-defined property (e. g. connectivity) and the overall energy consumption is minimized are called Range Assignment Problems (RAPs) [1]. There are many interesting RAPs. For example in the Minimum

T. Spyropoulos and K.A. Hummel (Eds.): IWSOS 2009, LNCS 5918, pp. 226–231, 2009.
© IFIP International Federation for Information Processing 2009

Energy Broadcast Problem (MEB), one node has to transmit a message to all other nodes of the network. These nodes can relay the message, so the topology is a directed tree rooted at the first node spanning the whole network.

In this paper, we want to concentrate on topologies that are symmetrically connected, i.e. a link is established only if both nodes have a transmission range high enough to reach the other node. This RAP is called the Weakly Symmetric Range Assignment Problem (WSRAP). It is NP-hard [2,3], and can be formulated as follows:

Given a network $G = (V, V \times V)$ of wireless nodes $V$ and a cost function $c : V \times V \to \mathbb{R}$, we search for the spanning tree $T \subseteq V \times V$ that minimizes the overall power consumption

$$C(T) = \sum_{v \in V} \underbrace{\max_{u \in V : \{u,v\} \in T} c(u, v)}_{\text{transmission power of node } v} \tag{1}$$

Usually, the cost function $c$ is modelled as a polynomial function of the Euclidean distance $d$ using a distance-power gradient $\alpha$: $c(u, v) = d(u, v)^\alpha$. Measurements have shown that this distance-power gradient $\alpha$ can take values between 1 and 6. In an ideal obstacle-free environment, it takes the value of $\alpha = 2$.

In the same way, the transmission power assigned to a node translates to a transmission range. In the WSRAP, the transmission power of a wireless node is set such that the node can reach its farthest neighbour. Note that the formulation can be changed to a formulation for the Minimum Spanning Tree (MST) problem just by exchanging the max with a sum in (1). However, while the MST can be solved in polynomial time, the WSRAP is NP-hard. The MST can still be used as an approximation of the WSRAP. It has been shown that the MST already gives a 2-approximation [2,3].

A valuable side effect of the overall power minimization is a reduction of interference in the wireless network, as the lower transmission ranges translate to smaller interference areas. However, actually minimizing interference is a different optimization problem, and is not part of our analysis.

## 3    Related Work

The WSRAP was first introduced by Călinescu et al. in [2] as a variant of another Range Assignment Problem searching for strong connectivity only. They show that the Minimum Spanning Tree (MST) is a 2-approximation for the WSRAP. This approximation ratio is also shown to be tight.

Cheng et al. rediscover the problem in [3]. They emphasize that the problem is different than the strong connectivity range assignment problem and provide a new proof of NP-hardness. They also rediscover the MST as a 2-approximation, and present a new greedy heuristic. This heuristic builds the tree in a way that resembles Prim's algorithm for building MSTs, but it needs global knowledge and can only be transformed to a distributed algorithm with a high level of cooperation between the nodes.

Improving upon these global heuristics, Althaus *et al.* [4] present two local search heuristics. In the edge-switching (ES) heuristic, edges from the tree are replaced by non-tree edges re-establishing the connectivity. In the edge-and-fork-switching (EFS) heuristic, pairs of edges, so-called forks, are inserted in the tree, and the resulting cycles are cut again by removing other edges. In both heuristics, the process is repeated until a local optimum is reached. The authors also try to limit the number of hops for the ES heuristic, as this translates to the level of cooperation in a distributed algorithm. However, they do not proceed in this direction and do not model the wireless network as a self-organizing system.

Optimal topologies can be found using Mixed Integer Programming formulations (MIP) by Althaus *et al.* [4] or Montemanni *et al.* [5]. In both papers, problem instances with up to 40 nodes were solved. While Althaus *et al.* take more than an hour for each of these instances, Montemanni *et al.* can solve an instance of this size in less than two minutes. However, calculation times increase exponentially with the network size for both algorithms.

Near optimal topologies can be found in shorter time using our Iterated Local Search (ILS) based on the described local search heuristics [6]. Solutions for problem instances with up to 1000 nodes are presented. The largest network size for which optimal topologies are known is 100 (using the MIP from Montemanni it takes a couple of days), and the strongest ILS finds topologies with average costs of only 0.002 % above the optimum in less than three minutes.

To the best of our knowledge, no fully developed distributed algorithm for the WSRAP has been proposed so far. There are distributed algorithms for the MST which could be used to construct a 2-approximation to the WSRAP. But the challenge of designing a distributed algorithm for the local search heuristics that includes propagation of routing information and cycle detection, and that works without a high level of cooperation, has not yet been tackled.

## 4    A Distributed Algorithm

In our distributed algorithm, we assume that each node maintains a list of neighbours, a list of links, and a routing table. Links are established and revoked using a set of control messages. Nodes relay messages for other nodes and react on incoming control messages. The nodes also become active from time to time to try and establish new links or improve the current network topology.

When a node joins a network, it sends out a broadcast ping message to make itself known. Using the signal strength of the response messages, it can calculate the necessary transmission power to reach its neighbouring nodes.

In a first phase, shown in Fig. 1, the node actively sends out `AddLink` messages to the closest not yet reached neighbour. The usual response is a `ConfirmLink` message, containing the routing information of the other node. The link is established and the routing information is updated. Both nodes then propagate the new routing information to their part of the network.

In a second phase, shown in Fig. 2, the node intentionally sends `AddLink` messages to neighbouring nodes that can already be reached. Using the routing

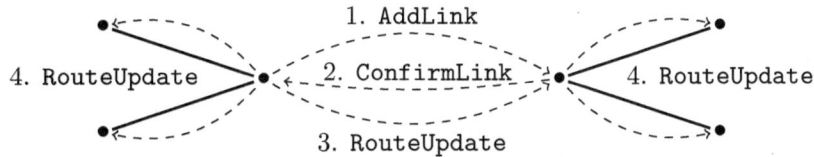

**Fig. 1.** Adding a new link

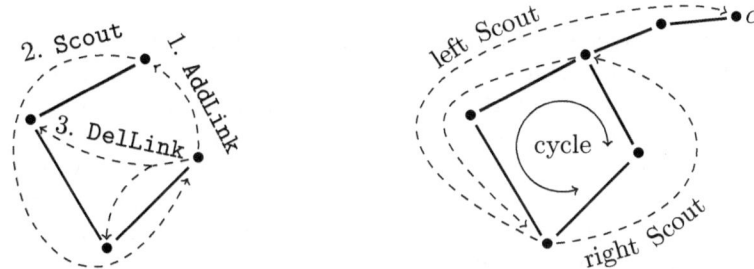

**Fig. 2.** Replacing a link                **Fig. 3.** Cycle detection

information, these nodes send a Scout back to the origin node. Each node that relays the Scout adds information on how much energy it could save if it were to remove the links the Scout is taking. When the Scout reaches the origin node, it decides based on this information whether it is favourable to add the new link and which remote link should be deleted. If so, DelLink messages are sent to the nodes adjoining the remote link, and an AddLink message is sent again to the neighbouring node. Using the new routing information, the neighbouring node should respond with a ConfirmLink and the link is established. This phase of the protocol can be seen as a distributed version of the ES heuristic.

Since our aim is to create a self-organizing wireless network, we do not synchronize the nodes. So it can and will happen, that establishing a link creates a cycle in the topology. When the routing information is propagated after adding the link, some remote nodes will recognize the cycle, because they receive contradicting routing information for another node $c$. As can be seen in Fig. 3, this node $c$ is not necessarily part of the cycle, it can also be several hops away from a node of the cycle. The algorithm has to take care that only links of the actual cycle are considered to be removed. The path from the cycle to node $c$ must not be severed, otherwise the network would be split and the cycle would remain. If a cycle is detected, two Scouts are sent to the node $c$, one for each route. If the Scout passes a node that knows a different route back to the origin node, it sends the Scout back over this route. If the Scout reaches the node $c$, but this node $c$ does not know a different route back, it ignores the Scout. So, only one Scout will return to the origin node, this Scout has only traversed the links of the cycle, and the origin node can decide which link should be removed and send the necessary DelLink messages.

## 5    Experiments

For our experiments, we used nine sets of well established test instances [6,7,8]. Each set contains 30 instances, where $n$ nodes are randomly located in a $10\,000 \times 10\,000$ grid, using a uniform distribution. Euclidean distance was used and the distance-power gradient was set to $\alpha = 2$ as in an ideal environment. The sets use $n \in \{20, 50, 100, 200, 500, 1000\}$ nodes, and there are three sets of clustered instances ($n \in \{100, 200, 500\}$, with $c = \lfloor \sqrt{n} \rfloor$ clusters of size $1000 \times 1000$, the cluster centres placed at random locations in the $10\,000 \times 10\,000$ grid).

Each experiment was repeated 30 times and average values are used for the following discussion. Our simulator was written in C++, and the simulations were conducted on a 2.5 GHz Intel Xeon running 64 bit Linux 2.6. The nodes became active after random waiting intervals between 1 and 10 seconds. To allow for actual concurrency, we simulated a message delay by waiting for 50 ms before processing a message after receiving it. In our experiments we limited the number of links to be added in phase 2 to five. We will show that this already gives quite good results. In normal operation, the nodes could continue trying to improve the topology, possibly piggy-backing the control messages on actual transmissions in the network.

Figure 4 shows the results for the uniformly distributed networks. The excess is plotted over time, averaged over all 30 runs for each of the 30 instances of the same size. The excess defines how much higher the overall energy consumption is compared to the optimal topology (or the best known topology for $n \geq 100$). An excess of 0 % means that an optimal topology has been found. In phase 1, which lasts about 20 seconds, the overall transmission energy rises as new links are established to connect the network. Then, the excess is between 17 % and 40 %. By this time, some nodes have already entered phase 2. The results of the topology changes in phase 2 can be seen in the figures. For larger networks, the transformation is slower, but they also reach the same final excess over the best known topologies: between 1.4 % and 3.6 %.

For the clustered networks (Fig. 5), both the initial and final excess is higher. The initial excess lies between 39 % and 76 %, and the final excess between

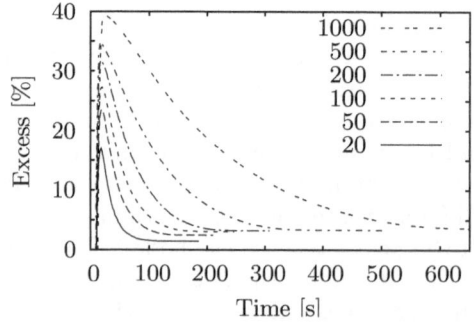

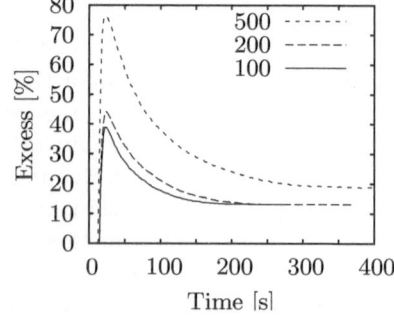

**Fig. 4.** Results for uniformly distributed random networks

**Fig. 5.** Results for clustered random networks

13 % and 19 %. In clustered instances, often only one node from each cluster maintains connections to other clusters. If the wrong node is selected to maintain these connections, our distributed algorithm, which is based on the ES heuristic, cannot transfer this task to another node. On the other hand, the amount of energy that actually is saved in these clustered networks is higher.

# 6   Conclusion

We have presented a self-organizing distributed algorithm to reduce the transmission energy in wireless networks. In experiments with well established test instances, both uniformly distributed and clustered, we have shown that the algorithm can reach close-to-optimal topologies. The energy that is saved using the improved topologies translates to a longer network lifespan without increasing the batteries of the nodes.

In future work, we want to improve the handling of clustered networks, e. g. by allowing the nodes to add more links in phase 2. The increased protocol overhead will have to be weighed against the expected improvements. We also want to expand our algorithm to other RAPs, such as the MEB.

# References

1. Santi, P.: Topology Control in Wireless Ad Hoc and Sensor Networks. John Wiley & Sons, Chichester (2005)
2. Călinescu, G., Măndoiu, I.I., Zelikovsky, A.: Symmetric connectivity with minimum power consumption in radio networks. In: Baeza-Yates, R.A., Montanari, U., Santoro, N. (eds.) Proc. 2nd IFIP International Conference on Theoretical Computer Science. IFIP Conference Proceedings, vol. 223, pp. 119–130. Kluwer, Dordrecht (2002)
3. Cheng, X., Narahari, B., Simha, R., Cheng, M.X., Liu, D.: Strong minimum energy topology in wireless sensor networks: NP-completeness and heuristics. IEEE Transactions on Mobile Computing 2(3), 248–256 (2003)
4. Althaus, E., Călinescu, G., Măndoiu, I.I., Prasad, S.K., Tchervenski, N., Zelikovsky, A.: Power efficient range assignment for symmetric connectivity in static ad hoc wireless networks. Wireless Networks 12(3), 287–299 (2006)
5. Montemanni, R., Gambardella, L.M.: Exact algorithms for the minimum power symmetric connectivity problem in wireless networks. Computers & Operations Research 32(11), 2891–2904 (2005)
6. Wolf, S., Merz, P.: Iterated local search for minimum power symmetric connectivity in wireless networks. In: Cotta, C., Cowling, P. (eds.) EvoCOP 2009. LNCS, vol. 5482, pp. 192–203. Springer, Heidelberg (2009)
7. Al-Shihabi, S., Merz, P., Wolf, S.: Nested partitioning for the minimum energy broadcast problem. In: Maniezzo, V., Battiti, R., Watson, J.-P. (eds.) LION 2007 II. LNCS, vol. 5313, pp. 1–11. Springer, Heidelberg (2008)
8. Hernández, H., Blum, C., Francès, G.: Ant colony optimization for energy-efficient broadcasting in ad-hoc networks. In: Dorigo, M., Birattari, M., Blum, C., Clerc, M., Stützle, T., Winfield, A.F.T. (eds.) ANTS 2008. LNCS, vol. 5217, pp. 25–36. Springer, Heidelberg (2008)

# A Distributed Power Saving Algorithm for Cellular Networks

Ingo Viering[1], Matti Peltomäki[2], Olav Tirkkonen[3,4], Mikko Alava[2],
and Richard Waldhauser[5]

[1] Nomor Research GmbH, Munich, Germany
[2] Department of Applied Physics, Helsinki University of Technology (TKK), Finland
[3] Department of Communications and Networking, TKK, Finland
[4] Nokia Research Center, Helsinki, Finland
[5] Nokia Siemens Networks, Munich, Germany

**Abstract.** We consider power saving in a cellular network. Subject to constraints generated by network planning and dynamic Radio Resource Management algorithms, there is room for reducing Base Station (BS) transmit powers. We suggest that power is reduced in a way that does not move cell boundaries significantly. For this, there is a power difference constraint $d_{\max}$ that upper limits the *difference* in power reductions in neighboring BSs. The BSs exchange information about their current level of power. Each BS has a target maximum power reduction. We propose a distributed algorithm, where the BSs take turns to reduce their power, respecting their neighbors current power settings and the power difference constraint. We show that this algorithm converges to a global optimum.

## 1 Introduction

Self-organization is a wide ranging research trend in modern networking. In the scope of wireless networking, research on Self-Organized Networks (SON) ranges from general principles of cognitive and ad hoc networks to concrete problems in standardization and implementation of near future mobile networks [1,2]. Here, we concentrate on a specific SON problem of current interest for the standardization of the next release of the Evolved Universal Terrestrial Radio Access Network (E-UTRAN), a.k.a. Long Term Evolution (LTE). The SON-related requirements for LTE-Advanced [3] are:

- energy efficiency of NW and terminals,
- SON in heterogeneous and mass deployments,
- avoiding drive tests.

This paper discusses a distributed algorithm for realizing energy savings in cellular networks. As such, it deals directly with energy efficiency, and with heterogeneous deployments—the only truly scalable way of implementing networking algorithms for heterogeneous mass deployments is to distribute the algorithms to the deployed network elements themselves.

The distributed algorithm for power saving is designed to respect cell boundaries up to a degree. It can be proven to converge in linear time to a global optimum.

T. Spyropoulos and K.A. Hummel (Eds.): IWSOS 2009, LNCS 5918, pp. 232–237, 2009.

## 2   Motivation for Power Saving

Power efficiency has not been an issue in network planning, and modern cellular communication networks such as LTE have been designed to operate in an interference limited manner. As a consequence adapting the transmit power will be an important SON feature. The target is to minimize output power without endangering coverage.

As an example we take the evaluation scenario "Case 1" defined in 3GPP [4]. This is a heavily interference limited scenario with the Inter-Site Distance 500m, and a Tx power of 46dBm. There is a significant potential for power saving. Figure 1 shows the average sector throughput in this scenario with a full buffer traffic model, and different Tx powers at the BSs. A reduction of 12dB (to 34dBm) leads to a negligible loss of 1.9% in throughput—the network remains interference limited. Based on this we argue that especially in small cells, there may be significant room for optimizing the transmit power of base stations.

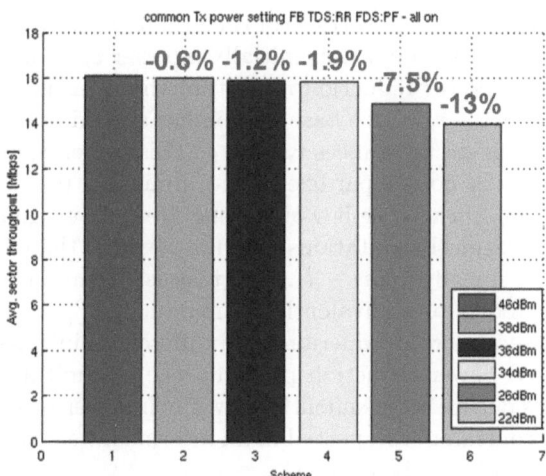

**Fig. 1.** Loss in throughput when reducing base station transmit power

## 3   Distributed Network Power Saving

We propose an algorithm for reducing the power in cellular networks, where the base stations decide the reductions in the cells, based on local information such as load, and reference signal receive power (RSRP) measurements received from User Equipments (UEs).

As discussed in [5], an important requirement of power adaptation is that the coverage areas of neighboring cells shall not be affected, at least not by an individual BS. In other words, power adaptation of an individual cell shall not change the cell boundaries. The cell boundaries are carefully desinged in the network planning phase, and there will be other SON algorithms which may dynamically optimize the boundaries for traffic reasons etc. To avoid clashes, we allow an individual cell to reduce its power only if all neighbors reduce their power in a rather similar way. For this, we assume a maximum power reduction difference $d_{\max}$, which is set by a central entity. An intial power difference may exist for two neighboring cells, designed in network planning and by other SON algorithms. For energy saving purposes, a cell may save power in a way that increases the power difference to neighbors with at most $d_{\max}$. As an example, assume that there are two cells, and in normal operation (i.e. given by network planning) the transmit power in cell A would be 46dBm, and in cell B 43dBm. In a power saving phase, cell A would like to reduce its power by 12dB to 34dBm and cell B by 21dB to 22dBm. If $d_{\max} = 2$dB, cell B could not reduce its power by more than 12dB+2dB, i.e. not to less than 29dBm.

### 3.1   System Model

The power reduction problem is defined as follows. There is a graph $G = G(V, E)$ of interference couplings between base stations, based on e.g. Neighbor Cell Lists. There are $N$ BSs, which are the vertices of the graph. The set of neighbors of BS $n$ is denoted by $\mathcal{N}_n$. Each node $n$ uses initially a power $p_{n,0}$, and makes periodic decisions related to a power reduction $r_n$ so that the transmit power used after the decision is $p_n = p_{n,0} - r_n$. Each base station has a maximum power reduction $m_n$ such that under all circumstances $r_n \leq m_n$. This power reduction is selected so that cell coverage is not jeopardized—the minimum power base station $n$ could use is $p_{n,0} - m_n$, otherwise it would generate a coverage hole. The power reductions of two adjacent base stations may not be more than $d_{\max}$ units apart, or $|r_n - r_m| \leq d_{\max}$ for all $(n, m) \in E$. The task is to minimize $\sum_v p_n$ subject to the constraints above, or equivalently, to maximize $\sum_v r_n$. For convenience, we measure all powers, power reductions and differences in dB-scale.

In the distributed power reduction problem, the minimization is to be done such that (a) nodes decide independently how much power to use, (b) power reduction difference contraint is respected in all intermediate stages of the process, and (c) the global minimum is achieved.

### 3.2   Suggested Algorithm

The Distributed Power Saving (DiPoSa) algorithm works as follows:

1. There is a predetermined periodicity determining the moments when each BS makes a decision related to its power usage.
2. For simplicity, the algorithm is assumed to be iterative and asynchronous—the decision period is assumed to be the same for all BSs, and it is assumed that the BSs make decisions in an unsynchronized manner. This translates to the situation that the distributed algorithm sweeps through the BSs in a random order, and then sweeps through again in the same order. One sweep is called an iteration.[1]

---

[1] Note that this is readily generalized to different periods for different BSs, or even periods changing from decision to decision.

3. At the moment BS $n$ makes a decision related the power to use, it knows the power reductions $\mathcal{R}_n = \{r_m\}_{m \in \mathcal{N}_n}$ used by all its neighbors $m$ at that moment.

4. The BS selects the power reduction $p_n = \min\left(\{m_n\} \cup \{r_m + d_{\max}\}_{m \in \mathcal{N}_n}\right)$

In the last step the BS computes the maximum power reduction allowed by the powers used by each neighbour, by adding the power reduction of the neighbour and the maximum power reduction difference $d_{\max}$. If the smallest power reduction allowed by the neighbours is larger than the target maximum reduction $m_n$, the BS uses the smallest power reduction allowed by the neighbours. Otherwise, it uses the target maximum reduction.

## 3.3 System Requirements

DiPoSa is distributed so that the decision of the power used is made at the BS. In order to make the decisions, the BS needs to know the maximum power reduction $d_{\max}$, and the power reductions at use by its neighbors $\mathcal{R}_n$. The former is assumed to be signaled by Operations and Maintenance, and it is time-invariant. Note that it could even be different for different cells, but for the sake of illustration we have assumed that it is the same in the network. The latter are regularly exchanged directly between the BSs. In LTE, the network of BSs is meshed by direct BS-to-BS X2-interfaces, which is the natural candidate for exchanging $\mathcal{R}_n$ information.

## 3.4 Convergence to Global Optimum

The objective function that each node minimizes is completely local, but it is connected by the power reduction constraint to the neighbors. In principle, connections may link up to generate long-distance connections between nodes. Nevertheless, the distributed algorithm converges to a global optimum, and in a time that is linearly proportional to $\max_n m_n/d_{\min}$.

**Proposition 1.** In a finite network, the DiPoSa algorithm described in Section 3.2 converges to a global optimum.

*Proof:* First we have to prove that DiPoSa converges to some state. This is easiest to do by observing that the BSs participate in an exact potential game [6]. There is a simple global potential function which is a sum of the power reductions achieved in the network. The utility function that each BS tries to maximize at each update is the individual power reduction. At each update, the increase in the updating nodes utility corresponds exactly to the increase in the global potential. Thus the settings for a exact potential game are fulfilled and DiPoSa converges. Next, we prove by contradiction that the converged state of DiPoSa is the global optimum. Assume that the algorithm has converged to a state $\mathcal{S} = \{r_n\}$, and that the global optimum corresponds to $\mathcal{O} = \{o_n\}$. If $\mathcal{S} \neq \mathcal{O}$, $o_n > r_n$ for at least a $n = n_1$. Due to Step 4 in DiPoSa, the reason for this must be that for at least for one neighbor of $n_1$, say $n_2$, we have $o_{n_2} > r_{n_2}$, and that

$$r_{n_1} = r_{n_2} + d_{\max} < o_{n_1} \tag{1}$$

Due to Step 4 in DiPoSa, a neighbor of $n_2$ has to be away from optimum. This neighbor of $n_2$ cannot be $n_1$, as from (1) we have $r_{n_1} > r_{n_2}$. Thus there exists at least another neighbor of $n_2$, say $n_3$, for which $o_{n_3} > r_{n_3}$. Repeating this induction step $N$ times we observe that the last BS is also away from optimum, $o_{n_N} > r_{n_N}$, and from equations induced from (1) we know that $r_{n_N} < r_n + d_{\max}$ for all $n \neq n_N$. Thus in Step 4 for $n_N$ the reduction $r_{n_N}$ can be increased. This contradicts the assumption that $\mathcal{S}$ is a converged state.     ∎

A slightly more involved argument would show that DiPoSa converges in linear time, which is upper bounded by a linear function $t_{\mathrm{conv}} \propto \frac{\max_n m_n}{d_{\max}}$.

## 4   An Example

Figure 2 gives an example for a heterogeneous cell layout with different cell sizes, modeling the cells covering the center and outskirts of a medium-size city. The network setting is generated as a Voronoi diagram, where the x- and y-coordinates of the underlying point set is Laplacian distributed. Every cell bears two numbers. The first contains the individual (uncoordinated) maximum power reduction $m_n$ of every cell. These are determined in an ad hoc manner, the larger the cell the smaller $m_n$. The second number is the $r_n$ in the converged state of

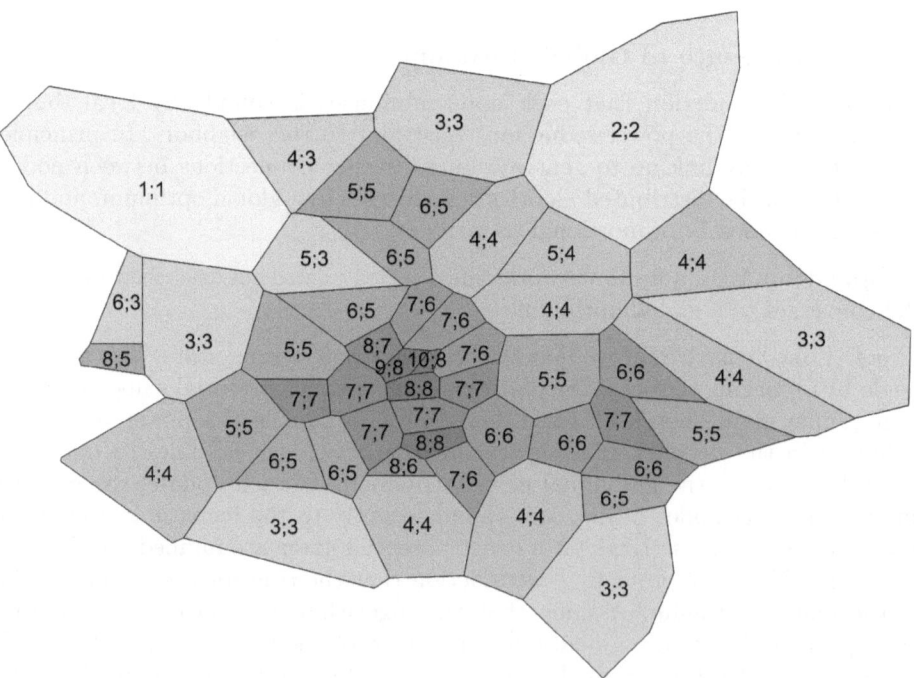

**Fig. 2.** Example network with maximum power reductions and reached optimum power reductions

DiPoSa, how much every cell is allowed to actually reduce the power. Here, we have used $d_{max} = 2$dB. If the original transmit power is the same for all cells, i.e. $p_{n,0} = p_0$, the converged solution saves 67% of the overall output power in the example network. Without the $d_{max}$ constraint, i.e. if every BS $n$ reduced by $m_n$, the savings would be 70%.

## 5   Conclusion

In this paper, we proposed a distributed power reduction algorithm, where each BS has a local target power reduction, and the reductions at use at the BSs should not differ between neighbors by more than a maximum $d_{max}$. The objective functions for each BS are decoupled, but the constraint $d_{max}$ reduces the action space of the BSs, and induces long-range correlations. We observed that these long-range correlations do not prevent a distributed solution. The proposed algorithm is a straight forward implementation of the limitations of the local action spaces by which these long range correlations become realized in the converged solution. Due to the locality of the objective functions, the algorithm was proven to converge to the global optimum.

**Acknowledgments.** This research is partially funded by Nokia.

## References

1. 3GPP: Self-configuring and self-optimizing network use cases and solutions. Technical Report TR 36.902 (2008), http://www.3gpp.org
2. Döttling, M., Viering, I.: Challenges in mobile network operation: towards self-optimizing networks. In: Proc. IEEE ICASSP (April 2009)
3. 3GPP: Requirements for further advancements for E-UTRA (LTE-Advanced). Technical Report TR 36.913 (2008), http://www.3gpp.org
4. 3GPP: Physical layer aspects for evolved universal terrestrial radio access (UTRA). Technical Report TR 25.814 (2009), http://www.3gpp.org
5. 3GPP: De-centralized optimization of downlink transmit power. Technical Report R3-091929 (2009), http://www.3gpp.org
6. Monderer, D., Shapley, L.: Potential games. Games Econom. Behav. 14, 124–143 (1996)

# Local Optimum Based Power Allocation Approach for Spectrum Sharing in Unlicensed Bands

Furqan Ahmed and Olav Tirkkonen

Helsinki University of Technology
Espoo, Finland
{furqan.ahmed,olav.tirkkonen}@tkk.fi

**Abstract.** We present a novel local optimum based power allocation approach for spectrum sharing in unlicensed frequency bands. The proposed technique is based on the idea of dividing the network in a number of smaller sub-networks or clusters. Sum capacity of each cluster is maximized subject to constraint on total power of each user in a cluster. On its turn each user in a cluster maximizes the sum capacity by calculating power allocations that correspond to a local optimum. Total power constraint of each user and effect of interference from other users in the network is taken into account for finding local optimum solution. Comparison of achieved network sum capacity is done with the well known iterative water filling method. Numerical results show that the proposed cluster based local optimum method achieves higher capacity than selfish iterative water filling and is therefore suitable for geographically distributed networks.

## 1 Introduction

Resource allocation for devices working in unlicensed bands has gained significant research interest because of its impact on the performance. An efficient resource allocation is the one in which it is not possible to improve the performance of one system without causing degradation in some other systems performance.

Our focus in this paper is on efficient power allocation for devices in unlicensed bands. We discuss a scenario where a number of users are sharing spectrum in an unlicensed band. The main aim is to find power allocation for each node that maximizes the sum capacity of entire network. Given the importance of this problem, a number of authors have addressed it using different analysis techniques. The well known selfish iterative water filling (IWF) power allocation method was proposed in [1] using a game theoretical approach. In [2] it has been extended and comparison of different power allocation approaches is also given. Most studies have been done considering flat fading case, however flat fading results have been generalized for frequency selective fading channels in [3]. Some other recent works related to distributed power allocation problem include [4], which discusses maximization of a logarithmic utility function or capacity, jointly for all the links. Method discussed by [4] assumes that all the distributed decision makers have information of the price of

T. Spyropoulos and K.A. Hummel (Eds.): IWSOS 2009, LNCS 5918, pp. 238–243, 2009.
© IFIP International Federation for Information Processing 2009

interference that is caused by them to all receivers of the network. Same problem has recently been addressed in [5] for cognitive radio networks.

In this paper we examine the performance of similar power allocation scheme as in [2] but the network model is changed to a more random one. A concept of distributed power allocation is presented for capacity maximization within different clusters. The rest of the paper is organized as follows: in section 2 system model is developed and parameters used throughout the paper are introduced. Section 3 presents the explanation of our proposed local optimum based power allocation scheme. In section 4, performance is analyzed with the help of a numerical example and simulation results are presented.

## 2   Network Model

In this section we describe the system model that includes network architecture, power constraints on transmitter-receiver pairs and expressions for calculating the sum capacity of the network. The network model used in this study is similar to the network examined in [2], but the locations of transmitters and receivers are more random. Figure 1 shows an example layout of the network with 16 links.

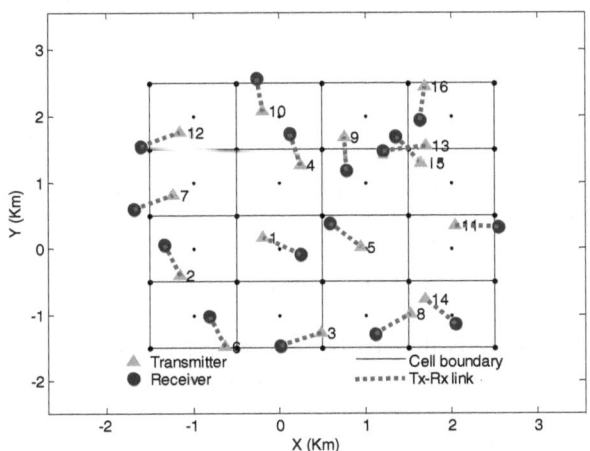

**Fig. 1.** Network architecture

Network consists of square shaped cells with a transmitter receiver pair (also called link) in each cell. The alignment and directions of links are random with middle point of each link lying randomly in square shaped cell. Numbered triangles indicate the transmitters and circles indicate locations of receivers in network. The links in network share the same frequency band having bandwidth equal to B. Bandwidth is divided into N number of channels and transmitters of all links can allocate their transmit power freely over these N channels. All the transmitters have fixed maximum transmit power which they cannot exceed. The Shannon capacity achieved by one link is given by the expression:

$$C_r = \sum_{j=1}^{N} \log_2(1 + \frac{p_{r,j} g_{r,r,j}}{I_{r,j} + N_0}) \cdot \tag{1}$$

Where $N$ is the number of channels, $p_{r,j}$ is the power allocated by link $r$ on channel $j$ and $g_{t,r,j}$ is the gain from transmitter $t$ to receiver on channel $j$. The spectral density of additive white Gaussian noise is $N_0$ and $I_{r,j}$ is the interference from other transmitters in network that receiver $r$ experiences on channel $j$. The interference can be obtained as follows:

$$I_{r,j} = \sum_{\substack{t=1 \\ t \neq r}}^{N} p_{t,j} g_{t,r,j} \cdot \tag{2}$$

To simplify we define:

$$\tilde{I}_{r,j} = \frac{I_{r,j} + N_0}{g_{r,r,j}} \cdot \tag{3}$$

Using this notation an equivalent expression for the capacity of link $r$ denoted by $C_r$ is given by:

$$C_r = \sum_{j} \log_2(1 + \frac{p_{r,j}}{\tilde{I}_{r,j}}) \cdot \tag{4}$$

The sum capacity of the network is the sum of capacities of all links and can be determined as follows:

$$C_{sum} = \sum_{r=1}^{M} C_r \cdot \tag{5}$$

Power allocation of link $r$ is defined by vector $p_r = [ p_{r,1} \, p_{r,2}.....p_{r,N} ]$ and maximum power constraint that has to be followed by all links in network is characterized as:

$$\sum_{j=1}^{N} p_{r,j} = P_{\max} \cdot \tag{6}$$

We do not consider the interior points of power constraint equation here. The assumption is that each transmitter uses full power and follows the maximum power constraint. Our aim is to allocate that power efficiently across all channels in an optimum way in order to maximize the network sum capacity.

## 3 Local Optimum Based Power Allocation

In the local optimum based cooperative power allocation scheme, each link is aware of the links within a certain area. The updating link calculates its power to maximize the sum capacity of sub-network and will start using this new allocation.

To the best of our knowledge, cooperative distributed power allocation based on cluster (as opposed to full network) interference information has not been addressed in literature before. When the cluster becomes large enough to encompass the whole network of all updating links, this scheme is close in spirit to distributed multichannel asynchronous pricing scheme studied in [4]. In [4], the interference to other users is abstracted by a price function but in our case we consider the utility function (capacity) with an aim to directly employ cooperative optimization. In order to optimize power allocation on each link in an asynchronous way, let us consider an updating link r with set of neighbors denoted by $\mu_r$. The index of optimizing link is outside this set i.e. $r \notin \mu_r$. It is assumed that updating link is aware of channel gains between its transmitters and all receivers in cluster as well as interference powers across all channels at all receivers that are included in a cluster. The sum capacity of the cluster is the objective function for optimization and can be expressed as:

$$C_{sum,r} = \sum_j (\log_2(1 + \frac{p_{r,j}}{\tilde{I}_{r,j}}) + \sum_{n \in \mu_r} \log_2(1 + \frac{f_{n,j}}{p_{r,j} + J_{n,j}})) \cdot \quad (7)$$

Where $f_{nj}$ is the scaled transmit power of neighbor $n$ on channel $j$, and $J_{nj}$ is the scaled total interference and noise experienced at channel $j$ of receiver $n$ except the interference caused by $r$. The expressions are given by:

$$f_{n,j} = p_{n,j} / g_{r,n,j} \cdot \quad (8)$$

$$J_{n,j} = \tilde{I}_{r,j} / g_{r,n,j} - p_{r,j} \cdot \quad (9)$$

It is assumed that a signaling protocol exists based on which transmitter of link $r$ acquires information of the effect of its transmission to neighbors from which $f_{n,j}$ and $J_{n,j}$ are computed. The objective function is the sum capacity given by equation 7, and will be optimized under power constraint equation given by 6. Resulting power allocations will maximize the capacity of cluster. Like iterative water filling approach, the cooperative local optimum needs to be iterated over several asynchronous power allocation updates by all links.

A numerical example of cooperative local optimal power allocation is presented in next section in which we compare the cumulative density functions of the capacities obtained using both techniques. The capacity cumulative density function of cluster based approach is compared with selfish iterative water filling and random power allocation methods. Simulation results show that sum capacity achieved by optimizing cluster capacities is higher than the one achieved by using distributed selfish iterative water filling approach.

## 4  Simulation Results

In this section, we present simulation results to evaluate the performance of proposed scheme. To achieve numerical results, more specific model assumptions have been made. Details of parameters and values used in the simulation are given in table 1.

**Table 1.** Simulation parameters

| Parameter | Symbol | Value |
|---|---|---|
| Total bandwidth | $B$ | 10 MHz |
| No. of links | $M$ | 16 |
| No. of channels | $N$ | 4 |
| Max. Tx power | $P_{max}$ | 16 dBm |
| Thermal noise level | $N_T$ | -174 dBm/Hz |
| Noise figure | $N_F$ | 6 dB |
| Path loss exponent | $\alpha$ | 3.76 |

Figure 2 shows the clusters used in simulations of local optimum approach; the entire network is divided into 4 sub-networks or clusters $L_1$, $L_2$, $L_3$ and $L_4$. To optimize objective functions under given power constraints we have used built-in MATLAB optimization function called fmincon. Performance is evaluated by comparing capacity cumulative density functions obtained by maximized capacities calculated using different power allocation schemes. Using the simulation parameters specified in table 1, Monte Carlo simulation method is used for calculating capacities for local optimum and selfish iterative water filling schemes. Using the iterative water filling solution as starting point, on its turn, transmitter of a link selected randomly from cluster updates its power allocation to maximize the sum capacity of the cluster on the basis of most recent interference and power situation signaled by neighbors. Randomly ordered optimizations are performed by links of all four sub-networks, selected one by one. We compare the capacity CDF of local optimum based cluster capacity maximization power allocation scheme with selfish iterative water filling method as well as random power allocation. The network sum capacity CDFs are compared in figure 3.

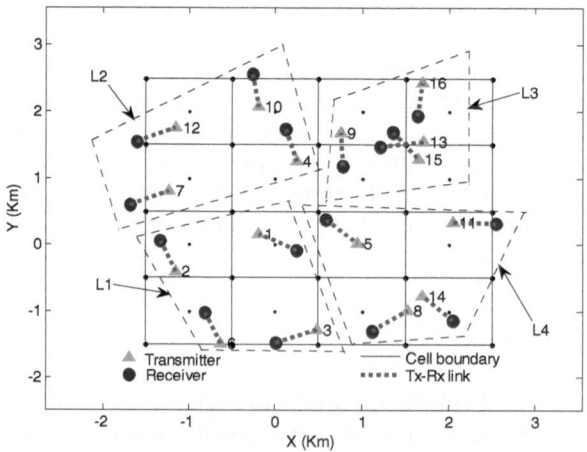

**Fig. 2.** Network architecture with four clusters

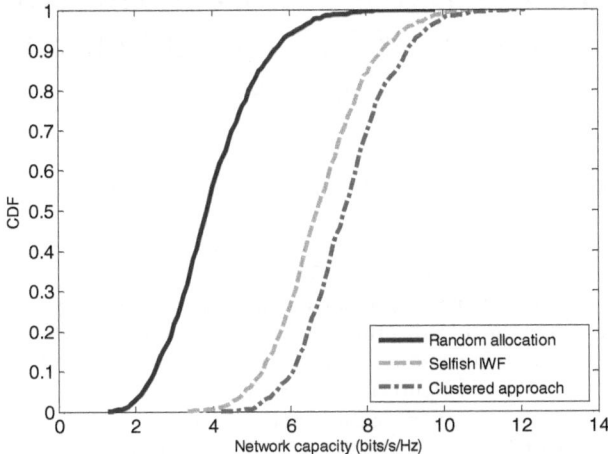

**Fig. 3.** Comparison of network sum capacity CDFs

Network sum capacity for proposed cooperative cluster based distributed power allocation was found to be around 0.7 bits/s/Hz greater than the mean network sum capacity achieved using selfish iterative water filling. We conclude that cluster based locally optimal power allocations is an effective distributive power allocation strategy which can achieve higher network sum capacity than selfish iterative water filling scheme.

# References

1. Chung, S.T., Kim, S.J., Lee, J., Cioffi, J.M.: A Game-Theoretic Approach to Power Allocation in Frequency-Selective Gaussian Interference Channels. In: Proc. IEEE International Symposium on Information Theory, p. 316 (2003)
2. Etkin, R., Parekh, A., Tse, D.: Spectrum Sharing for Unlicensed Bands. IEEE J. Sel. Areas Comm. 25(3), 517–528 (2007)
3. Xu, Y., Chen, W., Cao, Z.: Optimal Power Allocation for Spectrum Sharing in Unlicensed Bands. IEEE Comm. Letters 12(7), 511–513 (2008)
4. Huang, J., Berry, R.A., Honig, M.L.: Distributed Interference Compensation for Wireless Networks. IEEE J. Sel. Areas Comm. 24(5), 1074–1084 (2006)
5. Lee, W.Y., Akyildiz, I.F.: Joint Spectrum and Power Allocation for Inter-Cell Spectrum Sharing in Cognitive Radio Networks. In: Proc. DySPAN, October 2008, pp. 1–12 (2008)

# Economics-Driven Short-Term Traffic Management in MPLS-Based Self-adaptive Networks

Paola Iovanna[1], Maurizio Naldi[2], Roberto Sabella[1], and Cristiano Zema[3]

[1] Ericsson Telecomunicazioni S.p.a.
Via Moruzzi 1, 56124 Pisa, Italy
{paola.iovanna,roberto.sabella}@ericsson.com
[2] Dipartimento di Informatica, Sistemi e Produzione
Università di Roma "Tor Vergata"
Via del Politecnico 1, 00133 Rome, Italy
naldi@disp.uniroma2.it
[3] CoRiTeL c/o Ericsson Telecomunicazioni S.p.a.
Via Anagnina 203 00118 Roma, Italy
cristiano.zema@ericsson.com

**Abstract.** Today's networking environment exhibits significant traffic variability and squeezing profit margins. An adaptive and economics-aware traffic management approach, needed to cope with such environment, is proposed that acts on short timescales (from minutes to hours) and employs an economics-based figure of merit to reallocate bandwidth in an MPLS context. Both underload and overload deviations from the optimal bandwidth allocation are sanctioned through the economical evaluation of the consequences of such non-optimality. A description of the traffic management system is provided together with some simulation results to show its operations.

## 1  Introduction

Traffic on the Internet is affected by an ever growing variability, reflected both in its patterns and in its statistical characteristics, which require traffic management solutions to rely on online traffic monitoring. Cognitive packet networks (CPN) are a pioneer example of self-aware networks [1], since they adaptively select paths to offer a best-effort QoS to end-users. That concept has been advanced in [2] through self-adaptive networks, which employ a traffic management system acting on two timescales in an MPLS infrastructure to achieve QoS goals (with constraints on the blocking probability for connection-oriented networks, and on packet loss, average delay, and jitter for connectionless networks).

However, network design and management procedures can't be based on QoS considerations alone, since the economical issue is of paramount importance for any company. Even the QoS obligations, embodied in a Service Level Agreement (SLA), have an associated economical value, under the form of penalties or compensations when those obligations are violated. Though QoS goals can be

T. Spyropoulos and K.A. Hummel (Eds.): IWSOS 2009, LNCS 5918, pp. 244–249, 2009.

met by overprovisioning, network operations could result expensive in the long run. Even with limited overprovisioning the currently unused bandwidth could be assigned otherwise, providing additional revenues: its careless management is an opportunity cost and a source of potential economical losses. An effective traffic management system should implement a trade-off between the contrasting goals of delivering the required QoS (driving towards overprovisioning) and exploiting the available bandwidth (driving towards efficiency). Deviations in either way are amenable to an economical evaluation, so that traffic management economics appear as the natural common framework to manage network operations.

In this paper we propose a novel engine for the traffic management system envisaged for self-adaptive networks in [2], using economics as the single driver, to cater both for QoS violations and for bandwidth wastage. In particular, we focus on its inner feedback cycle, i.e., that acting on shorter timescales. We describe its architecture in Section 2 and its forecasting engine in Section 3. We introduce a new economics-based figure of merit to drive traffic management decisions in Section 4. We finally report in Section 5 some early results showing the dynamics of such figure of merit in a simulated scenario.

## 2   The Traffic Management System: Overview

We consider a traffic management system in an MPLS context, where the traffic is channelled on LSPs (Label Switched Path), in turn accomodated on traffic tunnels. We have to allocate bandwidth to LSPs to achieve an effective use of the network resources. We resume the traffic management system acting on two timescales put forward in [2] and focus on the short timescale subsystem. In this section we describe in detail that subsystem.

A schematic diagram of the Short Term Management Subsystem (STMS) is reported in Fig. 1. The traffic measurements block monitors each traffic tunnel and forecasts the evolution of traffic for the next time interval (the domain of the SMTS is on timescales of the order of hours, hence the *nowcasting* name). The nowcasting engine (block B in Fig. 1) employs the Exponential Smoothing technique in the versions considered in [3] to build a time series of traffic. This time series is in turn fed as an input to the cost function block, which evaluates the cost associated to the current combination of traffic and allocated

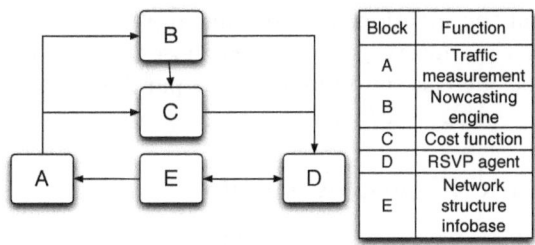

**Fig. 1.** Short Term Management System

capacity. Rather than minimizing deviations from the QoS objectives (which is the common approach to bandwidth management, as in [3]), bandwidth allocation is here driven by the willingess to maximize the provider's revenues. The correcting actions taken by the STMS (not considered here) on the basis of the trend observed are: Modification of LSP attributes (e.g., their bandwidth); Rerouting of LSPs; Termination of LSPs, in particular of the lower priority ones (pre-emption); Dynamic routing of new unprecedented requests.

## 3   Traffic Nowcasting

Our system includes a traffic measurement subsystem (Block A in Fig. 1), which feeds a traffic prediction subsystem (block B in the same picture). The measurements are conducted on each LSP, through a counter measuring the cumulative number of bytes transferred on that LSP during a given period of time (typically of 5 minutes, in agreement with what SNMP-based devices provide, and to be chosen as a trade-off between readiness of reaction and accuracy); at the end of each period the byte count is transferred to the nowcasting block and the counter is reset. The byte count divided by the period length provides the average bandwidth employed during that period.

Two forecasting methods are considered, both based on the Exponential Smoothing (ES) approach and analysed in [3]:

1. ES with linear extrapolation (ESLE);
2. ES with predicted increments (ESPI).

In both methods the classic Exponential Smoothing recursive formula is adopted unless when both underestimation ($F_j < M_j$, where $M_j$ is the traffic measurement at time $j$ and $F_j$ is the traffic forecast for the same time) and a growing trend ($M_j > M_{j-1}$) are observed at the same time. In that case different forecasting algorithms are used in the two methods, as follows.

**ESLE method.** If both underestimation and a growing trend take place the forecast is equal to the latest measurement ($M_j$) plus the latest measured increase ($M_j - M_{j-1}$).

**ESPI method.** When both underestimation and a growing trend take place, the forecast is equal to the latest forecast plus a specified increment. equal to: a) a fixed fraction of the latest measured increment $z > \alpha(M_j - M_{j-1})$ on the first interval the mentioned conditions apply; b) the estimated increase $\Delta_{j+1}$ on following time intervals as long as those conditions apply. In case b) the estimate of the increase is obtained by a parallel basic ES approach, i.e. $\Delta_{j+1} = \alpha\Delta_j + (1 - \alpha)(M_j - M_{j-1})$.

## 4   An Economic Figure of Merit

In the past the figure of merit for a traffic management system was chosen to achieve the maximum efficiency of transmission resources subject to QoS constraints [4], but such approach fails to consider the economic value associated to the usage of bandwidth (not simply the capital cost incurred in building the transmission infrastructure, but also the costs associated to alternative uses of the same bandwidth). In this section we propose a new figure of merit for traffic management, that takes a wider view of the monetary value of bandwidth allocation decisions.

An improper bandwidth allocation impacts on the provider's economics in two opposite ways. If the LSP is overused, congestion takes place, with failed delivery of packets and possible SLA violations. If the LSP is underused, chunks of bandwidth are wasted that could be sold to other users (the provider incurs an opportunity cost). Common approaches to bandwidth management either focus on just the first issue, overlooking bandwidth waste, or lack to provide an economics-related metric valid for both phenomena. A first attempt to overcome these limitations has been made by Tran and Ziegler [3] through the introduction of the Goodness Factor (GF), which employs the load factor $X$ on the transmission link (the LSP in our case), i.e., the ratio between the expected traffic and the allocated bandwidth, whose optimal value is the maximum value that meets QoS constraints $X_{opt}$. The GF is then defined as

$$GF = \begin{cases} X/X_{opt} & \text{if } X < X_{opt} \\ X_{opt}/X & \text{if } X_{opt} \leq X < 1 \\ (1/X - 1)/X_{opt} & \text{if } X \geq 1 \end{cases} \tag{1}$$

The relationship between the GF and the load factor is shown in Fig. 2 (dotted curve) for $X_{opt} = 0.7$. Over- and under-utilization bear different signs and can be distinguished from each other. The value of the GF in the optimal situation is 1. The GF takes into account both underloading and overloading, but fails to put them on a common scale, since it doesn't consider the monetary losses associated to the two phenomena: the worst case due to under-utilization bears $GF = 0$, while the worst case due to over-utilization leads to the asymptotic value $GF = -1/X_{opt}$. In addition, the GF as defined by expr. 1 is discontinuous when going to severe congestion ($X > 1$). We have also developed a continuous version of the Goodness Factor, where the function behaviour when the load factor falls in the $X_{opt} \leq X \leq 1$ range is described by a quadratic function; the modified version of the Goodness Factor (used for the simulation analysis reported in Section 5) is given by expr. (2) and shown in Fig. 2 (solid curve).

$$GF_{\text{mod}} = \begin{cases} X/X_{opt} & \text{if } X < X_{opt} \\ 1 - \left(\frac{X - X_{opt}}{1 - X_{opt}}\right)^2 & \text{if } X_{opt} \leq X < 1 \\ (1/X - 1)/X_{opt} & \text{if } X \geq 1 \end{cases} \tag{2}$$

In our approach we introduce a cost function whose value depends on the current level of LSP utilization, putting on a common ground both under- and over-utilization. The minimum of the cost function is set to zero when the LSP utilization is optimal. As we deviate from the optimal utilization level the cost function grows. The exact shape of the function can be defined by the provider, since it depends on its commercial commitments. However we can set some general principles and provide a simple instance. If a SLA is violated due to insufficient bandwidth allocation, the provider faces a cost due to the penalty defined in the SLA itself. On the other hand, an opportunity cost may be associated to the bandwidth unused on an LSP; the exact value of the cost may be obtained by considering, e.g., the market price of leased lines. A very simple example of the resulting cost function is shown in Fig. 3. The under-utilization portion takes into account that leased bandwidth is typically sold in chunks (hence the function is piecewise constant), e.g., we can consider the typical steps of 64 kbit/s, 2 Mbit/s, 34 Mbit/s, and so on. The over-utilization portion instead follows a logistic curve, that asymptotically leads to the complete violation of all SLAs acting on that LSP, and therefore to the payment of all the associated penalties.

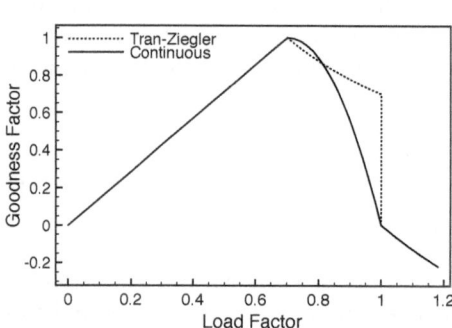

**Fig. 2.** Goodness Factor                    **Fig. 3.** Cost function of STMS

# 5   Simulation Analysis

After introducing in Section 4 the cost function to replace the Goodness Factor, we now show how the two metrics behave in a simulated context, through the use of the Network Simulator (ns2).

The simulation scenario considers a single LSP on which we have generated traffic over an interval of 6 hours with a sampling window size of 5 minutes. The traffic was a mix resembling the UMTS service composition, with the following services (and the pertaining percentages on the overall volume): Voice (50%); SMS (17.7%); WAP (10.9%); HTTP (7.8%); MMS (5.7%); Streaming (4.1%); E-mail (3.8%). This traffic mix was simulated at the application layer [5].

The STMS described in Section 2 readjusts the LSP bandwidth after the load factor (which provides the direction to follow) and the Cost Function (which provides a measure of the adequacy of bandwidth readjustments). The optimal load factor was set at 0.82; whenever this threshold is exceeded the bandwidth is increased (the reverse action takes place when the load factor falls below 0.82).

In Fig. 4 the observed rate and the load factor are shown together during the 6 hours interval. Though the rate exhibits significant peaks, the load factor is kept tightly around the optimal value by the bandwidth readjustment operations.

Both performance indicators are shown in Fig. 5. The Cost Function oscillates between two values since for most of the time the load factor falls in the stair-like under-utilization area. This is due to the granularity of sold bandwidth, which may make small changes in the load factor not relevant for the opportunity cost. On the other hand, the continuous changes of the Goodness Factor would induce readjustments when there's nothing to gain by reallocating bandwidth.

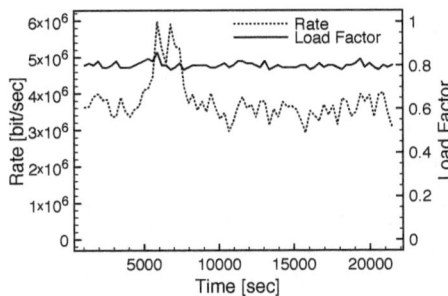
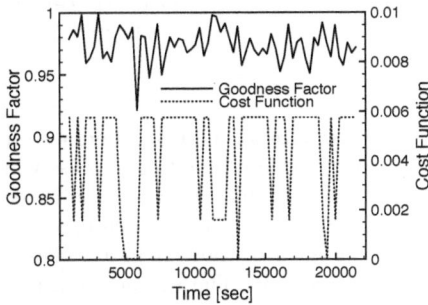

**Fig. 4.** Load on LSP                **Fig. 5.** Performance indicators

## 6   Conclusions

A traffic management system acting on short timescales and employing an economics-based figure of merit has been introduced to base traffic management on the consequences of bandwidth mis-allocation. Such figure of merit marks the deviations from the optimal allocation due to under- and over-utilization, and improves a previously defined Goodness Factor proposed by Tran and Ziegler. The traffic management system allows to adjust bandwidth allocation to achieve an economically efficient use of the network resources.

## References

1. Gelembe, E., Lent, R., Nu nez, A.: Self-aware networks and QoS. Proceedings of the IEEE 92(9), 1478 1489 (2004)
2. Sabella, R., Iovanna, P.: Self-Adaptation in Next-Generation Internet Networks: How to React to Traffic Changes While Respecting QoS? IEEE Transactions on Systems, Man, and Cybernetics - Part B: Cybernetics 36(6), 1218–1229 (2006)
3. Tran, H.T., Ziegler, T.: Adaptive bandwidth provisioning with explicit respect to QoS requirements. Computer Communications 28(16), 1862–1876 (2005)
4. Carter, S.F.: Quality of service in BT's MPLS-VPN platform. BT Technology Journal 23(2), 61–72 (2005)
5. Iovanna, P., Naldi, M., Sabella, R.: Models for services and related traffic in Ethernet-based mobile infrastructure. In: HET-NETs 2005 Performance Modelling and Evaluation of Heterogeneous Networks, Ilkley, UK, July 18-20 (2005)

# Self-organized Evacuation Based on LifeBelt

Kashif Zia and Alois Ferscha

Institute of Pervasive Computing,
Johannes Kepler University Linz,
Altenberger Strasse 69, A-4040 Linz, Austria
{kashif,ferscha}@pervasive.jku.at
http://www.pervasive.jku.at

**Abstract.** In this paper, we have investigated the feasibility of a self-organized evacuation process when compared with a centralized control. The evacuation strategy is based on 'predicted exit time' (a relation of 'estimated time to reach to an exit', 'exit capacity' and 'exit population') for each of the exit in a multi-exit environment, selecting the minimum value exit. The self-organized strategy is based on information propagation in a peer-to-peer fashion, initiated by a special agent in each of the exit area. The propagation range ('zone of influence') is dependent on intensity and direction of peers interaction. Based on the propagated dataset, each agent can make an autonomous decision, conceptually a converse of centralized strategy where each agent is directed by a server. The evacuation process in supported by a wearable device, i.e. LifeBelt. Through large scale simulations using cellular automata technique and a challenging airport terminal model, we have proved that an efficient evacuation based on principles of self-organization is a real possibility, even in an infrastructureless environment.

**Keywords:** Self-organized evacuation simulation.

## 1 Introduction

Several modeling approaches are applied to study crowd dynamics separately and in combination [1]. However, the most important approaches are: Cellular Automata (CA), Lattice Gas (LG), Social Force (SF), and Fluid Dynamics (FD) models. For the sake of this paper, the reason for choosing CA as a modleing choice its intrinsic capability to model basic structures and hence the global space, in conjunction with behavioral rules, along with its simplicity. A CA model is based on a regular grid of the cells in which each cell is occupied by a single individual. The space (grid occupancy) and local (neigborhood state) rules in combination, describe the next cells an agent needs to occupy at each time stamp. Considering that an individual knows his direction of motion (towards an exit), the strategy of choosing the next cell to move to depends on the intensity and scope of his perception in the direction of the exit. In more technical terms, intensity is defined as individual's ability to interact with other individuals in all directions, whereas scope is defined as individual's ability to interact with a

T. Spyropoulos and K.A. Hummel (Eds.): IWSOS 2009, LNCS 5918, pp. 250–255, 2009.

only a subset of it. For example, it may be assumed that a normal vision allows an individual to see the state of the cells within an intensity equal to three cells (all cells within three cells radius). Moreover it can be assumed that, due to limitations incurred by human vision, an individual can only see what lies reasonably ahead (in the line of sight), hence limiting the scope of cells he can view. Combining intensity and scope, we have termed the interaction range of an individual as his Zone of Influence (ZoI).

In contrast with most of the CA mechanism, we focus on three aspects of an evacuation situation, overlooked by most of the research community. The first difference lies in not considering an evacuation environment as a normal structure with no breakdowns. For example, in an emergency situation it is hard to assume that there would be no power failure due to structural collapse. If that is the case in reality, there would be no vision (particularly at night) to view any thing within ZoI. Secondly there is little effort in designing a strategy for optimal exit selection (in a multi-exit environment) based on factors other than distances to the exits. For example, authors in [2] have presented a strategy of selecting an exit based on occupants' density (Exit Population (EP)) in an area of interest (Exit Area (EA)) around each of the exit, in conjunction the with distance. There are two issues in this model. (i) The model is based on human perceptions and does not account for limited or no visibility. It is 'assumed' that individuals are aware of the exit area activities, irrespective of the exit distance. (ii) For optimal exit selection, the model does not consider the Exit Capacity (EC) the width or capacity an exit has. We already designed a more efficient strategy [3] and used it in this paper (an implementation in a more complex environment). Our model is based on EP, EC and Time to reach to an Exit Area (TEA). It is designed for environments with no visibility and hence rely on two-way communication mechanism between server and individuals. Still it can be useful in unlimited special cases in an environment where normal human perceptions are active, for example, assisting handicapped people in evacuation, or assisting in a very large unknown environment. Lastly, there are self-organized data dissemination strategies [4] but there is no evidence of its usage in crowd evacuation simulations. Through this paper, we have introduced a novelty by extending Ferscha's model [3] for a truly distributed setting with no centralized server.

For all the three principles highlighted in the paragraph above, there is a need to equip each individual with a helping device having following capabilities:

- It should provide the neighborhood sensing to assist individuals with taking microscopic next cell decision in the dark. The aspects of neighborhood sensing absolutely desirable in this context are distance and orientation, so that it can fill in the requirements for intensity and scope respectively.
- It should provide location coordinates to the centralized server to be used in optimal exit selection mechanism.
- It should provide a peer-to-peer communication mechanism to help disseminating the information to be used in a self-organized evacuation strategy.

We have designed a wearable helping device, LifeBelt [3], fulfilling all these requirements. Using LifeBelt, optimal exit selection can either be applied in a centralized or distributed manner. At local level, a decision to move to the next cell can also be made based on orientation and distance information provided by LifeBelt, keeping the individual aggregatively inline with the directional guidance provided by the server or extracted from the self-organized data dissemination. Additionally, LifeBelt provides vibro tactile stimulus in the hip area thus keeping the evacuee attentive throughout the evacuation process or to guarantee a successful evacuation from areas with no visibility.

## 2    Optimal Exit Selection Centralized (OESC) Mechanism

The optimal exit selection strategy chooses the minimum value of Predicted Exit Time (PET) between the PET values for all the exits. PET is based on three parameters - the Exit Population (EP), the Exit Capacity (EC) and the Time to reach to an Exit Area (TEA) and is calculated as:

$$PET_i = TEA_i + (Ep_i/EC_i) \qquad (1)$$

where subscript i represents a typical exit. The following pseudo code represents the mechanism of evacuation executed in each time stamp, in a centralized setting.

```
For each individual
   Let position = cell-type
   if position = 'exit' -> SAFE
   else if position = 'EA'
       turn-towards 'exit-in-sight' and move-forward
   else
       get-optimal-exit
       turn-towards optimal-exit and move-forward
```

For each of the individual 'position' is the variable which represents the cell type on which an individual is residing. If position is 'exit', it means that the individual has reached to an exit and is safe. If position describes as if individual being in an exit area (EA), the individual can see the 'exit-in-sight' and has to direct towards the corresponding exit by turning towards the exit in sight and moving forward. Finally, if individual is neither at exit or within an EA, it relies on guidance for optimal-exit provided by the central server. After application of eq 1 for each of the exit, the cental server points towards 'optimal-exit' (get-optimal-exit) thus enabling the individual to perform a turn towards corresponding exit and then moving forward. A more detailed discussion of local cell selection mechanism adopted for 'move-forward' is presented in [5] and [3].

## 3    Optimal Exit Selection Distributed (OESD) Mechanism

The mechanism needs a special agent in each EA, which collects the presence of individuals in an exit area. This information is propagated towards the exterior

individuals in a peer-to-peer fashion. After each round of propagation initiated
by the special agents, each individual can perform a single step motion. The
following pseudo code represents the mechanism of evacuation executed in each
time stamp, in a distributed setting:

```
//Propagation: recursive call
    For each individual 'i'
        if dataset (i) is updated
            for each individual 'ni' in the neighborhood
                if ni is not in EA
                    update // create dataset of ni
                    set update (ni) = true
//motion: executed in each iteration after propagation
    ... // first six lines of OESC pseudocode
        find-optimum-exit
        find-route-neighbor
        turn-towards route-neighbor and move-forward
```

In the propagation phase, each individual recieves a dataset from neighbors. If
the dataset recieved is updated, it is propagated to the neighbors within ZoI of
the propagating individual if and only if the target neighbror is not in EA. The
process terminates when number of individuals updated in a recursive step are
zero. The pseudocode for motion is same as that of centrallized case, except that
the optimal-exit is calculated by each individual from the information recieved
from the neighorhood (find-optimum-exit) and is not directed by central server.
After 'finding' the optimal exit, the individual gets the direction of the neighbor which points towards the optimal exit (find-route-neighbor), turns towards
corresponding neighbor and then moves forward.

## 4    Simulation

The simulation environment (groud terminal of Ibiza airport, not shown to save
the space) has overall cellular dimension of 120*40 including the obstacles (es-
calators, elevators and counters), exits, baggage recalim areas, check-in area and
pathways. There were two purposes of the simulation. The first purpose was
to investigate the potential of self-organized evacuation, following an optimal
exit selection strategy. Whereas the second purpose was to establish the effect
of ZoI on the efficiency of self-organized evacuation. To achieve these goals, we
performed a large scale simulation for the two evacuation strategies (OESC and
OESD) with the same randomly placed set of 500 individuals. Each strategy was
tried with four possible values of ZoI: 2.5, 5.0, 7.5 and 10.0. A comparison of
time required to evacuate (all realistic individuals) and number of individuals
saved between two strategies is shown in Fig. 1 which helps in drawing useful
conclusions. Before jumping into conclusions, there are few terms (phenomenan)
which require definitions. (i) *Migration:* an individual changing its destination
from one exit to another. Essentially, this happens due to emergence of new

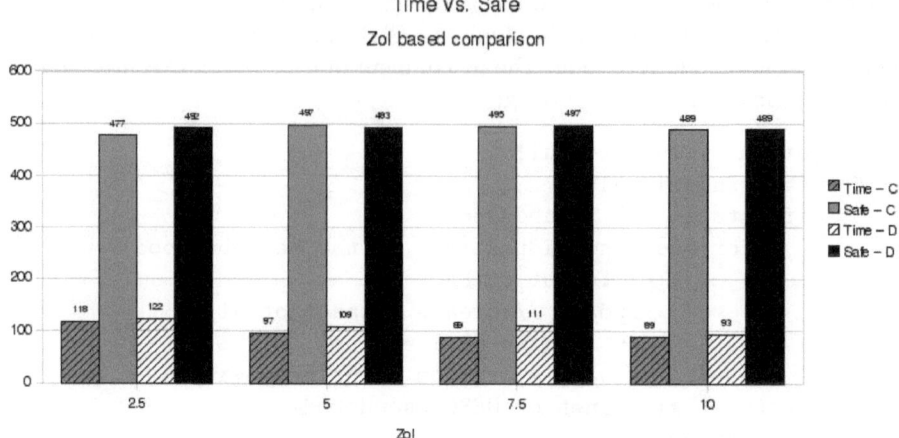

**Fig. 1.** ZoI based comparison. C = centralized, D = distributed.

situations in a surrounding. For example a migration is necessary to balance the load from exits with low EC to exist with higher EC. (ii) *Exaustion:* a situation in which there is no more individuals in EA of an exit. Following results are drawn:

1. In case of very small ZoI (2.5) with OESC, the migration starts very late (after 40th iteration), in fact after exaustion of low EC exits. The low EP (due to less ZoI) does not allow the factor $Ep_i$ / $EC_i$ to be significant unless the EP becomes zero. That is the reason, with the similar directional guidance and due to limitation in local navigational capabilities, the individuals stuck around obstacles (particularly the ramp beside exit 4). This decreases the number of individuals saved (477). In OESD, the migration phenomena remains the same, but there is an increase in saved quantity (492) due to the reason that the individuals are not provided with absolute direction towards optimal exit. They follow the connected individuals which are connected to the exit by a chain of individuals, hence by passing the obstacles.

2. In case of ZoI equal to 5 with OESC, the migration starts early (around iteration 10). This helps in decreasing the evacuation time and an increase in number of individuals saved. With greater ZoI, it is obvious that there would be late migration as well. In OESD, the pattern is almost the same, except that there is no possibility of late migration due to disconnections. Obviously this decreses the saved quantity from 497 (OESC) to 483 (OESD).

3. In case of ZoI equal to 7.5, both OESC and OESD strategies ensure same number of saved individuals, which is on higher side. Obviously, the migration starts very early and there is alot of fluctuation throughout. In case of OESC, the exaustion is very balanced for all exits. The difference between evacuation time (89 vs. 111) is due to very intense migrations in OESD, which in fact, let the exist with smaller EC finsih first.

4. A further increase in ZoI (ZoI = 10) balances the difference in evacuation time (between two strategies), but there is a slight decrement in number of individuals saved (for both strategies). This decrease is environment specific and may not be there in other environments. Typically, a huge ZoI, let the EA of exit 4 extend itself on the other side of the ramp, leaving few individuals there in later stages of simulation.

On top of results presented here (related with effect of ZoI value on the efficiency), it is obvious that self-organized evacuation is not impossible. Although slightly on the lower side, the evacuation efficiency of OESD is certainly comparable with OESC. Additionally, self-organized evacuation based on LifeBelt is the only possible strategy in an infrastructureless environment where human perceptions are not active or overwhelmed.

## 5    Conclusion

In this paper, we have investigated the effects of using LifeBelt on the efficiency of crowd evacuation from a harazdous environment. The factors making an environment hazardous can be a complete blackout, collapse of infrastructure or lack of evacuee attention. In such a situation, a self-organized evacuation strategy remains the only possibility. LifeBelt being an enabling technology provides us with essential features required for performing self-organized evacuation process. Through simulation, We found evacuation efficiency of self-organized strategy comparable with the centralized strategy in a multi-exit environment.

**Acknowledgments.** This work is supported under the FP7 ICT Future Enabling Technologies pro- gramme of the European Commission under grant agreement No 231288 (SOCIONICAL).

## References

1. Zheng, X., Zhonga, T., Liua, M.: Modeling crowd evacuation of a building based on seven methodological approaches. Building and Environment 44(3), 437–445 (2009)
2. Weifeng, Y., Hai, T.K.: A nove algorithm of simulating multi-velocity evacuation based on cellular automata modeling and tenability conditions. Physica A: Statistical Mechanics and its Applications 379(1), 250–262 (2007)
3. Ferscha, A., Zia, K.: On the Efficiency of LifeBelt based Crowd Evacuation In: 13th IEEE/ACM International Symposium on Distributed Simulation and Real Time Applications, Singapore, October 25-28, IEEE Computer Society Press, Los Alamitos (2009)
4. Adler, C., Eigner, R., Schroth, C., Strassberger, M.: Context-adaptive Information Dissemination in VANETsMaximizing the Global Benefit. In: Proc. 5th IASTED Int. Conf. on Communication Systems and Networks, p. 712 (2005)
5. Ferscha, A., Zia, K.: LifeBelt: Silent Directional Guidance for Crowd Evacuation. In: Proceedings of the 13th International Symposium on Wearable Computers (ISWC 2009). IEEE Computer Society Press, Los Alamitos (2009)

# On the Role of Self-organisation as a Model for Networked Systems

Hajo Greif

Dept. of Science and Technology Studies, University of Klagenfurt, Sterneckstr. 15,
9020 Klagenfurt, Austria
hajo.greif@uni-klu.ac.at

**Abstract.** In this paper, the role of the concept of self-organisation as
a model in the analysis and design of advanced networked systems is
investigated. In a first step, criteria for the definition of scientific models
and their explanatory roles are introduced on the background of theories
of models in the philosophy of science: intended scope, selection of the
properties modelled, type of analogy, and levels of formalisation, abstraction and idealisation. In a second step, the applicability of these criteria
to model-building in engineering is discussed, in order to assess some of
the implications and limitations of modelling networked systems as self-
organised systems, with particular attention to the role of the systems'
environments in these models.

## 1 Introduction

It has become a common practice in computer science and related fields to invoke the functional and organisational principles of natural systems, such as
organisms, populations or ecosystems, as models for the analysis and design of
technological artefacts. Those natural systems are adapted to their respective
environments by way of processes of random variation and natural selection;
and they are adaptive to variations in these conditions by virtue of their self-
organising properties, in which higher-level, complex behaviours are produced
from sets of reiterated interactions between more basic elements and their functions (e.g. individuals or organs) [1,2].

The concept of self-organisation has achieved particular popularity in models
for advanced information and communication networks, as in [3,4], with early
references dating back the 1980s, e.g. [5]. Among the plethora of definitions that
have been proposed in this field, common conceptual ground is found, firstly,
in the mode of adaptation of the systems, which occurs in ad-hoc, situated
and dynamic fashion rather than along predetermined routines, and, secondly,
in their adaptive qualities, which lie in their distributed and localised structure,
from whose operations more complex structures and functions emerge in bottom-
up fashion. One may expect that these properties of the model translate to the
level of the target system, i.e. that which is to be designed. If the model is
successful, these properties of the network will enable real-world applications
that are embedded in, and adaptive to, the ever-changing and sometimes hardly
predictable environment of human actions and interactions. Or so it seems.

T. Spyropoulos and K.A. Hummel (Eds.): IWSOS 2009, LNCS 5918, pp. 256–261, 2009.

The purpose of this essay, which argues from a philosophy of science perspective, is to raise awareness to some possible implications of the use of the concept of self-organisation in the design of information and communication networks. It will do so by first generally discussing the epistemic role of models in science, in order to apply the outcomes of that discussion to the present case.

## 2   Models in Science

In the natural sciences, the target system of a model, in most cases, is the explanandum of a theory. The question to be answered by introducing a model normally is: What properties of an entity or what kind of causal interchange with its surroundings that are inaccessible to observation by currently available means make it behave in a certain way? In the absence of direct evidence of the physical and causal structures in question, an answer is sought in the selective ascription to the target system of properties known to pertain to entities from other domains. Thus, a relation of analogy is established.

The establishment of such analogies is an important step in the construction and application of scientific theories. Analogies may be applied in an informal, 'psychological' way, making the domain of the theory intelligible by using the structure and behaviour of some fairly well-explored system to describe the predicted structure and behaviour of the target system. The importance of models even on this informal level can be observed in quantum theory, which arguably is so notoriously hard to comprehend because there is no better-known system available to stand in for that theory's highly abstract notions [6, p. 45 f].

In a more systematic fashion, analogies may also serve to establish correspondence rules between theoretical and empirical concepts, which otherwise would remain detached from each other and empirically unproductive, and they may serve to systematically expand the domain of application of a theory. To do so, clear definitions of the extension of both the target system and the model are required. Only a certain part of the model can be expected to correlate with the target system, and only in certain ways.

On these grounds, several criteria for scientific models are articulated in the philosophical literature:

(I) It should be spelled out precisely which properties are deemed analogous between model and target system [6,7]. The selection should be such that the roles played by the chosen properties in the different systems should be *isomorphic* to a certain extent [8]. In computer models of the human mind, the functions of mental traits are explained by ascribing certain functional properties of computers to them, but processor architecture or storage methods are not among these properties.

(II) It should be monitored what character the relation of analogy is meant to have when introducing the model, and what character it turns out to have after testing the model. A distinction can be made between *positive*, *negative* and *neutral* analogies [7]: relations between model and target system that are already known to hold, relations that are already known not to hold, and, most interestingly, relations that, at the current stage of inquiry, cannot yet be proven to hold, but will help to evaluate the present theory when it is eventually found to hold or not to hold.

(III) It should be explicated in which way the analogy is meant to hold. Models may represent their target systems in a variety of ways [9]: They may *abstract* from the concrete properties of the target system in focusing on analogies in effects, functions, or behaviours in general (e.g. in being a computer model of a biological system); they may be *idealisations* of the target system, in having certain properties that the target system does not, and probably cannot, have (e.g. noise-less transmission or point masses); and they may be chosen to *approximate* the target system's behaviour (e.g. in assuming a mean value for a variable effect).

(IV) It should be defined what kind of structure the analogy is to have: The most fundamental distinction is to be found between *substantive* or "material" and *formal* analogies [6,7]. The former may be comprised of physical objects, such as the wire-and-plastic structure of a DNA model, or of descriptions in natural language. Stricter conditions apply to formal analogies, in that they only hold if an identical formal structure can be proven to underlie the behaviour of different systems. If, for example, a behavioural pattern within some population can be represented by a certain algorithm, this pattern may serve as a formal model of the interactions of network nodes only if the same algorithm is to be applied to the latter.

These criteria, in different constellations, cater for all kinds of models in the natural sciences. They serve to define both the extension of the model (what target domain it refers to) and its intension (what it says about that domain). I will now discuss the applicability of these criteria to models in the realm of technology, and then proceed to their possible implications for the present case.

## 3    Models and Technology

The most obvious distinction between models in science and models in the realm of technology lies in their direction of fit. Models in science are meant to help explaining a given phenomenon, being generally models *of* a certain target system. Any model that is found to misrepresent the properties of its target system, e.g. in mistaking a negative analogy for a positive one (see III above), or in missing out one important property of the target system (see I above), is either falsified or must be restricted in its domain of application. If however the design of artefacts is concerned, and if the model is meant to be a model *for* the target system, i.e. the artefact to be designed, such misrepresentation may also be a case of faulty implementation, and thus of lacking world-on-model fit.

Moreover, what is represented by a technological model is not something that has a history of existing and being observed within the real world in the same way as a natural system. What actually has to be included in the model, and in precisely which way it needs to fit onto the target system, is thus less clear from the outset. In the worst case, it is only found out on the real-world implementation of the model. Most prominently, choosing the wrong properties or choosing the wrong level of abstraction or idealisation may result in the overall failure of the technological system so modelled [10,11].

On the positive side of the balance, approximations in formal models, unlike in the natural sciences, need not be analytical in order to count as a solution, as their criteria of success are pragmatical: If the model is found to work in engineering practice, it is sufficiently confirmed [11]. In the present case, these observations on the role of technological models are of particular importance to the question of how the environment of the target system is taken into account.

## 4    Self-organisation as a Model for Networks

Advanced contemporary networked systems, when modelled as self-organising systems, are related to their environments in two particular ways that distinguish them from other technological systems:

**A. The environment is part of the model.** If the systems in question are modelled on natural systems and their behaviour in natural environments, abstraction and idealisation will become particularly challenging tasks. First and foremost, self-organisation in natural systems is itself a highly abstract and idealised scientific model of the formal kind, with the purpose of revealing one common trait among a plethora of physical, chemical and biological phenomena. It may be decently articulated mathematically, but in its implications, the concept of self-organisation is not fully explored to date [2]. Consequently, the mode of abstraction and idealisation will be different in each case of application to new domains, requiring attention to a variety of additional factors, since the specific kinds of interaction of the functional elements of each system with and within its respective environment are the *topic* of the model, not something that could be omitted from the picture in any way. Algorithms of ant behaviour will have to take different system-environment relations into account than, e.g., models based on cellular automata, and they will be applicable on different levels of abstraction, idealisation and approximation. The models themselves, as they stand in biology, may be found not to provide sufficient guidance to this task.

Moreover, the more precise a model is to be, and the richer its repertoire of analogies is to be, the more difficult it will be to address the right level of model-building. If the analogy is very informal, there will be few restrictions on its application, and there will be little risk of failure beyond having picked a bad metaphor, but its epistemical value will be limited to phenomenological similarities with inspirational function, as in [3, contribution by Flake et al.]. If the analogy is highly formal, it will be applicable in a straightforward and systematic fashion, but its scope will be limited, too, since it only covers one or a few systems, and since the set of properties it includes will be small, as in [3, contribution by Dousse / Thiran]. Yet, unlike for scientific models, all that precision might ultimately be in vain, as the system so modelled may turn out to be faultily implemented in its environment. If however one chooses the middle ground of a systematic heuristic, as in [4,5,12], model-building might prove both a very difficult and very rewarding endeavour. If one uses natural self-organising systems as a (material) analogy that shall be both of comprehensive scope *and* of systematic value – if, for example, it is to provide us with neutral analogies that can be tested for becoming positive ones, so as to advance our inquiry –, the selection of the right set of properties to be modelled and of the correct level of abstraction and idealisation, *with particular respect to the system's environment*, rather than formal precision, assume superior importance.

**B. The modelled system is embedded in its environment.** The environment of a system, it is argued in some corners of evolutionary biology, does not reduce to the spatio-temporal surroundings of that system; instead, it is coextensive with those conditions in its surroundings that are relevant to its further behaviours – which may vary significantly between different systems even if placed in the same surroundings [13]. The specific organisation of the system and its interactions with and within its surroundings thus define which conditions are the relevant ones. If networked systems are designed as self-organised, they will, at least implicitly, incorporate models of their environment, and, if successful, they will be adaptive to a certain set of conditions and variations therein.

However, although the behaviours of their human counterparts belong to the conditions relevant to the systems, and thus to their environments, it does not follow that these systems are adaptive towards human actions and purposes from the latter's perspective, so as to be perceived by them as embedded in *their* environments. First of all, the property of self-organisation does not necessarily apply to all levels of the system. As the analogy proposed by the model is of partial nature *by definition*, other parts or other levels of the system's organisation may be of a different kind, and they may well be so perceived. There is no usable body of evidence to date that could tell us how self-organising networked systems actually fit into human environments, and how they are actually being perceived, as they have no history of being part of such environments. In order to achieve a systematic match between self-organising properties modelled into the systems and their intended perception, the inclusion of models of human beliefs, desires and actions with regard to these systems seems recommended.

For example, communication networks may be modelled along the principles of self-organisation in fairly detailed and formal fashion on a certain level, as in [5,12]. The aim is a coherent and stable, yet flexible infrastructure for all varieties of uses under all varieties of circumstances. However, on this first, infrastructural level, the systems' adaptivity to human behaviours is limited to the latter's movements in space-time and to the transition between different usage contexts. No attempt is made to anticipate, and adapt to, the purposes of human beings in terms of what world affairs and accomplishments they are directed at. Yet these purposes define what kinds of services are actually required and used, and what contents are communicated. Addressing these contents may be facilitated by a self-organising systems architecture, but if its model indeed contains a neutral analogy that might capture human purposes proper, this analogy will have to be independently validated – on pragmatical grounds.

# 5    Conclusion

The concept of self-organisation not only is a difficult concept in itself, for its combination of complexity, abstraction and intended scope. In the analysis and design of advanced information and communication networks, it poses particular challenges, as its function is not to provide explanatory models of the structure and behaviour of natural systems in their specific environments, but to provide design models for the structure and behaviour of technological systems in their environments – whose conditions are difficult to predict. Still, there may be constructive uses for the concept of self-organisation – in spite of the criticism it received even from some of its protagonists [1,2], and in spite of the inverse correlation between its popularity and the agreement on its definition.

# References

1. Ashby, W.R.: Principles of the Self-Organizing System. In: von Foerster, H., Zopf, G.J. (eds.) Principles of Self-Organization: Transactions of the University of Illinois Symposium, London, pp. 255–278. Pergamon Press, Oxford (1962)
2. Sumpter, D.: The Principles of Collective Animal Behaviour. Phil. Trans. R. Soc. B 361, 5–22 (2006)
3. Staab, S., Heylighen, F., Gershenson, C., Flake, G.W., Pennock, D.M., Roure, D.D., Aberer, K., Shen, W.M., Dousse, O., Thiran, P.: Neurons, Viscose Fluids, Freshwater Polyp Hydra – and Self-Organizing Information Systems. IEEE Intell. Sys. 18(4), 72–86 (2003)
4. Prehofer, C., Bettstetter, C.: Self-Organization in Communication Networks: Principles and Design Paradigms. IEEE Commun. Mag. 43(7), 78–85 (2005)
5. Robertazzi, T., Sarachik, P.: Self-Organizing Communication Networks. IEEE Commun. Mag. 24(1), 28–33 (1986)
6. Nagel, E.: The Structure of Science. Harcourt, Brace & World, New York (1961)
7. Hesse, M.B.: Models and Analogies in Science. University of Notre Dame Press, Notre Dame (1966)
8. da Costa, N., French, S.: Science and Partial Truth: A Unitary Approach to Models and Scientific Reasoning. Oxford University Press, Oxford (2003)
9. McMullin, E.: Galilean Idealization. Stud. Hist. Phil. Sc. 16, 247–273 (1985)
10. Laymon, R.: Applying Idealized Scientific Theories to Engineering. Synthese 81, 353–371 (1989)
11. Hansson, S.O.: What Is Technological Science? Stud. Hist. Phil. Sc. 38, 523–527 (2007)
12. Di Caro, G.A., Dorigo, M.: Ant Colonies for Adaptive Routing in Packet-Switched Communications Networks. In: Eiben, A.E., Bäck, T., Schoenauer, M., Schwefel, H.-P. (eds.) PPSN 1998. LNCS, vol. 1498, pp. 673–682. Springer, Heidelberg (1998)
13. Lewontin, R.C.: Organism and Environment. In: Plotkin, H.C. (ed.) Learning, Development, and Culture, pp. 151–170. Wiley & Sons, Chichester (1982)

# Addressing Stability of Control-Loops in the Context of the GANA Architecture: Synchronization of Actions and Policies

Nikolay Tcholtchev, Ranganai Chaparadza, and Arun Prakash

FOKUS Fraunhofer Institute for Open Communication Systems, Berlin, Germany
{Nikolay.Tcholtchev,Ranganai.Chaparadza,
Arun.Prakash}@fokus.fraunhofer.de

**Abstract.** As research on self-managing networks is on the rise, a very important question requiring answers is: How to ensure that all autonomic entities/elements in a network or in a single node work together harmoniously with the aim of maintaining service availability and good levels of quality of service (QoS) provided to the end users? In this paper, we propose to enhance the *Generic Autonomic Network Architecture* (GANA) – a recently emerged architectural Reference-Model for Autonomic Networking, such that actions, policy enforcements and/or (re-) configurations, issued by different autonomic (decision making) entities, are synchronized in such a way that they lead to the best possible (long-term) reaction of the system to the challenging conditions the network is exposed to.

**Keywords:** autonomic behaviors of a control-loop, stability of a control-loop(s), self-managing networks, GANA.

## 1 Introduction

*Autonomicity* – realized through control-loop structures is an enabler for advanced and enriched self-manageability of network devices and networks. As research on self-managing networks is on the rise, a very important question requiring answers is: How to ensure that all autonomic entities/elements in a network or a single node, work together harmoniously with the aim of maintaining service availability and good levels of quality of service (QoS) provided to the end users? Autonomic entities/elements normally control only some of the diverse aspects that are significant for self-manageability of networks. In a complex environment or architecture involving several control-loops that need to execute in parallel, the challenge is to ensure that the autonomic elements' behaviors are synchronized towards a global network goal. That is required in order to avoid a situation whereby each autonomic element is working towards its own goal but, the overall set of actions/policies degrades drastically the performance and dependability of the network. Such a situation could even result in unwanted oscillations and instabilities in the control-loops. These issues raise the question of how stable and efficient are the control-loops of an autonomic node or network. Therefore, methodologies and mechanisms are

T. Spyropoulos and K.A. Hummel (Eds.): IWSOS 2009, LNCS 5918, pp. 262–268, 2009.

required for synchronizing the actions undertaken by diverse autonomic elements in a self-managing node and the network as a whole. One applicable approach is the use of Formal Description Techniques (FDTs) during the design of an autonomic node or network [2]. On the other hand, the approach that is presented in this paper aims at ensuring the stability of control-loops during the run-time/operation of an autonomic node/network.

The GANA *(Generic Autonomic Network Architecture)* [1] is evolving in the course of the EFIPSANS [3] project. We propose to extend the GANA architecture in a way ensuring that the set of actions/policies issued by diverse autonomic entities in a particular time slice are analyzed such that only those actions that contribute to the overall global goal of the network, are allowed to be executed.

The rest of this paper is organized as follows: Section 2 deals with the architectural extensions to GANA for the purposes of addressing run-time stability of control-loops of an autonomic node and network. Section 3 defines an optimization problem that is at the heart of our approach. Section 4 provides the results of some experiments that were conducted in order to evaluate of our approach. Finally, section 6 draws conclusions and outlines some future research directions.

## 2   Extending the GANA in Order to Address Control Loop Stability

The *GANA* specifies a *control loop* by introducing two basic concepts – a *Managed Entity (ME)* and a *Decision Element (DE)*. Conceptually a DE is an autonomic element responsible for the management of concrete MEs (e.g. protocol modules), and realizes a control loop based on its continuous learning cycle. Through a specified control loop a DE is able to exercise specific algorithmic schemes and/or policies on its ME(s). Taking into account that control loops can be implemented at different levels of abstraction – e.g. on node or network level, GANA provides the *Hierarchical Control Loops (HCLs)* framework consisting of four levels at which DEs and associated control loops can be introduced – *Protocol Level* (e.g. a control loop inside a single protocol), *Functions Level* (in-built management of basic networking functions such as Forwarding), *Node Level* and *Network Level*. Thereby DEs on a higher level manage DEs on a lower level. For more information we refer the reader to [1].

In this work, we borrow some ideas from the area of Control Theory and apply them inside the GANA architecture. Firstly, the layered structure of the *HCL* should be used for implementing stability hierarchically. Secondly, every set of DEs which have some overlapping and possible contradictions in the actions, policies or configurations they are responsible for, should refer to an arbiter that allows only those management actions to be executed, which are beneficiary for the overall fitness of the network. We call this arbiter an **Actions Synchronization Module (ASM)**. An ASM is considered as part of a dedicated DE that has been elected or is by design the most appropriate one for acting as an arbiter (enabling the negotiation over the tentative actions). We consider plain actions, policy enforcements, and the deployment of complex configuration profiles as (management) actions. Therefore, we require that every DE should keep a list (catalogue) of the actions it is allowed to issue without having to consult an upper level DE. As illustrated in Figure 1, if a DE (e.g. DE1.1, DE1.2 or DE1.3) faces a problem that is beyond its local scope, i.e. the action to be issued as a response to some challenging conditions is not inside the aforementioned catalogue, it should consult its upper level DE (DE2.1 or DE2.2). The upper level DE should in turn consult the corresponding ASM (hosted by DE2.2) that is expected to solve an optimization problem and to respond back with a set of actions that are allowed to be executed.

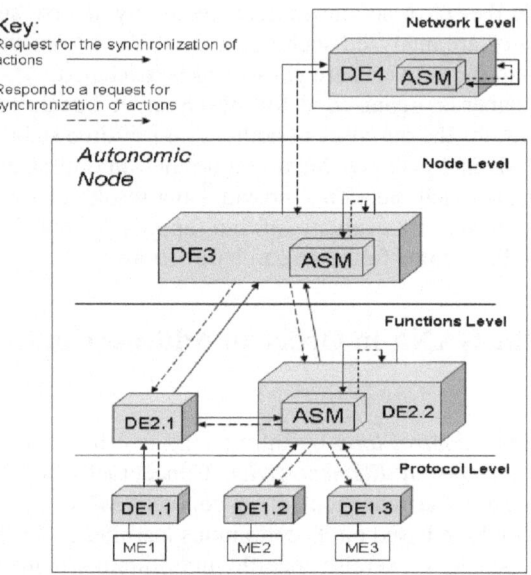

**Fig. 1.** Hierarchically addressing stability within GANA

After the allowed actions have been selected, the upper level DE informs the lower level DEs whether or not they are allowed to "fire" some of the actions on their corresponding MEs (ME1, ME2, and ME3). On the other hand if the upper level DE "recognizes" that the actions it has been requested to synchronize are beyond its competence, it should further consult its corresponding upper DE (e.g. DE3 in Figure 1) on the higher level.

## 3   A Binary Integer Program for Addressing Run-Time Stability in GANA

In this section we define an optimization problem that should be solved by an ASM whenever a set of actions needs to be synchronized.

Let $Q$ be a set of independent QoS metrics ($|Q| = n \in N^+$). *Independent* in this case means that the metrics do not have co-relations with each other. Let $W$ be a set of weights, each of which is assigned to one of the metrics contained in $Q$ ($|W| = n \in N^+$, $w_i \in W$, $w_i \in R^+$, $i \in \{1, ..., n\}$). These weights represent the importance of each metric for the overall health of the network. Considering a particular point in time $t_0$, the values of the metrics in $Q$ are represented by a vector $Q(t_0)=(q_1(t_0), ..., q_n(t_0))^T \in R^n$. Suppose that the higher the values of $q_1(t_0), ..., q_n(t_0)$ the better the performance of the system (for metrics which require to be minimized one can take the reciprocal value), then the following expression gives the network fitness at $t_0$.

$$NF(t_0) = \sum_{i=1}^{n} w_i q_i(t_0) = \langle w, Q(t_0) \rangle \qquad (1)$$

Hence the goal of the autonomic mechanisms in the node/device and the network is to maximize $NF(t)$ throughout the operation of the system. Additionally let $A$ ($|A|=M \in N^+$) be the set of actions that the ASM is responsible for. By $a_j \in A$ with $j \in \{1, ..., M\}$ we denote a single action. Further, the domain relation of an action $d:A \rightarrow \{0,1\}^n$ is introduced. The relation takes as an input an action and returns a (0-1) binary vector. If the $i^{th}$ component of the vector is 1 then the $i^{th}$ metric is influenced by the input action, and respectively if its 0 then the metric is not influenced. We define the *domain matrix* of $A$ as $D = (d(a_1) \cdots d(a_M)) \in \{0,1\}^{n \times M}$. Further, we introduce the impact relation of $A - I:Q \times A \rightarrow R$. The output value of $I(i,j)$ stands for the impact the $j^{th}$ action has on the $i^{th}$ metric. Thus, if only action $a_j$ is executed at point in time $t_0$, then the new value of $q_i$ will be $q_i(t_0)+I(i,j)$. Based on the definitions given above, and assuming that in a particular time slice a total number of $m \in N^+$ ($m \leq M$) actions have been requested for synchronization, the following optimization problem is defined

$$p = \arg\max_{p \in \{0,1\}^m} \sum_{i=1}^{n} w_i (q_i(t_0) + \sum_{j=1}^{m} p_j I(i,j)) = \arg\max_{p \in \{0,1\}^m} w^T I_m p + w^T Q(t_0) \quad (2)$$

$$s.t. \ D_m p \leq c$$

$I_m$ stands for an $n \times m$ real-valued matrix where $I_m(i.j)$ represents the impact of the requested $j^{th}$ action on the $i^{th}$ metric. Moreover, in (2), $p$ stands for a (0-1) vector which gives whether a particular action has been allowed to be executed or not. If the $j^{th}$ position of $p$ is 1 then the corresponding action has been selected for execution, otherwise it has to be dropped. Hence, an optimization with respect to $p$ is equivalent to selecting the optimal set of actions requested in a particular time frame. Moreover, the matrix $D_m$ is a sub-matrix of the domain matrix $D$ of $A$ consisting only of the columns which represent the domains of the requested actions. The vector $c \in N^n$ determines the extent to which actions with overlapping domains are allowed. For example, if $n=4$ and $c=(1,1,1,1)^T$, i.e. only four QoS metrics are considered, then the additional constraint says that only one action influencing a single metric is allowed. That is, the additional constraint can be used to enforce the resolution of conflicts between actions with overlapping domain regions.

Since the term $w^T Q(t_0)$ in (2) is just a constant that reflects the current state of the network, it can be dropped, which is very good news because it means that the values of the QoS metrics are not needed for the optimization. Thus, no interaction with the monitoring functions measuring the metrics is needed. The final optimization problem takes the following form:

$$p = \arg\max_{p \in \{0,1\}^m} w^T I_m p$$

$$s.t. \quad D_m p \leq c$$

While we are aware of the fact that linear mixed integer programs are NP-hard in general, modern solvers are quite advanced and would provide a solution that is the best possible based on the underlying optimization algorithm.

## 4   Evaluations

In order to conduct some initial evaluations of the proposed methodology, we implemented an ASM on a Linux Ubuntu Machine. The prototype is based on Multi-Threading and the Coin-OR solver [4][5].The focus of the evaluations was set on the issue of overhead produced by solving the previously derived optimization problem.

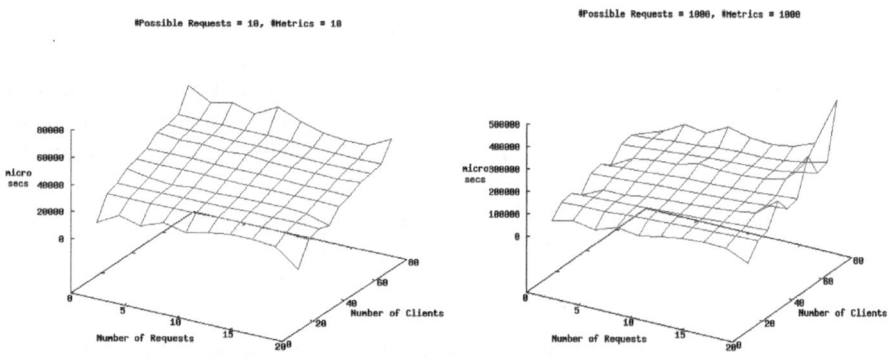

**Fig. 2.** Performance evaluations of the proposed approach

The evaluations were conducted on a single machine - Intel(R) Pentium(R) M processor 1.60GH, 509.2 MB RAM. Our goal was to examine the performance of the overall processes implemented by the ASM component. Thus several experiments were carried out with different numbers of metrics and actions, different numbers of clients simultaneously requesting for synchronization, and different number of requests issued by every client. The values of the impact matrices and the weights reflecting the significance of the metrics were randomly generated for each experiment. Figure 2 presents the results of our performance evaluations. The time measurements that are plotted on the Z-axis correspond to the maximum response time of an ASM component, after each of the simultaneously started ASM clients have issued a number of requests (shown on the X-axis). In the above described environment it was impossible to complete the experiments with more than 70 clients started simultaneously. This could be due to the prototype design or an overload of the operating system. Despite this drawback, one can observe (Figure 2) that for 70 and fewer clients the response times were quite reasonable. Hence, depending on the number of the possible and relevant requests, we deduce that one could use this approach presented here for both fast decisions, e.g. on routing, forwarding level or decisions related to the resilience of the system, and for synchronisations which are not strongly time constrained.

## 5   Conclusions and Future Work

This paper proposed an approach for addressing run-time stability in the context of the GANA. GANA is a recently emerged Reference Model for Autonomic Network engineering. However, the methodology proposed here is applicable not only to GANA but also to any other architectural model for autonomic systems engineering. The idea presented here is to introduce special components called Actions Synchronization Module (ASM), which should be part of a Decision Element (DE). DEs are responsible for managing their assigned Managed Entities (MEs). The aforementioned ASMs provide the functionality of "gathering" (from the DEs) management actions, policies and configuration profiles to be deployed,   and synchronizing them based on a binary linear program that selects the best subset of actions allowed to execute at a particular time slice. By doing that issues such as contradicting actions and policies, oscillations and chaos caused by various DEs trying to optimize the performance of their own MEs, without considering the overall network fitness are avoided.

Our evaluations showed that the prototype is applicable inside a GANA node and scales well (for the purposes of GANA) with respect to the number of clients as well as the size of the requests and the optimization problem parameters. Thus, our methodology could be applied for tasks with real-time requirements such as improving the resilience of the autonomic systems.

In general, we conclude that it is important to apply all the methodologies that guarantee stability of control-loops, ranging from the application of model-driven/FDTs-driven analysis, simulation and validation of control-loop structures and behaviors at design phase, to the application of optimization techniques at run-time.

Our future work in that direction will mainly target the application of the presented methodology in a real GANA network, as the one that is developed in the course of EFIPSANS [3]. As a short paper focusing mainly on the importance and applicability of Synchronization techniques in addressing some aspects of stability of interacting control-loops, we have left out the illustration with real-world scenarios (a subject to be covered by the future long-version of this paper). Also as part of further work, scalability issues will be further examined and addressed.

**Acknowledgement.** This work has been partially supported by EC FP7 EFIPSANS project (INFSO-ICT-215549).

## References

1. Chaparadza, R.: Requirements for a Generic Autonomic Network Architecture (GANA). suitable for Standardizable Autonomic Behavior Specifications for Diverse Networking Environments, IEC Annual Review of Communications 61 (2008)

2. Bai, J., et al.: A Model Integrated Framework for Designing Self-managing Computing Systems. In: The proceedings of FeBid (2008)
3. EC funded- FP7-EFIPSANS Project: http://efipsans.org/
4. Thorncraft, S.R.: Evaluation of Open-Source LP Optimization Codes in Solving Electricity Spot Market Optimization Problems. In: The proceedings of ORMMES 2006 (2006)
5. Coin-OR: as of date (October 1, 2009), http://www.coin-or.org/

# Author Index